120 Advances in Polymer Science

Physical Properties
of Polymers

With contributions by
I. Capek, J. Casas-Vázquez, M. Criado-Sancho,
L. J. Fetters, S. Förster, N. Hadjichristidis,
D. Jou, J. W. Mays, M. Schmidt, Z. Xu

With 45 Figures and 21 Tables

 Springer

ISBN 978-3-662-14863-1   ISBN 978-3-540-49049-4 (eBook)
DOI 10.1007/978-3-540-49049-4

© Springer-Verlag Berlin Heidelberg 1995
Library of Congress Catalog Card Number 61-642

Originally published by Springer-Verlag Berlin Heidelberg New York in 1995.

Softcover reprint of the hardcover 1st edition 1995

Typesetting: Macmillan India Ltd., Bangalore-25
SPIN: 10470574   02/3020 - 5 4 3 2 1 0 Printed on acid-free paper

Editors

Table of Contents

Structure/Chain-Flexibility Relationships of Polymers

Zhongde Xu*[1], N. Hadjichristidis[2], L.J. Fetters[3], and J.W. Mays[4]
[1] Department of Polymer Science and Engineering, East China University
of Science and Technology, Shanghai, 200237, People's Republic of China
[2] Department of Chemistry, University of Athens, Panepistimiopolis,
Zographou, Athens (157 71) Greece
[3] Exxon Research and Engineering Company, Corporate Research
Laboratories, Annandale, NJ 08801, USA
[4] Department of Chemistry, University of Alabama at Birmingham,
Birmingham, AL 35294, USA

The relationships between polymer conformational characteristics, i.e., unperturbed chain dimensions and their variation with temperature, and polymer structure are reviewed and critically discussed. Emphasis is placed on structure/conformation relationships for structurally well-defined polymethacrylates, polydienes, and polyolefins. Stiff chain polymers are also discussed. It is shown that conformational characteristics measured in bulk systems by small-angle neutron scattering are sometimes quite different than those measured for the same chains in theta solvents. Finally, relationships between parameters of chain flexibility and bulk properties, such as plateau modulus, molecular weight between entanglements, and tube diameter, are demonstrated.

---

*Also affiliated with Structure and Analysis Research Laboratory, University of Science and Technology of China, Hefei, 230026, People's Republic of China

Advances in Polymer Science, Vol. 120
© Springer-Verlag Berlin Heidelberg 1995

List of Symbols and Abbreviations

A_2	second virial coefficient
A_o	parameter in Bohdanecky wormlike cylinder model
A_η	parameter in Bohdanecky wormlike cylinder model
A'	constant
B	excluded volume parameter
B_o	constant in Bohdanecky wormlike cylinder model
B_η	parameter in Bohdanecky wormlike cylinder model
BSF	Burchard–Stockmayer–Fixman
c	polymer concentration
C	constant
C_∞	characteristic ratio
d	cylinder diameter
dn/dc	specific refractive index increment
d_r	reduced cylinder diameter (d/l')
d_t	tube diameter
f	force applied in thermoelasticity experiments
G_N^o	plateau modulus
HHPP	head-to-head polypropylene
K	light scattering optical constant
K_Θ	unperturbed parameter
ℓ	bond length
L	contour length
l'	Kuhn statistical segment length
L_r	reduced contour length (L/l')
l_u	projected length of the repeating unit along the molecular axis
m_o	average mass per main chain bond
M	molecular weight
M_c	molecular weight at which break in log η_o–log M plot occurs
M_e	molecular weight between entanglements
M_L	molecular weight per unit contour length
$\bar{M}_n$	number-average molecular weight
Mu	molecular weight of a repeating unit
$\bar{M}_w$	weight-average molecular weight

n_o	solvent refractive index
N	number of main chain bonds
N_A	Avogadro constant
N	molecular weight of polymer divided by molecular weight of the repeating unit in the Bohdanecky wormlike cylinder model
PB	poly(1-butene)
PBD	polybutadiene
PDMB	poly(2,3-dimethyl butadiene)
PDMS	poly(dimethyl siloxane)
PE	polyethylene
PEB	ethylene-butene *copolymers*
PEE	poly(ethyl ethylene)
PEP	ethylene-propylene *copolymers*
PI	polyisoprene
PIB	polyisobutylene
PMMA	poly(methyl methacrylate)
PMPD	poly(2-methyl pentadiene)
PMYRC	polymyrcene
PP	polypropylene
PS	polystyrene
PVCH	poly(vinyl cyclohexane)
q	persistence length
$\langle R^2 \rangle$	mean-square end-to-end distance
$\langle R^2 \rangle_o$	unperturbed mean-square end-to-end distance
$\langle R_g^2 \rangle$	mean-square radius of gyration
$\langle R_g^2 \rangle_o$	unperturbed mean-square radius of gyration
$\langle R_g^2 \rangle_z$	z-average of mean square radius of gyration
$\langle R_g^2 \rangle_w$	weight average of mean square radius of gyration
RIS	rotational isomeric state
R	gas constant
SANS	small angle neutron scattering
T	temperature
u	polydispersity ratio
V_2^*	critical volume fraction of polymer
V_1	molar volume of solvent
V	volume
V_{sp}	sphere volume
X	aspect ratio (length/diameter)
α	chain expansion factor
β	binary cluster integral
ΔL	subsegment length of a wormlike chain
ΔR_Θ	excess Rayleigh ratio
$[\eta]$	intrinsic viscosity
$[\eta]_o$	intrinsic viscosity under theta conditions

θ	$180°$ minus bond angle
Θ	light scattering angle
κ	temperature dependence of $\ln \langle R^2 \rangle_0$
λ_0	wavelength of light in a vacuum
$\bar{v}$	partial specific volume of polymer
σ	conformation factor
ϕ	average rotation angle about main chain bonds
Φ_0	Flory hydrodynamic parameter for a flexible linear unperturbed coil
Φ	Flory hydrodynamic parameter
χ_1	polymer/solvent interaction parameter
ψ	supplementary angle

1 Introduction

An enormous number of investigations conducted over the past half-century have focused on development and implementation of methods to probe macromolecular features such as chain dimensions, chain dynamics, molecular weight, and molecular shape. The intense interest in this area has been fueled by the recognition that the unique and versatile properties of polymers are a direct consequence of their being endowed with an extreme range of accessible conformations. From a practical standpoint, this means that bulk mechanical and thermal properties, morphology, and processing behavior reflect both the average equilibrium chain dimensions and the chain flexibility or ease with which changes in size and shape can occur. Thus, the development of an understanding of how polymer structure affects chain flexibility has been a major focus of polymer science since its inception. Much progress has been made in this area on both theoretical and experimental fronts. Flory [1] reviewed developments in this area up to 1969 in a classic monograph. Establishing correlations between polymer conformation and mechanical and thermal properties has proved more difficult. Recently, however, much progress has been made in terms of correlating the plateau modulus and other rheological parameters [2–9] with parameters of chain flexibility.

In this article, we will review major methods for the evaluation of unperturbed dimensions of both highly flexible and less flexible (wormlike) chains. Methods for extracting the temperature dependence of unperturbed dimensions will also be described. Experimental results are tabulated and discussed. The emphasis here is on developing structure/chain-flexibility relationships for a few series of carefully chosen polymers rather than providing an exhaustive tabulation. Furthermore, new developments and unresolved issues are stressed in an attempt to identify areas needing further attention. Finally, interrelationships between structure, flexibility, and mechanical properties will be discussed.

2 Characterization of Chain Flexibility: Theory

Polymers exhibit conformations ranging from tight coils to highly extended structures such as rigid rods and rigid helical chains. The parameters and, to some extent, the methods utilized for evaluating the degree of chain flexibility are usually different for "flexible" versus "stiff" chains.

For flexible polymers the most common parameter in use is Flory's [1, 10] characteristic ratio C_∞, defined as

$$C_\infty \equiv \lim_{N \to \infty} \frac{\langle R^2 \rangle_o}{N\ell^2} = \frac{[\langle R^2 \rangle_o / M] m_o}{\ell^2} \tag{1}$$

where $\langle R^2 \rangle_o$ is the unperturbed (theta condition) mean-square end-to-end distance, N is the number of main chain bonds of length ℓ, M is the molecular weight of the polymer, and m_o is the average mass per main chain bond. The value of C_∞ is a quantitative measure of the impact of hindered rotation about main chain bonds and rather fixed bond angles on $\langle R^2 \rangle$. For a freely jointed chain, which has neither rotational hindrances nor bond angle restrictions, $\langle R^2 \rangle$ is equal to $N\ell^2$ [11, 12]. Thus, a value of C_∞ is equal to 1 for the freely jointed chain, and larger C_∞ values are indicative of greater departures from freely jointed character, i.e., diminished flexibility.

The relative contributions of fixed bond angles and hindered rotation can be elucidated by modification of the freely jointed chain model. Eyring [13] showed that when N is very large $\langle R^2 \rangle$ can be calculated as

$$\langle R^2 \rangle = (N\ell^2)\left(\frac{1 + \cos \theta}{1 - \cos \theta}\right) \tag{2}$$

where θ is equal to 180° minus the fixed bond angle. Thus, for a polyethylene backbone $\cos \theta = 0.333$ and $\langle R^2 \rangle = 2N\ell^2$ with fixed tetrahedral bond angles for a saturated hydrocarbon backbone being expected to double $\langle R^2 \rangle$.

In light of the above, an alternative chain flexibility parameter, which is commonly used, is the conformation (steric) factor σ

$$\sigma = (\langle R^2 \rangle_o^{1/2})(\langle R^2 \rangle_f^{1/2})^{-1} \tag{3}$$

which is obtained from the experimental $\langle R^2 \rangle_o$ value and the $\langle R^2 \rangle$ value calculated for the freely rotating chain i.e., $\langle R^2 \rangle_f$. σ provides a measure of the relative increase in the end-to-end distance brought about by hindrances to rotation only. Obviously, C_∞ and σ are related for tetrahedral hydrocarbon backbones by

$$C_\infty = 2\sigma^2 \tag{4}$$

As Flory has pointed out [14], the use of C_∞ is to be preferred over σ. This is because small changes in bond angles can lead to large changes in σ, and bond angles are rarely known to within more than a few degrees.

The effect of restricted rotation can be taken into account by introducing an additional term into Eq. 2. If the hindering potentials are mutually independent for neighboring bonds and symmetrical [15–17], $\langle R^2 \rangle$ becomes

$$\langle R^2 \rangle = (N\ell^2)\left(\frac{1 + \cos \theta}{1 - \cos \theta}\right)\left(\frac{1 + \langle \cos \phi \rangle}{1 - \langle \cos \phi \rangle}\right). \tag{5}$$

Here $\langle \cos \phi \rangle$ is the average rotation angle ($\phi = 0°$ for *trans*).

The assumptions upon which Eq. 5 is based are invalid for many macromolecules. Of great significance in this regard has been the development of rotational isomeric state (RIS) models [1, 10, 18, 19]. These models generally consider only a few discrete conformers of relatively low energies. Interdependence of bond rotations, asymmetric rotational potential curves, and finite chain length effects can all be dealt with using modern RIS methods [1, 10].

The Kratky–Porod wormlike chain model [20, 21] is widely used for describing conformational characteristics of less flexible chains. The polymer is viewed as a semi-flexible string (or worm) of overall "contour length" L with a continuous curvature. The chain is subdivided into N segments of length ΔL, which are linked at a supplementary angle ψ. The persistence length q (Fig. 1) is defined as

$$q = \lim_{\substack{\psi \to 0 \\ \Delta L \to 0}} \frac{\Delta L}{1 - \cos \psi} \tag{6}$$

[22] and is thus a measure of the tendency for segments in the polymer chain to "remember" the orientation of adjoining (and other) segments in the chain. Wormlike chains will exhibit conformations ranging between random coils and rigid rods depending on the value of the ratio L/q. Therefore, q provides a measure of chain stiffness. Furthermore, it can be shown [23] that at large L a wormlike chain becomes Gaussian and q is related to the length of the Kuhn statistical segment l' as

$$q = l'/2 \tag{7}$$

Thus, the value of l' is also commonly taken as a measure of chain stiffness. Flory [24] has shown that there is also a simple relationship between the persistence length and the characteristic ratio

$$C_\infty = \frac{2q}{\ell} - 1. \tag{8}$$

Here ℓ is the average bond length and ∞ indicates, as usual, the limiting value for infinite chains. Hence, C_∞ can also be employed in comparing relative

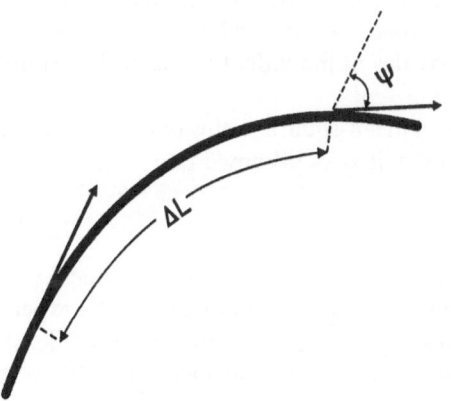

Fig. 1. A segment of a persistent chain (parameters are defined in the text)

stiffness of wormlike chains. In this article, we will follow the usual convention of using C_∞ to describe highly flexible chains and q to describe less flexible ones.

Finally, some rather recent developments must be noted. Several years ago, Yamakawa and co-workers [25–27] developed the wormlike continuous cylinder model. This approach models the polymer as a continuous cylinder of hydrodynamic diameter d, contour length L, and persistence length q (or Kuhn length l'). The axis of the cylinder conforms to wormlike chain statistics. More recently, Yamakawa and co-workers [28] have developed the helical wormlike chain model. This is a more complicated and detailed model, which requires a total of five chain parameters to be evaluated as compared to only two, q and L, for the wormlike chain model and three for a wormlike cylinder. Conversely, the helical wormlike chain model allows a more rigorous description of properties, and especially of local dynamics of semi-flexible chains. In large part due to the complexity of this model, it has not yet gained widespread use among experimentalists. Yamakawa and co-workers [29–31] have interpreted experimental data for several polymers in terms of this model.

3 Experimental Determination of Polymer Chain Flexibility

3.1 Flexible Chains

From Eq. 1, the evaluation of the characteristic ratio requires measurement of the *unperturbed dimensions* of the polymer chain. Polymers exhibit their characteristic unperturbed dimensions in the bulk amorphous state, i.e., chain dimensions under these conditions reflect the influence of short range, rotational isomeric state, effects only. Prior to the availability of small-angle neutron scattering (SANS) in the mid-1970s [32–34], no method was available that allowed chain dimensions in polymer melts to be measured directly. SANS utilizes the fact that different isotopes result in different scattering amplitudes for neutrons. Thus, selective deuterium labelling of some chains, followed by dispersing these chains in a "solvent" of otherwise identical but non-deuterated chains, allows the conformational properties of individual chains to be probed in the melt.

SANS allows $\langle R_g^2 \rangle_o$, the unperturbed mean-square radius of gyration, to be measured. For a linear unperturbed chain, it is well-known that

$$\langle R^2 \rangle_o = 6 \langle R_g^2 \rangle_o. \tag{9}$$

Hence, the value of $\langle R^2 \rangle_o$ may be computed. A detailed discussion of the SANS technique is beyond the scope of the present review, and details of the experimental protocol and data analysis have been reviewed previously [35]. Although unperturbed dimensions from SANS studies on melts are still very scarce in comparison to the huge number of studies reported from the more accessible dilute solution measurements, important recent SANS results [9, 36–39] suggest that some of the prevailing views on polymer chain flexibility and dilute solution behavior must be reconsidered (see Sect. 5).

The two dominant techniques currently in use for measuring unperturbed dimensions are light scattering and viscometric measurements on dilute polymer solutions. Macromolecular dimensions in solution are affected by both long range (excluded volume) and short range, rotational isomeric state, interactions [40]. In thermodynamically good solvents, where solvent/polymer-segment interactions are favored, the chain will expand to enhance these interactions and minimize polymer/polymer interactions ("excluded volume effect"). Conversely, in a thermodynamically poor solvent the chain will contract to minimize unfavorable polymer/solvent interactions. Flory [40] predicted that for a polymer chain in dilute solution there would exist an ideal or unperturbed state at the so-called theta (θ) condition. Here, through appropriate choice of solvent and temperature, the tendency for chain segments to interact with other chain segments or with solvent is balanced (the second virial coefficient A_2 is equal to zero). Flory [40, 41] suggested that polymer chains in solution at the theta condition and those in the bulk amorphous state should exhibit, at least approximately, identical conformations.

In practice, there is a growing body of experimental evidence that shows that the choice of theta solvent can have an impact on the magnitude of measured "unperturbed" dimensions [39, 42–44] and/or the Flory hydrodynamic parameter Φ_o under theta conditions [30, 45]. Theory anticipated [46] and has rationalized [45, 47] such effects. In fact, today, polymer chains are viewed by theory as being only quasi-ideal at the theta state [48–50].

The most direct dilute solution technique for measuring unperturbed dimensions is light scattering. From a light scattering experiment on dilute solutions of monodisperse polymer, the molecular weight M, the mean-square radius of gyration $\langle R_g^2 \rangle$, and thermodynamic interactions (A_2) may all be probed using the classical Zimm analysis [51]

$$\frac{Kc}{\Delta R_\Theta} = \frac{1}{M}\left[1 + \frac{16\pi^2 n_o^2}{3\lambda_o^2} \langle R_g^2 \rangle \sin^2(\theta/2) \right] + 2A_2c + \cdots \tag{10}$$

where $K = [2\pi n_o]^2 (dn/dc)^2 (N_A \lambda_o^4)^{-1}$, c is the polymer concentration, ΔR_Θ is the excess Rayleigh ratio, Θ is the scattering angle, n_o is the solvent refractive index, λ_o is the wavelength of light in vacuum, dn/dc is the specific refractive index increment, and N_A is the Avogadro constant. The value of $\langle R_g^2 \rangle$ will be $\langle R_g^2 \rangle_o$, the unperturbed value, when A_2 is equal to zero. Molecular weight heterogeneity is an important consideration here, since use of Eq. 10 for a polydisperse material will give a weight-average molecular weight $\bar{M}_w$ but a z-average size $\langle R_g^2 \rangle_z$. Thus, in working with polydisperse materials it is important to correct for polydispersity effects. The most common approach is to compute the weight-average mean-square radius of gyration $\langle R_g^2 \rangle_w$ using the equation

$$\langle R_g^2 \rangle_w = \frac{u+1}{2u+1} \langle R_g^2 \rangle_z \tag{11}$$

[52] were u is the polydispersity parameter equal to $\bar{M}_w/\bar{M}_n - 1$ with $\bar{M}_n$ being the number-average molecular weight. Equation 11 assumes a Schulz–Zimm distribution of molecular weight. To minimize the impact of approximate corrections such as Eq. 11 it is important to work with well-fractionated, narrow molecular weight distribution materials whenever possible.

Frequently, light scattering experiments are not conducted under theta conditions. This may be because theta conditions are not known or because of the greater ease associated with conducting light scattering measurements in good solvents. Baumann [53] proposed an equation that is equivalent to the Burchard–Stockmayer–Fixman extrapolation procedure for intrinsic viscosity (see below)

$$(\langle R_g^2 \rangle/M)^{3/2} = (\langle R_g^2 \rangle_o/M)^{3/2} + BM^{1/2} \tag{12}$$

where B, the excluded volume parameter, is related to the binary cluster integral β [60]. Equation 12 suggests that $\langle R_g^2 \rangle_o/M$ may be evaluated by extrapolating good solvent values $(\langle R_g^2 \rangle/M)^{3/2}$ versus $M^{1/2}$ to $M^{1/2} \Rightarrow 0$. In our experience, procedures that attempt to correct for excluded volume effects should be used with caution for chains with large excluded volumes [54].

Intrinsic viscosity approaches for evaluating unperturbed dimensions are by far the most popular techniques. The intrinsic viscosity $[\eta]$ is easy to measure with great accuracy using simple, inexpensive equipment. Under theta conditions, the intrinsic viscosity $[\eta]_o$ is related to M by the Flory-Fox equation [55, 56]

$$[\eta]_o = K_\Theta M^{1/2} \tag{13}$$

with

$$K_\Theta = \Phi_o(\langle R^2 \rangle_o/M)^{3/2}. \tag{14}$$

Flory [57] predicted many years ago that Φ_o should be a constant for linear random coils, independent of the nature of the polymer and solvent. It is now known that good solvent values of $\Phi = [\eta] M \langle R^2 \rangle^{-3/2}$ are smaller than values measured at theta conditions. A large number of experimental studies on different well-defined linear flexible polymers performed under theta conditions have yielded values of $\Phi_o = 2.5(\pm 0.1) \times 10^{21}$ mol^{-1} when $[\eta]$ is in dL g^{-1} [58]. Zimm [59] has computed a value of 2.51×10^{21} mol^{-1} using a numerical simulation whereby the usual pre-averaging of the Oseen tensor, a necessary limitation of theoretical approaches to date, was avoided. Thus, there is ample reason to believe that Φ_o exhibits a constant (or near constant) value for most flexible polymers. As mentioned above, however, Yamakawa and co-workers

[30, 45] have recently reported carefully derived Φ_o values for poly(methyl methacrylate) (PMMA) samples in two different theta solvents: acetonitrile and 1-chlorobutane. Taking into account experimental errors, apparently slightly lower values were obtained in acetonitrile. Values of Φ_o measured by Konishi et al. [45] for four other polymer/solvent systems are constant within experimental errors and agree with Φ_o measured for the PMMA/1-chlorobutane system. It was shown that the different Φ_o values for PMMA could be attributed to differences in $[\eta]_o$ and not $\langle R_g^2 \rangle_o$ [30]. Further work appears to be necessary, but it seems clear that Φ_o shows at most a mild variation with changing nature of the flexible polymer and the solvent. Thus, unperturbed dimensions can safely be determined using Eqs. 13 and 14.

As in the case of light scattering, fractions of low polydispersity are preferred since $[\eta]$ is sensitive to the viscosity-average molecular weight $\bar{M}_v$ instead of to the usually measured $\bar{M}_w$ or $\bar{M}_n$. In utilizing $[\eta]_o$ to evaluate $\langle R^2 \rangle_o$ it is preferable to use $\bar{M}_w$ (instead of $\bar{M}_n$) since this value is closer to $\bar{M}_v$. Correction factors for polydispersity effects [60] should be applied to K_Θ prior to computing C_∞, although the corrections are much smaller for $[\eta]$ measurements than for $\langle R_g^2 \rangle$.

A large number of methods have been developed for estimating K_Θ from $[\eta]$ measurements in thermodynamically good solvents. These techniques have been periodically reviewed [54, 61, 62], so only a few key methods will be discussed here. The most widely used method, in part because of its simplicity, is the method of Burchard [63] and Stockmayer and Fixman [64]. The BSF equation, which was based on earlier theoretical work by Kurata and Yamakawa [65]

$$[\eta] = K_\Theta M^{1/2} + 0.51 \Phi_o BM \tag{15}$$

suggests a plot of $[\eta]M^{-1/2}$ versus $M^{1/2}$ to obtain K_Θ as the intercept (Fig. 2A).

The ellipsoid polymer model upon which Eq. 15 is based [65] has been abandoned because it predicts a molecular weight dependence of the coil expansion factor that does not agree with experimental data. Thus, Eq. 15 may be regarded as purely empirical and rewritten as

$$[\eta] = K_\Theta M^{1/2} + CM \tag{16}$$

where C is a constant. The BSF equation is therefore of no use for estimating the excluded volume parameter B. It is well-known, however, that the BSF plot provides reliable estimates of K_Θ as long as chain expansion is not too great, i.e., at modest, but not too low molecular weights and in solvents that are not too good [66].

More recently, Tanaka [67] proposed

$$([\eta]M^{-1/2})^{5/3} = K_\Theta^{5/3} + 0.667 \Phi_o^{5/3}(\langle R^2 \rangle_o M^{-1}) BM^{1/2} \tag{17}$$

using a Padé approximation. Stickler and co-workers [68] demonstrated that for PMMA solutions linear plots are obtained – even at very large excluded

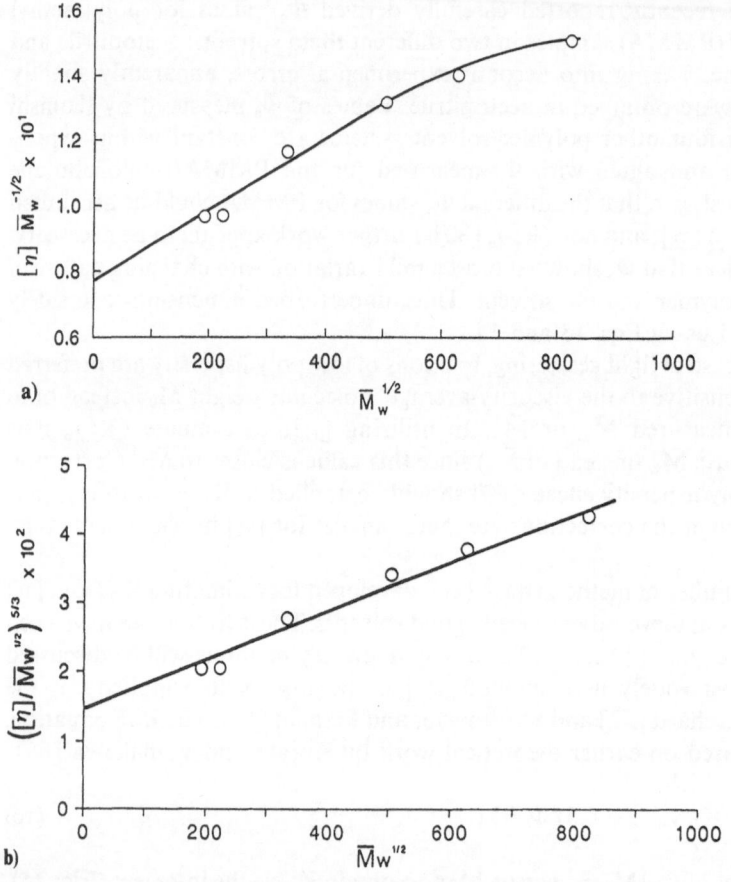

Fig. 2. a. BSF plot for estimation of unperturbed dimensions of polystyrene in n-butyl chloride at 25 °C. **b** Tanaka plot for the data of Fig. 2a. Notice that this plot is linear over the entire range of excluded volumes investigated. Reprinted with permission from Macromolecules, 24: 199 (1991). Copyright 1991 American Chemical Society

volumes – using the Tanaka plot. Similar findings were reported for the polystyrene/1-chlorobutane system [69]. Hence, compared to the graphical procedure described above, Tanaka's method appears to be superior in correcting for large excluded volumes (Fig. 2b). It is to be preferred for estimating K_Θ in the case of good solvent data. Rarely it is used for evaluating B.

It should also be mentioned that the validity of a number of the intrinsic viscosity extrapolations has been established [70, 71] using data obtained under extremely poor – worse than theta solvent – conditions. All of the studied methods, including that of Tanaka [67], deviated from linearity at large chain contraction ratios. This is not surprising, as extrapolation procedures based on two parameter theory cannot be expected to apply under conditions where ternary (3 body) interactions are strong. Nevertheless, reliable extrapolations to obtain unperturbed dimensions could be carried out [70, 71]. Consequently, [η] data obtained in very poor solvents can be used to estimate unperturbed dimensions.

3.2 Stiff Chains

As with flexible chains, most studies of conformational behavior of stiff chains have involved light scattering and intrinsic viscosity studies of dilute polymer solutions. Excluded volume effects are of much diminished significance for stiffer chains, so measured persistence lengths usually show only a mild dependence on the nature of the solvent. In fact, Norisuye and Fujita [72] have shown that excluded volume effects become measurable only when chains are very long ($L \geq 100\,q$). Thus, the choice of solvent is normally of less importance in studying the conformation of stiff chains than it is for flexible ones. Temperature, however, frequently has a pronounced impact on the value of q for stiff chain polymers [73].

Benoit and Doty [74] have related q and $\langle R_g^2 \rangle$ by the equation

$$\langle R_g^2 \rangle = q^2 [1/3(L/q) - 1 + (2q/L) - (2q^2/L^2)[1 - \exp(-L/q)]]. \quad (18)$$

Hence, measurement of $\langle R_g^2 \rangle$ by light scattering leads directly to q. However, light scattering measurements on stiff polymers are often difficult, and conformational studies are most frequently conducted using intrinsic viscosity measurements.

The methods described above for estimating unperturbed dimensions of flexible polymers from intrinsic viscosity measurements are not applicable to stiff chain materials. As Stockmayer [66] has pointed out, if the Mark–Houwink–Sakurada exponent exceeds 0.8 the previously cited methods are certain to underestimate unperturbed dimensions because of the non-Gaussian chain stiffness and/or the hydrodynamic draining exhibited by such chains. As an example, Stockmayer [66] cited the work by Helminiak et al. [75] on the ladder polymer poly(phenylsilsesquioxane). Fractions of this polymer gave an A_2 of zero in chloroethane at 50.5 °C, suggesting θ conditions. Conversely, the Mark–Houwink–Sakurada exponent measured under these conditions was 0.89 as expected for an intrinsically stiff chain. Application of Eq. 15, while nicely linear, yielded $\langle R^2 \rangle_o/M = 3.2 \times 10^{-17}$ cm^2 mol g^{-1}. This is much smaller than the true unperturbed dimensions that were estimated as $\langle R^2 \rangle_o/M = 1.34 \times 10^{-16}$ cm^2 mol g^{-1} [66] using an approach applicable to wormlike chains [25, 26].

The Yamakawa-Fujii wormlike cylinder model [25, 26] has been widely used for estimation of l' (or q) of stiff chain materials from intrinsic viscosities. This approach is based on the equation

$$[\eta]_o = \Phi(l')^{3/2} L^{1/2} M_L^{-1} \quad (19)$$

[26] where $[\eta]_o$ denotes $[\eta]$ values unperturbed by excluded volume, Φ is the Flory hydrodynamic parameter, and M_L is equal to M/L. This model requires a knowledge of the cylinder diameter d. Also, the value of Φ depends on both $L_r = L/l'$ and $d_r = d/l'$, taking on the limiting value for flexible unperturbed coils in the limit $L_r \rightarrow \infty$. The experimental evaluation of all these parameters is difficult, and a number of approaches have been employed [76–82]. The best approach appears to be to plot $M^2[\eta]^{-1}$ versus $\ln(M)$ and evaluate M_L and d from the slope and intercept, respectively. The data are then fit to theoretical curves for different values of q on double logarithmic plots of $[\eta]$ versus M to see which q value gives the best fit [80, 82].

Bohdanecky [83] has recently described a simple graphical procedure that allows easy evaluation of the wormlike cylinder parameters. He showed that Φ is related to Φ_o by

$$\Phi = \Phi_o[B_o + A_o(l'/L)^{0.5}]^{-3} \qquad (20)$$

where B_o is a virtual constant and A_o is a known function of d_r. By integrating this approach for estimating Φ into the general framework of the Yama-kawa–Fujii model, the following equation was derived [83]

$$(M^2/[\eta]_o)^{1/3} = A_\eta + B_\eta M^{1/2}. \qquad (21)$$

Here

$$A_\eta = A_o\Phi_o^{-1/3}M/L \qquad (22)$$

and

$$B_\eta = B_o\Phi_o^{-1/3}(\langle R^2 \rangle_o/M)_\infty^{-1/2} \qquad (23)$$

The subscript ∞ denotes the random coil value. From this approach unperturbed dimensions can readily be evaluated. The Kuhn length l' may be computed as

$$l' = \langle R^2 \rangle_o/(Nl_u) \qquad (24)$$

where N is equal to M/M_u (M_u is the molecular weight of the repeating unit) and l_u is the projected length of the repeating unit along the molecular axis.

Equation 21 formally requires unperturbed values of $[\eta]$ but, as mentioned above, excluded volume effects will be negligible or small for most stiff chain polymers. In particular, the initial slope of the Bohdanecky plot, from which chain dimensions are derived, should generally be determined from data where excluded volume effects are negligible.

Reduced cylinder diameter d_r can be estimated from the Bohdanecky plot if the partial specific volume $\bar{v}$ of the polymer is known [83]. The pertinent equation is

$$d_r^2/A_o = (4\Phi_o/1.215\pi N_A)(\bar{v}/A_\eta)B_\eta^4. \qquad (25)$$

The quantity d_r^2/A_o depends empirically on d_r [83]

$$\log(d_r^2/A_o) = 0.173 + 2.158 \log d_r \tag{26}$$

provided d_r is smaller than 0.1.

4 Experimental Results for Polymer Chain Flexibility and Correlation with Structure

In this section, we will try to elucidate certain trends regarding the effect of macromolecular structure on polymer chain flexibility by examination of experimental data. There have been attempts to correlate chain flexibility parameters (q, C_∞, σ) with the molar volume of the substituent [84–86]. These empirical correlations have met with limited success but only when the correlation is restricted to a narrow subclass of macromolecules. The most successful attempts to predict conformational characteristics from structure are certainly those based on RIS models. Unfortunately, such models have appeared for only a limited number of macromolecules, and the development of reliable RIS models remains a job for the specialist. In addition, such models become quite involved for polymers with complex substituents or with complex backbone structures.

In light of the above, we will utilize only four families of polymers (polymethacrylates, polyolefins, polydienes, and stiff chain macromolecules) for trying to illustrate empirical trends. These polymers are chosen because they provide a broad spectrum of structures that have been carefully studied. No attempt is made to provide a comprehensive listing; the Polymer Handbook [87] serves this function.

4.1 Polymethacrylates

In Table 1 we present results for C_∞ of 50 different polymethacrylates having an exceptionally large range of substituents [88]. All of these C_∞ values are based on results from viscometry. Efforts were made to exclude data that appeared to be anomalous. In some instances data were available from several sources for the same polymer. In these cases, we have chosen the data for Table 1 by examining reported polydispersities, number of samples, the molecular weight range covered, methods used to determine molecular weights and molecular dimensions, and whether or not results were obtained directly in a theta solvent (preference being given to the data that was). In some instances, average C_∞ values are reported. A consistent value of the Flory hydrodynamic factor $(\Phi_o = 2.5 \times 10^{21} \text{ mol}^{-1})$ is employed throughout, and polydispersity corrections [60] are applied to K_Θ values prior to computing C_∞. Furthermore, nearly all

Table 1. Characteristic ratios of polymethacrylates

Substituent	Solvent	Temp. (°C)	Method[a]	C_∞	Ref(s).
methyl	4-heptanone	33.8	VT	7.3	89, 90
ethyl	2-propanol	36.9	VT	8.2	91
n-butyl	2-propanol	23.7	VT	8.6	92
n-hexyl	2-propanol	32.6	VT	11.1	93
n-decyl	ethyl acetate	11.0	VT	13.4	94
n-dodecyl	1-pentanol	29.5	VT	14.5	95, 96
n-tridecyl	ethyl acetate	27.0	VT	14.9	94
n-octadecyl	n-propyl acetate	36.0	VT	20.6	94
n-docosyl	amyl acetate	31.0	VM	24.1	94
isobutyl	various solvents	25.0	VM	10.2	97
tert-butyl	cyclohexane	10.0	VT	10.2	98
2-ethylbutyl	2-propanol	27.4	VT	9.8	99
cyclobutyl	1-butanol	37.5	VT	10.0	100
cyclopentyl	cyclohexane	36.0	VT	10.9	100
cyclohexyl	1-butanol	23.0	VT	11.6	101
cyclooctyl	2-butanol	45.0	VT	11.9	100
cyclododecyl	n-hexyl acetate	35.0	VT	14.2	100
2-decahydronaphthyl	dipropylketone	25.0	VM	14.8	102
isobornyl	1-octanol	36.9	VT	12.3	103
5-p-menthyl	methylpropylketone	25.0	VT	15.4	104
4-tert-butylcyclohexyl	1-butanol	25.0	VT	12.0	105, 106
1-adamantyl	cyclohexanone	30.0	VG	16.9	107
phenyl	acetone	25.0	VT	13.4	101
benzyl	various solvents	–	VT	ca. 10	108
diphenylmethyl	3-heptanone	45.0	VT	14.0	109
triphenylmethyl	hexamethylphosphoric triamide	25.0	VG	20.3	110
2-tert-butylphenyl	cyclohexane	18.4	VT	12.8	104
4-tert-butylphenyl	cyclohexane	25.0	VT	14.6	105
β-naphthyl	benzene	25.0	VM	16.9	102
4-biphenyl	benzene	25.0	VM	16.4	111
2-biphenyl	1,4-diozane	25.0	VM	17.8	112
4-1,1,3,3-tetramethylbutyl-phenyl	benzyl acetate	14.0	VM	18.7	113
2,6-dimethylphenyl	toluene	25.0	VT	18.3	114
2,6-diisopropylphenyl	toluene	25.0	VT	24.5	115
2-chloroethyl	o-dichlorobenzene	35.7	VT	10.5	116
2-chlorophenyl	various solvents	25.0	VM	11.8	117
4-chlorophenyl	carbon tetrachloride	25.0	VM	14.1	118
2,4,5-trichlorophenyl	benzene	24.0	VT	12.7	119
pentachlorophenyl	ethylbenzene	25.0	VT	26.4	120
tetrahydropyranylmethyl	isobutanol	30.5	VT	9.5	121
2-thiophenmethyl	chlorobenzene	25.0	VT	10.8	122
cyclohexyl thiolmethacrylate	cyclohexane	25.0	VM	9.6	123
phenyl thiolmethacrylate	methyl ethyl ketone	25.0	VT	11.2	124, 125
o-methylphenyl thiolmethacrylate	toluene	25.0	VG	11.1	125
triethyltin	tetrahydrofuran, toluene	25, 30	VG	9.4	126
tributyltin	cyclohexane, toluene	25.0	VG	14.3, 11.2	127
2-methoxyethyl	1-butanol	64.0	VT	11.1	128, 129
2-ethoxyethyl	1-butanol	38.0	VT	12.6	129
2-butoxyethyl	1-butanol	17.0	VT	15.5	129
2-(triphenylmethoxy)ethyl	mesitylene	47.0	VT	19.8	130

[a] VT, viscometry under theta or near theta conditions;
VM, viscometry under moderate solvent conditions (Mark–Houwink–Sakurada exponent larger than 0.5 but smaller than 0.6;
VG, viscometry in a thermodynamically good solvent

the polymethacrylates of Table 1 were made by free radical polymerization and, in the main, have similar tacticities; generally the probability of racemic triads ranges from 40–60%. We believe the C_∞ values of Table 1 to be accurate, generally, to $\pm 5\%$.

The effect of substituent size is seen on examining results for the *n*-alkyl series from methyl to docosyl (Fig. 3). A progressive increase in C_∞ from 7.3 to 24.1 is observed. Clearly, this increase in C_∞ reflects increased hindrance to rotation about main chain bonds as the substituent size increases. Likewise, C_∞ also increases with substituent size for polymethacrylates with alicyclic substituents ranging from cyclobutyl to cyclododecyl:

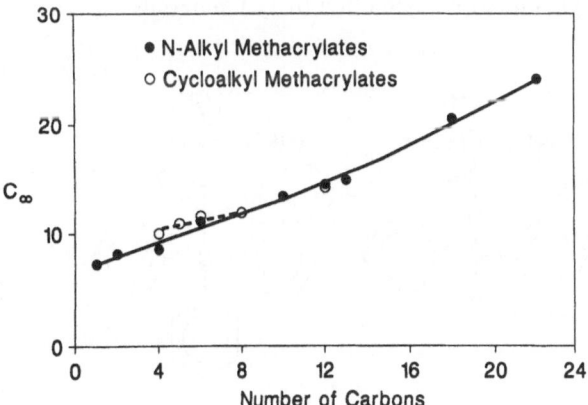

Results for the alicyclic series are also plotted in Fig. 3. Again, larger substituents impart larger C_∞ values. Other comparisons, which clearly demonstrate the influence of substituent size, include phenyl versus β-naphthyl,

Fig. 3. C_∞ as a function of the number of carbons in the alcohol used to make the ester for poly(*n*-alkyl methacrylates) and poly(cycloalkyl methacrylates). C_∞ increases with increasing side group size in each family. For small rings C_∞ is larger for cycloalkyl versus *n*-alkyl materials. Reprinted with permission from J. Macromol. Sci., Rev. Macromol. Chem. Phys., C28(3 & 4): 393 (1988). Copyright 1988 Marcel Dekker, Inc.

$$C_\infty = 13.4 \qquad C_\infty = 16.9$$

triethyltin versus tributyltin,

$$C_\infty = 9.4 \qquad C_\infty = 11.2 - 14.3$$

and benzyl, diphenylmethyl, and triphenylmethyl (trityl) materials.

$$C_\infty = 10.0 \qquad C_\infty = 14.0 \qquad C_\infty = 20.3$$

Flexibility of the substituent also plays a strong role in determining C_∞. This effect is illustrated by comparing n-alkyl and alicyclic materials.

$$\begin{array}{cc}
\text{CH}_3 & \text{CH}_3 \\
\text{~~~(CH}_2\text{—C)}_n\text{~~~} & \text{~~~(CH}_2\text{—C)}_m\text{~~~} \\
\text{C=O} & \text{C=O} \\
\text{O} & \text{O} \\
\text{(CH}_2\text{)}_3 & \square \\
\text{CH}_3 & \\
C_\infty = 8.6 & C_\infty = 10.0
\end{array}$$

$$\begin{array}{cc}
\text{CH}_3 & \text{CH}_3 \\
\text{~~~(CH}_2\text{—C)}_n\text{~~~} & \text{~~~(CH}_2\text{—C)}_m\text{~~~} \\
\text{C=O} & \text{C=O} \\
\text{O} & \text{O} \\
\text{(CH}_2\text{)}_5 & \bigcirc \\
\text{CH}_3 & \\
C_\infty = 11.1 & C_\infty = 11.6
\end{array}$$

The characteristic ratio of poly(cyclobutyl methacrylate) is greater than that of the corresponding n-alkyl material, despite the slightly larger size of the latter. This effect may be attributed primarily to the greater rigidity of the planar cyclobutyl group. However, the fact that the center of mass of the cyclobutyl substituent is closer to the main chain also contributes. This latter effect is discussed below. As the size of the groups increases to n-hexyl versus cyclohexyl this flexibility effect has diminished since cyclohexyl groups are quite flexible. A further increase in size to dodecyl groups results in virtually identical C_∞ values, again, because of increased flexibility of the large ring substituents (i.e., as the size of the side group increases the difference in flexibility between ring and linear groups is diminished).

$$\begin{array}{cc}
\text{CH}_3 & \text{CH}_3 \\
\text{~~~(CH}_2\text{—C)}_n\text{~~~} & \text{~~~(CH}_2\text{—C)}_m\text{~~~} \\
\text{C=O} & \text{C=O} \\
\text{O} & \text{O} \\
\text{(CH}_2\text{)}_{11} & \text{(cross-shaped ring)} \\
\text{CH}_3 & \\
C_\infty = 14.5 & C_\infty = 14.2
\end{array}$$

Other examples where side group flexibility is important include poly(cyclohexyl thiolmethacrylate) versus poly(cyclohexyl methacrylate) and poly(phenyl thiolmethacrylate) versus poly(phenyl methacrylate).

$C_\infty = 9.6$ $C_\infty = 11.6$

$C_\infty = 11.2$ $C_\infty = 13.4$

In thiolmethacrylates, where the oxygen linkage is replaced by sulfur, chain flexibility is substantially increased. Notice also that the magnitude of decrease in C_∞ for the two systems is roughly identical, as one might expect. These decreases in C_∞ mainly reflect the increased flexibility of the side group due to the decrease in the barrier to rotation about the C–S bond [124, 125]. Furthermore, the C–S bond is approximately 30% longer than the C–O bond and the bond angle is larger [131, 132]. This has the effect of moving the bulky substituents further away from the backbone. More over, sulfur compounds form much weaker hydrogen bonds than oxygen compounds. Results for the four polymers just mentioned also demonstrate that substitution of the rigid phenyl group for the flexible cyclohexyl group leads to an increase in C_∞.

Another example of how side group flexibility affects C_∞ is seen on comparing poly(tetrahydropyranylmethyl methacrylate) with poly(cyclohexylmethyl methacrylate) [133].

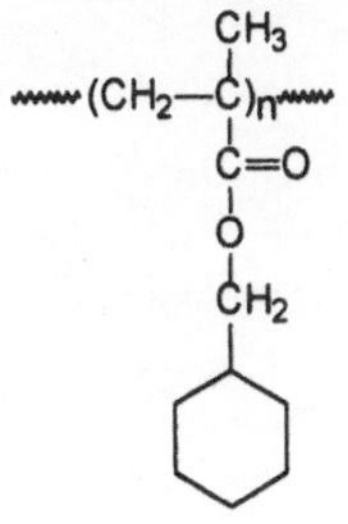

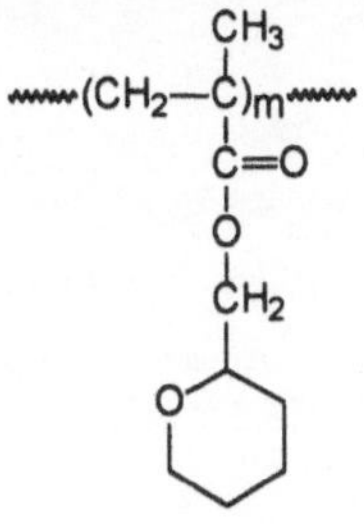

CH3
~~~(CH2—C)n~~~
         C=O
         O
         CH2

[cyclohexyl ring]

CH3
~~~(CH2—C)m~~~
         C=O
         O
         CH2

[tetrahydropyran ring with O]

$C_\infty = 11.9$ $C_\infty = 9.5$

Introduction of oxygen makes the ring more flexible because of the relative ease of rotation about C–O bonds [134]. Thus, it is clear that the rigidity of the substituent is an important factor controlling C_∞.

The influence of proximity of the center of mass of the side group to the backbone has been mentioned in prior examples and is especially important in the case of bulky substituents. This effect is clearly seen on comparing polymethacrylates with 4-biphenyl and 2-biphenyl substituents.

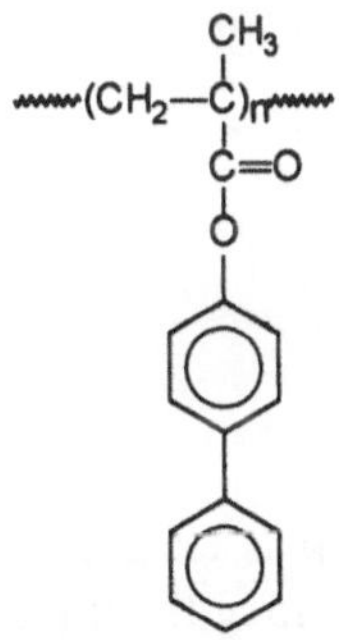

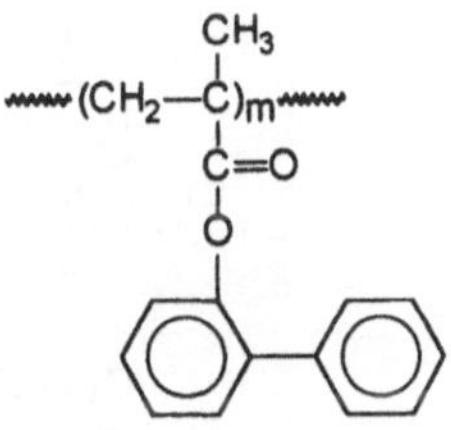

CH3
~~~(CH2—C)n~~~
         C=O
         O

[biphenyl - 4 position]

CH3
~~~(CH2—C)m~~~
         C=O
         O

[biphenyl - 2 position]

$C_\infty = 16.4$ $C_\infty = 17.8$

Here the substituents are identical, but the site of attachment has been varied. Attachment at the 2 position leads to larger hindrances to rotation about backbone bonds because the bulky and rigid substituent has been brought into closer proximity to the backbone. Likewise, a decrease in C_∞ is observed on going from a phenyl to a benzyl substituent.

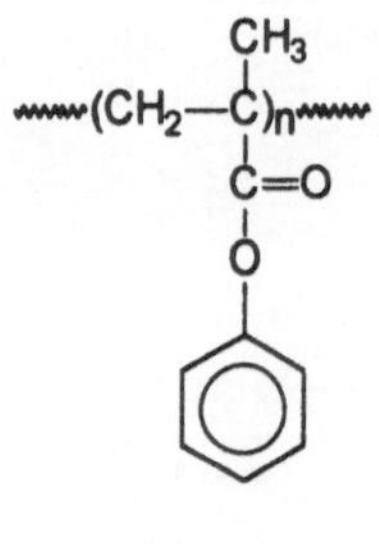

$C_\infty = 13.4$ 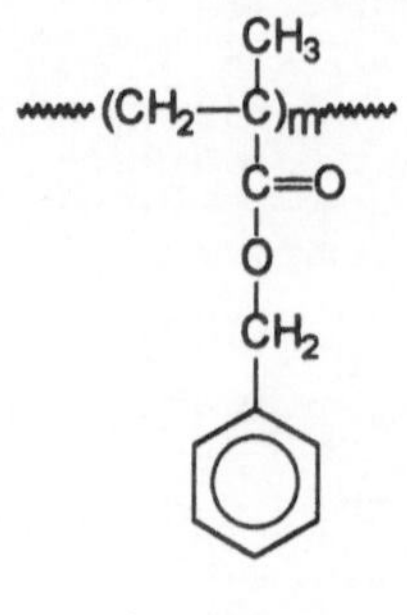 $C_\infty = 10$

Although the benzyl substituent is larger, steric hindrances to rotation are decreased substantially by moving the large and rigid aromatic ring further from the backbone. Also, the flexibility of the side group is enhanced by the presence of the methylene spacer.

The chemical nature of the substituent, which can lead to repulsions or attractions of a non-steric nature, can also affect C_∞. Consider the chlorine-substituted poly(phenyl methacrylates) shown below.

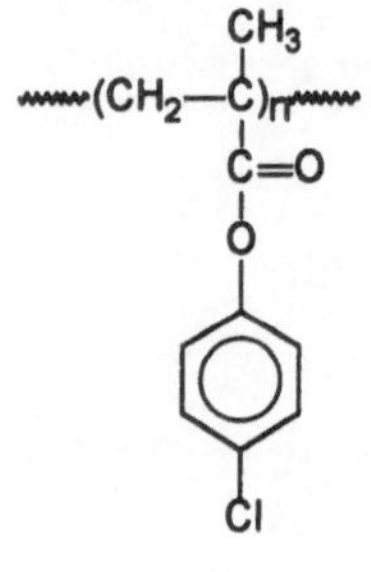

$C_\infty = 14.1$ 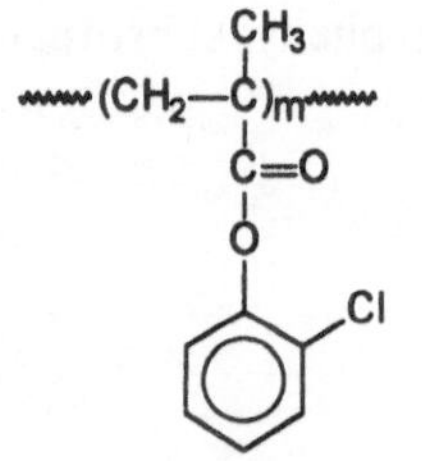 $C_\infty = 11.8$ 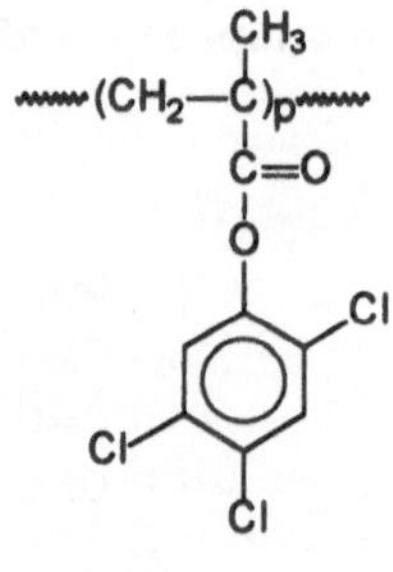 $C_\infty = 12.7$

Recalling the C_∞ value of 13.4 for poly(phenyl methacrylate), it is clear that replacing hydrogens with much larger chlorines leads – at least among the three materials shown – to at most a marginal increase in C_∞. In fact, two of the three chlorinated derivatives actually exhibit smaller C_∞ values than that observed for the unsubstituted material. A possible explanation for this effect is that substitution with chlorine disrupts specific interactions between neighboring aromatic groups [117, 119]. Similar trends are observed in chlorosubstituted polystyrenes including poly(2-chlorostyrene) ($C_\infty = 9.8$ [135]), poly(4-chlorostyrene) ($C_\infty = 10.1$–11.1 [61, 136–138]), poly(2,5-dichlorostyrene) ($C_\infty = 10.1$ [61]), and poly(3,4-dichlorostyrene) ($C_\infty = 10.5$ [139]). These values may be compared with the C_∞ value of about 10–11 for unsubstituted polystyrene [140–142]. As for polymethacrylates, the addition of a small number of chlorines to the ring seems to, at most, slightly reduce C_∞. Also, it should be noted that the lower C_∞ value observed for the 2-chloro derivative versus that for the 4-chloro material is paralleled in the styrene series where poly(2-chlorostyrene) exhibits the marginally lower C_∞ value. This is a surprising result if steric interactions are taken into account. It is relevant to note that for poly(pentachlorophenyl methacrylate) a very large C_∞ value of 26.4 has been reported [120].

In summary, we have identified four side group characteristics that influence C_∞. In general, C_∞ increases with an increase in size of the substituent, an increase in rigidity of the substituent, and with a decrease in distance between the center of mass of the substituent and the backbone. These effects are essentially steric in nature. Chain flexibility is also affected by specific interactions between substituents. It becomes quite difficult to empirically predict C_∞ when two or more effects oppose one another. As an example, the following series of polymers that all have hydrocarbon substituents with the same number of carbons is considered.

$$C_\infty = 15.4 \qquad C_\infty = 12.3 \qquad C_\infty = 16.9$$

Poly(5-p-menthyl methacrylate) exhibits an intermediate C_∞ value. It has the most flexible, yet most bulky substituent since no bridging is present. Poly(isobornyl methacrylate) has the lowest C_∞ value; its substituent, which is bridged, is more rigid but smaller than the menthyl group. Poly(1-adamantyl methacrylate) with a doubly bridged ring has the largest C_∞. Probably, the observed C_∞ variations could not be predicted.

4.2 Polydienes and Polyolefins

Over the past decade, Fetters and co-workers have synthesized and studied the chain flexibility of an extremely broad range of well-defined polydienes and polyolefins [36, 38, 39, 143–148]. The polyolefins were prepared by clean and quantitative hydrogenation [144–146] of polydienes of precisely controlled architecture. Thus, poly(1-butene) (a.k.a poly(ethyl ethylene)) of an atactic structure and a virtually monodisperse molecular weight distribution was synthesized by hydrogenation of 1,2-polybutadiene made via controlled ("living") anionic polymerization [144]. Likewise, "monodisperse" atactic polypropylene was prepared by hydrogenation of 1,4-poly(2-methyl-1,3-pentadiene) [146]. Because of the care taken both in synthesis and in characterization of these materials, they are ideal for studying structure/property relationships. These polymers (Table 2) range from structurally very simple materials, such as polypropylene and poly(1-butene), to complex *copolymers* or *ter*polymers. For some of the simpler structures, RIS models exist.

Polyethylene (PE) properties may be used as a basis for comparison, i.e., the polymers of Table 2 may be viewed as substituted PEs. The conformational properties of PE have been thoroughly studied by both theory and experiment. RIS models [1, 149] predict C_∞ to be 7.6 at 25 °C using a temperature coefficient for chain dimensions (dln $\langle R^2 \rangle_o$/dT = κ) of -1.1×10^{-3} K^{-1}. These values, based on a 3-state model (one *trans* and two *gauche* conformers), are in accord with those from experiments where C_∞ values at 25 °C range from 7.4 to 8.4 and κ ranges from -1.1 to -1.2×10^{-3} K^{-1} [37, 150–157]. The use of chain dimensions for PE at 25 °C facilitates the comparison with results from Table 2, where the reported chain dimensions are measured at ambient or near-ambient temperatures.

Flory, Mark, and Abe [158] used a 3-state RIS model to study the impact of placing small and/or flexible ("articulated") substituents on alternating carbons of PE. They noted that steric interactions between such substituents could be minimized if one or both of the backbone bonds adjacent to the methine carbon had a *gauche* conformation. This leads to adoption of a higher relative population of *gauche* conformers versus *trans* conformers than that observed for PE. Since *gauche* conformers are more compact than *trans* conformers, smaller C_∞ values are anticipated for chains with such substituents.

From Table 2, it can be seen that atactic polypropylene (PP) and atactic poly(1-butene) (PB) exhibit C_∞ values of 6.0 and 5.6, respectively. As expected from the above cited RIS results, these values are considerably smaller than the C_∞ value for PE of ca. 8. The decrease in C_∞ is slightly greater for ethyl versus methyl substituents. Additional comparisons of RIS predictions and experimental results for PP and PB will be presented below.

It is interesting to note that head-to-head polypropylene (HHPP) exhibits a virtually identical C_∞ value to that found for PP (5.8 vs. 6.0). To our knowledge, no RIS models exist for HHPP. However, Asakura and co-workers [159] have developed models that probe the influence of up to 50% head-to-head linkages on C_∞. Depending on the choice of model parameters, no change or a decrease in C_∞ with increased head-to-head content is predicted.

The chain dimensions of virtually alternating ethylene-propylene model polymers PEP have been studied experimentally [36, 38, 145, 147] and theoretically [160]. PEP is intermediate in structure between PE and PP with substituents on roughly every fourth backbone carbon. The C_∞ value of 6.8 for this polymer is also intermediate between those for PP and PE. It reflects the fact that, relative to PP, a lower level of *gauche* conformers is adequate to relieve steric effects caused by substituents. Similar to PEP, PEB represents *copolymers*

Table 2. Characteristic ratios of model polydienes and polyolefins at Ambient or near-ambient temperatures[a]

Abbreviation	Structure	C_∞	Ref(s).
1,4-PBD	$(CH_2-CH=CH-CH_2)_{93}-(CH_2-CH)_7$ $CH=CH_2$	5.1	143

1,4-PI	$(CH_2-CH=C-CH_2)_{93}-(CH_2-CH)_7$
	            CH_3 $C=CH_2$
	                                   CH_3
		5.1	143

1,4-PEBD	$(CH_2-CH=C-CH_2)_{90}-(CH_2-CH)_{10}$
	           CH_2 $C=CH_2$
	           CH_3 CH_2-CH_3
		5.4	148

1,4-PMYRC	$(CH_2-CH=C-CH_2)_{90}-(CH_2-CH)_{10}$
	           CH_2 $C=CH_2$
	           CH_2 CH_2
	         $CH=C(CH_3)_2$ $CH=C(CH_3)_2$
		6.3	148

1,4-PMPD	$(CH_2-C=CH-CH)_{100}$ CH_3 CH_3	5.5	146

1,4-PDMBD	CH_3
	 $(CH_2-C=C-CH_2)_{97}-(CH_2-C)_3$
	        $CH_3 CH_3$ $C=CH_2$
	                                  CH_3
		7.0	148

PP	$(CH_2-CH)_{100}$ CH_3	6.0	39, 146
HHPP	CH_3 $(CH_2-CH-CH-CH_2)_{97}-(CH_2-C)_3$ CH_3 CH_3 $CH_3-C=CH_2$	5.8	148
PB	$(CH_2-CH)_{99}-(CH_2-CH_2-CH_2-CH_2)_1$ CH_2-CH_3	5.6	39, 143, 148
PEP	$(CH_2-CH_2-CH-CH_2)_{93}-(CH_2-CH)_7$ CH_3 $CH(CH_3)_2$	6.8	36, 38, 147
PEB-73	$(CH_2-CH_2)_{73}-(CH_2-CH)_{27}$ CH_2-CH_3	6.5	145, 148

Table 2. Continued

Abbreviation	Structure	C_∞	Ref(s).
PEB-45	$(CH_2{-}CH_2{-}CH{-}CH_2)_{90}{-}(CH_2{-}C)_{10}$ with CH_3 on the first unit; the second unit bears CH_3, CH, $CH_2{-}CH_3$	6.0	148
PEB-20	$(CH_2{-}CH_2)_{20}{-}(CH_2{-}CH)_{80}$ with $CH_2{-}CH_3$	5.4	148
HPI-51	$(CH_2{-}CH_2{-}CH{-}CH_2)_{51}{-}(CH_2{-}CH)_{49}$ with CH_3 and $CH(CH_3)_2$	5.0	145
PBD-57	$(CH_2{-}CH{=}CH{-}CH_2)_{57}{-}(CH_2{-}CH)_{43}$ with CH, CH_2	6.0	145
PI 51	$(CH_2{-}CH{=}C{-}CH_2)_{51}{-}(CH_2{-}CH)_{49}$ with CH_3 and $C{=}CH_2$, CH_3	5.3	145
3,4-PI	$(CH_2{-}CH)_{74}{-}(CH_2{-}C)_{23}{-}(CH_2{-}CH{=}C{-}CH_2)_3$ with $C{=}CH_2$, CH_3; CH_3, CH, CH_2; CH_3	7.9	145
PMYRC-64	$(CH_2{-}CH{=}C{-}CH_2)_{64}{-}(CH_2{-}CH)_{36}$ with CH_2, CH_2, $CH{=}C(CH_3)_2$; CH_2, $C{=}CH_2$, CH_2, $CH{=}C(CH_3)_2$	7.6	148
PDMBD-55	$(CH_2{-}C{=}C{-}CH_2)_{55}{-}(CH_2{-}C)_{45}$ with CH_3 CH_3; CH_3, $CH_3{-}C{=}CH_2$	8.4	148
1,2-PBD	$(CH_2{-}CH)_{99}{-}(CH_2{-}CH{=}CH{-}CH_2)_1$ with $CH{=}CH_2$	7.0	144
H3,4-PI	$(CH_2{-}CH)_{74}{-}(CH_2{-}C)_{23}{-}(CH_2{-}CH_2{-}CH{-}CH_2)_3$ with $CH(CH_3)_2$; CH_3, $CHCH_3$; CH_3	7.2	145

Table 2. Continued

Abbreviation	Structure	C_∞	Ref(s).
H1,4-PMYRC	$(CH_2-CH_2-CH-CH_2)_{90}-(CH_2-CH)_{10}$ branch 1: $CH_2,\ CH_2,\ CH_2,\ CH(CH_3)_2$ branch 2: $CH-CH_3,\ CH_2,\ CH_2,\ CH_2,\ CH(CH_3)_2$	6.7	148
HPMYRC-64	$(CH_2-CH_2-CH-CH_2)_{64}-(CH_2-CH)_{36}$ branch 1: $CH_2,\ CH_2,\ CH_2,\ CH(CH_3)_2$ branch 2: $CH-CH_3,\ CH_2,\ CH_2,\ CH_2,\ CH(CH_3)_2$	8.3	148
HPDMBD-55	$(CH_2-CH-CH-CH_2)_{55}-(CH_2-C)_{45}$ branch 1: $CH_3,\ CH_3$ branch 2: $CH_3,\ CHCH_3,\ CH_3$	5.6	148

[a] All C_∞ values are based on intrinsic viscosity measurements under theta or near-theta conditions.

of ethylene and 1-butene made by hydrogenation of polydienes. Despite the larger size of ethyl versus methyl substituents, the C_∞ results for these materials are similar to those noted above for PEP. More specifically, C_∞ values of 5.4, 6.0, and 6.5 are found for copolymers with 20, 45, and 73% ethylene content, respectively. Thus, the C_∞ values are intermediate to those measured for the homopolymers, and a progressive decrease in C_∞ is observed as the 1-butene content increases. This trend has been predicted by Mattice [161] via a 3-state RIS model. Figures 4 and 5 show C_∞ as a function of ethylene content for PEP and PEB. With PEP, good agreement exists between theory and experiment up to ca. 80% propylene content. The theory of Mark [160] overestimates C_∞ for atactic PP. Similarly, for PEB agreement between theoretical and experimental data is good up to ca. 70% 1-butene content. Theoretical values of C_∞ [161] for materials containing more than 70% 1-butene are underestimated.

Results for structurally more complex hydrogenated mixed-microstructure polydienes are also listed in Table 2. HPI-51, with a substituent (methyl or isopropyl) on every third carbon, on average, exhibits the lowest C_∞ of any polymer in Table 2. The fact that the C_∞ value is lower than those of PP or PB

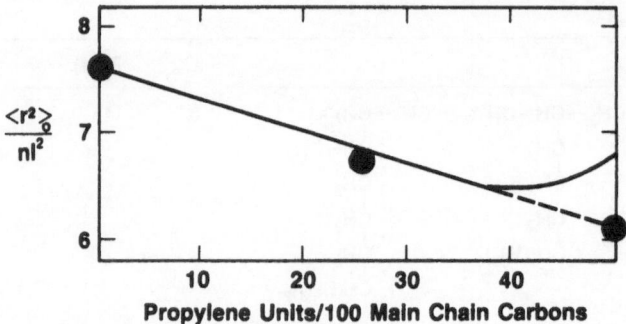

Fig. 4. Influence of propylene content on the characteristic ratio of *copolymers* of ethylene and propylene. The *solid curve* represents the RIS calculations of Mark [160]. Reprinted with permission from Macromolecules 22: 925 (1989). Copyright 1989 American Chemical Society

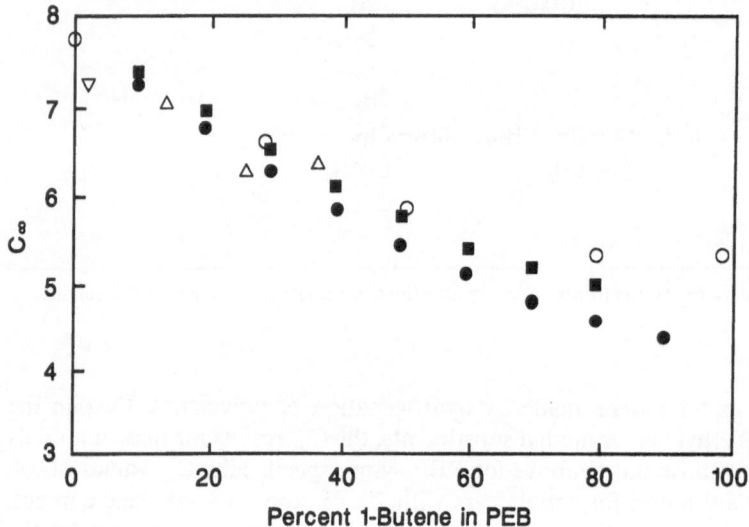

Fig. 5. Characteristic ratio of ethyl branched polyethylenes as a function of 1-butene content. Solid symbols are RIS values for statistic weights $\tau = 0$ (•) and 0.43 (■). Open symbols represent various experimental values from sources listed in [161]. Reprinted with permission from Macromolecules, 24: 6205 (1991). Copyright 1991 American Chemical Society

may well be attributable to the bulkiness of the isopropyl substituents. Conversely, for H3,4-PI, which generally has isopropyl groups on alternating carbons, the C_∞ value of 7.2 is only slightly smaller than that observed for PE. It should be noted, however, that H3,4-PI contains ca. 23% of disubstituted 2-methyl-1-butene units.

H1,4-PMYRC yields a C_∞ of 6.7. This value is quite similar to that obtained for the other materials with one flexible substituent per four backbone carbons

(PEP and PEB-45) despite the much larger size of the 4-methylpentyl substituent in the case of H1,4-PMYRC. It appears that for such materials the highly flexible nature of the substituent is probably more important than its overall size in terms of the impact on C_∞. Conversely, when the frequency of occurrence of the bulky substituent along the backbone is increased as in HPMYRC-64, a larger C_∞ of 8.3 is observed. The potential contribution of the interdependence of the conformations of neighboring substituents for long side-chain poly(1-alkenes) has been noted [162] and perhaps plays a role here. Finally, for the hydrogenated mixed microstructure poly(2,3-dimethyl-1,3-butadiene) (HPDMBD-55) a C_∞ value of 5.6 is found. Compared to HHPP ($C_\infty = 5.8$), the C_∞ value remains virtually unchanged despite incorporation of 29% poly(2,3-dimethyl-1-butene) units.

The study of model polydienes allows insight into the effect of unsaturation in the backbone or substituent on C_∞. Mark [163, 164] has published RIS models dealing with stereoregular (cis-1,4 or trans-1,4) polybutadiene (PBD) and polyisoprene (PI). Later, Abe and Flory [165, 166] developed RIS models applicable to stereoirregular PBD and PI having both cis and trans units present. Homogeneous anionic polymerization in hydrocarbon solvents leads to linear polydienes having extremely narrow molecular weight distributions and high 1,4 contents with both cis and trans isomers present [167]. In addition, a small amount ($\leq 10\%$) of 1,2 or 3,4 structures is present in these materials.

C_∞ values of 5.1 are found for both 1,4-PBD and 1,4-PI, which were made by anionic polymerization (Table 2). This value is much lower than that observed for PE. It might seem that polymers with double bonds in the backbone should be stiff, since they can have no rotation about such bonds. On the contrary, the double bond can lock in very compact (but sterically undesirable) cis conformers, thus diminishing unperturbed dimensions relative to that for PE. Also, there is enhanced ease of rotation about single bonds adjacent to double bonds [163, 164]. The result for PBD is in excellent accord with that predicted by Abe and Flory [165]; the result for PI is also in fair agreement with theory [166]. The experimental results clearly suggest that addition of a methyl substituent has no impact on C_∞. 1,4-Poly(ethyl butadiene) (1,4-PEBD) has a C_∞ of 5.4. Therefore, substitution of ethyl groups for methyl groups has only a marginal impact on C_∞. Conversely, 1,4-polymyrcene (1,4-PMYRC), a material with much larger substituents, exhibits a larger C_∞ value of 6.3. These findings are consistent with Mark's conclusion [164] that addition of small substituents (such as methyl) to the polybutadiene backbone will have a small impact on C_∞, as long as the backbone is not a high trans structure. Clearly, the distinction as to what constitutes large versus small substituents falls somewhere between ethyl and 4-methylpentenyl groups.

The C_∞ value of 5.5 for 1,4-poly(2-methyl-1,3-pentadiene) (1,4-PMPD) suggests that the addition of an extra methyl substituent at the 4 position of 1,4-polyisoprene has little influence on C_∞ of a mixed cis/trans material. Likewise, the C_∞ value of 7.0 for high trans-1,4-poly(2,3-dimethyl butadiene) is virtually identical to the value of 7.4 reported by Wagner and Flory [168] for *trans*-1,4-polyisoprene. Hence, at least in this instance, addition of an extra methyl group at the site of the double bond has little impact on C_∞.

At the opposite end of the polydiene spectrum are high 1,2-polybutadiene (1,2-PBD) and high 3,4-polyisoprene (3,4-PI). These materials have virtually all their double bonds within the substituents and, thus, may be viewed as substituted polyethylenes. Polar interactions may occur between side groups with carbon–carbon double bonds. The presence of double bonds in these substituents also diminishes their flexibility, i.e., these are not "articulated" side groups in the terminology of Flory, Mark, and Abe [158]. Therefore, steric interactions between these substituents are not expected to be eliminated merely by adopting higher relative populations of *gauche* conformations. Thus, in a simplistic sense, one would not expect to see diminished values of C_∞, relative to that of PE, for these materials. 1,2-PBD has a C_∞ of 7.0, which is similar to C_∞ for PE and much larger than the value of 5.6 found for its saturated derivative PB. 3,4-PI, with predominantly the bulkier isopropenyl substituent, has an even larger C_∞ of 7.9. These results are qualitatively consistent with theoretical expectations, and further comparison with theory requires further development of RIS models.

Results for the structurally more complex, mixed microstructure polydienes, which have substantial in-chain and side-chain unsaturation, are more difficult to interpret. Clearly, mixed microstructure polybutadiene (PBD-57) exhibits a C_∞ value (6.0) intermediate to that found for high 1,4 ($C_\infty = 5.1$) and high 1,2 ($C_\infty = 7.0$) materials. Conversely, a mixed microstructure polyisoprene (PI-51) of similar microstructure to PBD-57 yields a C_∞ of 5.3, a result virtually identical to that found for the high 1,4 material. Both mixed microstructure polymyrcene (PMYRC-64) and poly(2,3-dimethyl butadiene) (PDMB-55) yield C_∞ values (7.6 and 8.4, respectively) that are considerably larger than those for their high 1,4 counterparts. Interpretation of these results at the present level of theoretical development would be speculative.

4.3 Stiff Chain Polymers

As seen previously, the main feature determining the degree of flexibility of typical flexible polymers e.g., vinyl, acrylic, is the nature of interactions among their substituents. Even in extreme cases of very bulky substituents the chains remain quite flexible. Poly(triphenylmethyl methacrylate), poly(pentachlorophenyl methacrylate), poly(2,6-diisopropylphenyl methacrylate), and poly(*n*-docosyl methacrylate) may be considered as examples (Table 1). These chains have C_∞ values of approximately 20–26, which are the highest values among all polymethacrylates, polydienes, and polyolefins. Yet, calculated by using Eq. 8, they yield q values of only ca. 2 nm and, therefore, clearly are not stiff chain polymers.

Greater degrees of chain stiffness can be induced in several ways. Cellulose and cellulose derivatives are examples of semi-flexible chains where partial rigidity is caused by incorporation of β(1-4) linked glucopyranose rings into the backbone. In polypeptide and polynucleotide chains the molecules undergo "cyclization" through intramolecular hydrogen bonding, leading to highly ex-

Table 3. Persistence lengths of some stiff chain polymers

Polymer	Solvent	q (nm)	Method[c]	Ref(s).
cellulose	dimethylacetamide + 5% LiCl	11	LS, IV	171
hydroxypropylcellulose	dimethylacetamide	7	IV	172
hydroxypropylcellulose	dichloroacetic acid	10	IV	173
cellulose diacetate	dimethylacetamide	7	IV, LS	174
cyanoethylhydroxypropylcellulose	tetrahydrofuran	14	IV	175
poly(n-butyl isocyanate)	chloroform	30	IV	176
poly(n-hexyl isocyanate)	hexane	42	LS	177
poly(n-hexyl isocyanate)	dichloromethane	19	IV	178
poly(n-hexyl isocyanate)	toluene	38	IV	178
poly(1,4-benzamide)	dimethylacetamide + 3% LiCl	75	LS	179
PA-2[a]	dimethylacetamide + 1% LiCl	12	LS	180
PA-4[b]	dimethylacetamide + 4% LiCl	8.5	LS	180
poly(p-phenylenediamine-terephthalic acid)	sulfuric acid	29	LS	181
DNA	0.2 M NaCl	60	LS, IV, S	182, 183
xanthan	0.1 M NaCl	125	LS, IV, S	184, 185
collagen	aqueous systems	130–150	IV, S	186, 187
schizophyllan	water or 0.1 N NaOH	150–200	LS, IV, S	188, 189

(a)

PA-2

(b)

PA-4

[c] LS, light scattering;
IV, intrinsic viscosity;
S, sedimentation.

tended and stable helical conformations. Some poly(alkyl isocyanates) also exist in highly extended conformations due to partial double bond character in the main chain and steric interactions between carbonyl groups and n-alkyl substituents [169]. *Para*-linked aromatic polyamides and polyesters are also highly extended structures; because of partial double bond character rotation can only occur about bonds that do not allow the chain to bend back on itself [170]. The presence of the rigid aromatic rings contributes here as well.

Persistence lengths for a number of these stiff chain polymers are tabulated in Table 3. Cellulosics exhibit q values that are approximately one order of magnitude larger than those of the least flexible random coil materials of Table 1. Substitution of the cellulose appears to have an impact on flexibility, as does the nature of the solvent. Even less flexible are poly(n-hexyl isocyanate) and poly(n-butyl isocyanate); these polymers have q values of 19–42 nm. The aromatic polyamides listed in Table 3 cover a broad range of persistence lengths depending on the polymer structure. PA-2 and PA-4 polymers demonstrate that substitution of only one phenyl substituent per every four rings on either the diamine or terephthalic acid moiety substantially reduces the stiffness of the chain as compared to "pure" poly(1,4-benzamide) or poly(p-phenylenediamine terephthalic acid). This latter material is the polymer from which Kevlar fibers are produced. The reduction in chain flexibility is caused by the phenyl substituent, particularly if substituted on the terephthalic acid residue, forcing the adjacent carbonyl group out of the plane of the ring [180].

The q values for the biopolymers DNA, xanthan, collagen, and schizophyllan range between 60 and 200 nm. They are comprised of highly extended structures, with double-stranded DNA and xanthan exhibiting smaller q values than triple-stranded collagen and schizophyllan.

One very interesting aspect of stiff chain polymers is their tendency to form thermotropic or lyotropic liquid crystals [190]. This is not surprising, as Flory [191] has noted that "molecular asymmetry is a feature common to all liquid crystalline substances." Thus, random coils do not form liquid crystals, although less flexible polymers may. Regarding lyotropic liquid crystal formation, Flory [192] has demonstrated a direct dependence of the critical volume fraction V_2^* of polymer, which is necessary for mesophase formation, and the aspect ratio X (length over diameter ratio) for a true polymer rod

$$V_2^* \simeq \frac{8}{X}\left(1 - \frac{2}{X}\right). \tag{27}$$

Equation 27 is also known to reliably predict V_2^* for polymers that are not true rods if, as Flory [193] suggested, the aspect ratio of the Kuhn segment is utilized [175, 194, 195]. Furthermore, Eq. 27 provides some insight into the minimum aspect ratio for thermotropic liquid crystal formation. Most of the polymers of Table 3 are known to form liquid crystalline mesophases under appropriate conditions.

5 Influence of Temperature, Solvents, and Tacticity on Unperturbed Chain Dimensions

5.1 Temperature and Solvent Effects

It is well-known that temperature affects the conformational equilibria of macromolecules. This temperature dependence is most commonly expressed in terms of $\kappa = d\ln\langle R^2\rangle_0/dT$. A positive temperature coefficient indicates that more compact chain postures have lower energies; a negative temperature coefficient reveals that more compact conformations have higher energies. Both positive and negative temperature coefficients have been found for various polymers [1].

Several procedures are available for measuring κ. These methods include dilute solution techniques such as light scattering and viscometry, thermoelasticity measurements, and neutron scattering from isotopically labelled polymers. Solution methods have been broadly utilized for many years for measuring κ, with most of these studies involving viscometry. The reason for choosing viscometry is that values of κ are very small (usually of the order 10^{-4}–10^{-3} K^{-1}) and errors in measuring $\langle R_g^2\rangle_0$ by light scattering are usually too large to allow for reliable extraction of κ. Viscometric measurements involve much smaller experimental errors, but they yield chain dimensions indirectly [196]. Furthermore, the viscometric approach assumes that the hydrodynamic parameter Φ_0 does not vary with temperature. While, as discussed above, the various experimental determinations of Φ_0 are remarkably constant [30, 45, 58], its temperature dependence has, to our knowledge, not been directly examined. Particularly suitable for such investigations are polymer/solvent systems exhibiting both upper and lower critical solution temperatures, because through use of such systems the need to change solvents, and possibly encounter specific solvent effects, is avoided.

As mentioned in the Introduction and discussed below, a hazard inherent in any θ condition solution measurement of C_∞ (or q) or κ is specific solvent effects on the polymer conformation [42–44, 46, 47]. These effects are especially troublesome in attempting to evaluate κ since the influences of solvent and temperature can be comparable in magnitude, and thus can mask one another. This is illustrated by examining the intrinsic viscosity results of Table 4. At a given temperature, the C_∞ values are largest in the "cyclohexane series" (cyclohexane, methylcyclohexane, ethylcyclohexane). Within a given series of theta solvents (cyclohexanes, diesters, chloroalkanes), a progressive decrease in C_∞ is observed as temperature increases. The relative errors in K_Θ were minimized by conducting measurements on the same six polystyrene samples in each of the different solvents and then applying the BSF approach to determine K_Θ. Thus, errors in $\bar{M}_w$ cancel out and slight deviations from theta conditions are corrected for [43]. Furthermore, within experimental error consistent

Table 4. Unperturbed parameters for polystyrene from [43]

Solvent	$K_\Theta \times 10^2$ (mL/g)	C_∞	Temp. (°C)
1-chloro-*n*-decane	8.015	10.4	8.5
cycloheptane	8.099	10.5	19.0
cyclooctane	7.164	10.5	20.5
cyclopentane	8.402	10.7	20.5
1-chloro-*n*-undecane	7.748	10.2	32.8
cyclohexane	8.371	10.7	34.5
diethyl malonate	7.535	10.0	34.5
diethyl oxalate	7.353	9.8	58.2
1-chloro-*n*-dodecane	7.402	9.9	58.6
dimethyl succinate	7.301	9.8	67.6
methylcyclohexane	7.821	10.2	68.0
ethylcyclohexane	7.806	10.2	75.0

κ value of about -1×10^{-3} K^{-1} is derived from each series. This value of κ and the values of C_∞ are consistent with predictions of the RIS theory [197, 198]. The data of Table 4 suggests that cyclic theta solvents promote larger chain dimensions for polystyrene. Such a trend was also noted by Bohdanecky and Berek [44]. Similar results have been obtained for poly(*tert*-butylstyrene) in various cyclic and acyclic solvents [199]. Theory [46–48] has also considered the nature of specific solvent interactions.

As an additional example of how chain dimensions can be influenced by both the solvent and the method employed in their extraction, consider Table 5, where results on unperturbed dimensions of highly syndiotactic ("conventional") PMMA are collected. SANS results for C_∞ obtained in the melt are consistently higher than those values obtained from light scattering in solution (ca. 9.5 vs. 8.4). A still lower value of ca. 7.3–7.4 is derived from intrinsic viscosity measurements in four different theta solvents assuming $\Phi_o = 2.5 \times 10^{21}$ mol^{-1}.

Several interesting questions arise upon studying these data. Although all the samples used are consistent in tacticity, large differences in theta temperatures are noted in both butyl chloride and acetonitrile. For the former solvent, Quadrat and co-workers [200] reported a value of 35 °C, whereas Fujii and co-workers [30] employed a theta temperature of 40.8 °C. For acetonitrile an even large difference is noted: 28 °C was reported by Quadrat et al. [200], while 44.0 °C was measured by Fujii et al. [30]. These differences seem to lie well outside the experimental error that is likely in determining the theta temperatures and are probably too large to be attributed to variations in solvent purities. In both solvents the larger C_∞ values were obtained via light scattering measurements, which were conducted at the higher temperatures. The measurement of the theta temperatures may play a role in these differences.

Fujii and co-workers [30] also reported [η] values for PMMA in butyl chloride and acetonitrile at their measured theta temperatures. The Φ_o value computed for butyl chloride was consistent with Φ_o values obtained for other polymer/solvent pairs, but the Φ_o value computed for acetonitrile was slightly

Table 5. Solvent influence on the characteristic ratios of poly(methyl methacrylate)[a]

Solvent	Method	Temp. (°C)	$\langle R^2 \rangle_0/M$ Å^2 mol g^{-1})	C_∞	Ref(s).
various solvents	[η]$_\theta$	25–41	0.348[b]	7.43	200
4-heptanone	[η]$_\theta$	33.8	0.342[b]	7.30	90
butyl chloride	LS-θ	40.8	0.393	8.40	30
acetonitrile	LS-θ	44.0	0.393	8.40	30
melt	SANS	d	0.437–0.459	9.3–9.8	201–203

[a] ca. 75–80% racemic diads.
[b] Calculated using $\Phi = 2.5 \times 10^{21}$ mol^{-1}.
[c] m-Xylene (theta temp. 25 °C); butyl chloride (theta temp. 35 °C); acetonitrile (theta temp. 28 °C).
[d] Samples prepared in the melt with measurements at room temperature.

smaller than those observed for other systems. Thus, at least some of the C_∞ values calculated using $\Phi_0 = 2.5 \times 10^{21}$ mol^{-1} in Table 5 may be in error. No matter what the explanation is for the discrepancies in solution data, clearly SANS on bulk specimens gives larger C_∞ values. While we have no explanation for this trend, we will note that deuterium labelling is employed in the SANS work, and perhaps this labelling is at least partially responsible for these differences.

As an alternative to use of a theta solvent series for viscometrically evaluating κ, the athermal solvent approach, pioneered by Flory, Ciferri, and Chiang [151], is also employed sometimes. For solvents, which mix athermally with the polymer (frequently this is true for oligomeric structures with identical repeating units to that of the polymer being investigated), the dependence of [η] on temperature is determined for one or more samples. The temperature dependence of $\langle R^2 \rangle_0$ is then evaluated using the equation

$$[\eta] = \Phi_0 (\langle R^2 \rangle/M)^{3/2} M^{1/2} \alpha^3 \qquad (28)$$

[204] where α, the expansion factor, is given by

$$\alpha^5 - \alpha^3 = 27(2\pi)^{-3/2}(\bar{v}^2/N_A V_1) \cdot (\langle R^2 \rangle_0/M)^{-3/2} M^{1/2}(1/2 - \chi_1) \qquad (29)$$

and $\bar{v}$ is the partial specific volume of the polymer, V_1 is the molar volume of the solvent, N_A is the Avogadro constant, and χ_1 reflects polymer/solvent thermodynamic interactions. Flory and co-workers [151] derived

$$d\ln\langle R^2 \rangle_0/dT = (5/3 - \alpha^{-2})d\ln[\eta]/dT - (1 - \alpha^{-2})$$

$$\times [2\beta_2 - \beta_1 - (1/2 - \chi_1)^{-1}(d\chi_1/dT)] \qquad (30)$$

by solving Eq. 28 for the dependence of $\langle R^2 \rangle_0$ on temperature and using Eq. 29 to correct for the temperature dependence of α. By use of an athermal solvent $d\chi_1/dT$ becomes zero. This approach has been rather widely utilized for evaluating κ [147, 205–208]. However, it should be noted that validity of Eqs. 29 and 30 is questionable [209]. Thus, values of κ derived using the athermal approach should be viewed with caution. In our investigations anomalous results have sometimes been obtained using this approach [210].

As an alternative to dilute solution techniques, thermoelastic measurements on strained, amorphous polymer networks can also yield κ. Mark [211, 212] has reviewed this technique, which is based on the equation

$$\kappa = - [dln(f/T)/dT]_{L,v} \tag{31}$$

where f is the force required to maintain the sample at constant length L as temperature is varied. As Mark [211] has noted, values of κ obtained by this method appear to be insensitive to orientation, the presence of solvents during crosslinking or deformation, and the type of deformation employed. Therefore, specific solvent effects can be avoided.

McCrum [213, 214] recently suggested that the above approach is subject to large errors and based on an irrational premiss. He proposed a new method of "thermoviscoelasticity". Smith and Mark [215] have demonstrated McCrum's analysis to be flawed and have shown that the classical thermoelasticity approach is soundly based on theory. Indeed, there is excellent agreement between thermoelastic and viscometric results for poly(1-pentene) [206, 211], polyethylene [151, 154, 155, 211, 216], poly(dimethyl siloxane) [205, 211, 217], poly(ethylene oxide) [211, 218], poly(isobutylene) [207, 211, 216, 219] and poly(n-butyl methacrylate) [220, 221] (Table 6).

Agreement is not as good in the case of polystyrene (Table 7). Here, thermoelastic measurements yield small positive values of κ while viscometric results generally support negative values, especially when care is taken to minimize specific solvent influences, Interestingly, two of the earliest papers describing the measurement of polymer chain dimensions in the bulk via neutron scattering [33, 34] report temperature coefficients that are virtually zero for polystyrene. In an attempt to resolve this dilemma, additional investigations of temperature effects on polystyrene chain dimensions by neutron scattering are presently underway [233]. At this time, it seems safe to assume that the "true" value of κ for polystyrene is zero within experimental error, since the

Table 6. κ Values obtained by thermoelasticity and [η] measurements for some polymers

Polymer	$\kappa \times 10^3 (K^{-1})$		
	Thermoelasticity	Viscometry	Ref(s).
poly(1-pentene), isotactic	0.34	0.52	206, 211
polyethylene	− 1.05	− 0.8 to − 1.2	151, 154, 155, 211, 216
poly(dimethyl siloxane)	0.59	0.52	205, 211, 217
poly(ethylene oxide)	0.23	0.2	211, 218
poly(isobutylene)	− 0.19	− 0.1 to − 0.04	207, 211, 216, 219
poly(n-butyl methacrylate)	2.5	2.3	220, 221

Table 7. Experimental results for polystyrene temperature coefficient

Workers	$\dfrac{d\ln\langle R^2\rangle}{dT}$ (K^{-1})	Method[a]	Year	Ref(s).
Fox & Flory	-1.8×10^{-3}	$[\eta]_\theta - T$	1951	222
Bazuaye & Huglin	-0.3×10^{-3}	$[\eta]_\theta - T$	1979	223
Kotera, et al.	-1.0×10^{-3}	$[\eta]_\theta - T$	1971	224
Abdel-Azim & Huglin	-1.15×10^{-3}	$[\eta]_\theta - T$	1983	225
Kuwahara, et al.	-0.1×10^{-3}	$[\eta]_\theta - T$	1974	226
Mays, et al.	-1.1×10^{-3}	$[\eta]_\theta - T$	1985	43
Cotton, et al.	Negative	N.S.	1974	33
Abe & Fujita	Negative	$[\eta]_\theta - T$	1965	227
Schulz & Baumann	0	$[\eta]_\theta - T$	1963	228
Wignall, et al.	0 or slightly negative	N.S.	1974	34
Dusek	$0.4-0.7 \times 10^{-3}$	$F - T$	1967	229, 230
Orofino & Mickey	0.44×10^{-3}	$[\eta]_\theta - T$	1963	231
Orofino & Ciferri	0.37×10^{-3}	$F - T$	1964	232

[a] $[\eta]_\theta - T$ viscometry; N.S. neutron scattering; $F - T$ stress-temp. coefficient measurements.

positive values reported in Table 7 are very small and subject to finite errors. The reason for the large negative values obtained in most viscometric studies may be related somehow to local solvent effects on the conformational characteristics of the polymer. As discussed by Bahar et al. [47], asymmetric chains with bulky substituents are particularly susceptible to specific solvent effects. This is because the nature of side group interactions with the solvent differs depending on the local conformational characteristics of the chain, i.e., usually substituents attached to *gauche* conformers are more accessible for interactions with the solvent than are those attached to *trans* conformers. Hence, solvent molecules can influence the relative populations of conformers, affecting both C_∞ and κ (assuming solvent/substituent interactions are temperature dependent).

In Table 8 recent results obtained by neutron scattering in the melt for κ of various model polydienes and polyolefins are displayed. Also listed are κ values determined in theta solvents by viscometry and those predicted by RIS models. As mentioned earlier, the results for PE from thermoelastic experiments agree with results in Table 8, as do those for PEP and atactic PB [211]. Clearly, with the possible exception of PP, the neutron scattering results yield κ values in accord with RIS predictions. The early 3-state model of Flory, Mark, and Abe [158] predicts a κ of approximately zero, in agreement with neutron scattering from the atactic PP melt; other RIS models [160, 162, 236–240] predict larger temperature coefficients for this polymer. Furthermore, in most cases there also is agreement with results of theta solvent measurements. PP and PB are two instances where strong discrepancies clearly exist. These findings are perhaps reflective of the fact that polymers in a theta solvent are only in a quasi-ideal state [48–50], i.e., ideal chain statistics are not obtained at theta or any other temperature. Although A_2 is equal to zero at theta, ternary (3-body) interactions may be quite strong and indeed could be markedly different in different theta

Table 8. Temperature coefficients of model polydienes and polyolefins

Polymer[a]	$\kappa \times 10^3 (K^{-1})$			Ref(s).
	Melt	Theta	RIS Model	
1,4-PBD	0	-0.1[b]	ca. 0	165, 208, 234
1,4-PI	0.6	–	0.2	166, 234
1,2-PBD	-0.7[c]	-1.7	–	148, 235
PE	-1.2	-1.2	-1.1	37, 149, 151
PP	-0.1	-2.7	0 to -1.8	39, 147, 158, 160, 162, 236–240
PB	0.4	-2.3	ca. 0	39, 148, 158, 161
PEP	-1.2	-1.0	-1.1	38, 147, 160

[a] Structure are the same as in Table 2.
[b] Athermal solvent approach using intrinsic viscosities.
[c] Via rheological measurements; see [235]

solvents. In other works, although A_2 vanishes at theta, contributions from higher order interaction terms still influence the conformational properties of polymer chains. These so-called ternary interactions may possibly be strongly system dependent.

Based on the large body of experimental data studied in this review, it appears that theta solvents can usually provide a reasonable estimate of unperturbed dimensions but that neutron scattering is the preferred technique for extracting κ. Clearly, more work remains to be done on both experimental and theoretical fronts, if we are to be able to understand and predict solvent effects on polymer conformational characteristics.

5.2 Tacticity Effects

For vinyl, acrylic, and methacrylic polymers, RIS theory [1] predicts C_∞ and κ to depend on polymer tacticity. The unperturbed dimensions of stereoregular polymers have been reviewed by Jenkins and Porter [241]. These authors noted that while a large number of studies of unperturbed dimensions have been conducted for such materials, only in a few cases were tacticity or sequence distribution known with certainty. Poly(methyl methacrylate) is an exception where some careful results on chains of well-defined tacticity exist (Table 9).

As discussed in the previous section, the various methods used to measure C_∞ and κ can yield somewhat different results. Consequently, results from viscometry, SANS, and light scattering experiments are included in Table 9. For virtually 100% isotactic PMMA, viscometry and SANS give C_∞ values in reasonable agreement with one another, i.e., values of 10.2 and 10.7. Viscometry indicates a strongly negative values of κ for this material. These results are quite consistent with recent RIS calculations [244, 245].

For highly syndiotactic chains, C_∞ values range from 7.3–9.8, depending upon the method of measurement, as already noted. In addition, viscometric

Table 9. Characteristic ratios and temperature coefficients for PMMA chains of varying tacticity

Tacticity[a]	C_∞	Ref(s).	$\kappa \times 10^3 (K^{-1})$	Ref(s).
ca. 0	10.2[b]	241, 242	− 2.3[b]	243
ca. 0	10.7[c]	201	−	−
17	9.3[b]	242	−	−
65	7.5[b]	242	−	−
ca. 75–80	7.3–7.4[b]	90, 200, 242	1.6[b]	90
ca. 75–80	8.4[d]	30	0 to − 0.1[c]	246–248
ca. 75–80	9.3–9.8[c]	201–203	−	−
ca. 87	9.2[c]	201	−	−

[a] Percent racemic diads.
[b] Via viscometry.
[c] Via SANS.
[d] Via light scattering

measurements [90] give a large positive temperature coefficient, as expected based on the RIS model of Vacatello and Flory [244]. SANS results collected by three different groups, however, show conclusively that in the melt state κ is nearly zero for this material [246–248].

For PB and PP, the solution results for chain dimensions tabulated by Jenkins and Porter [241] also suggest tacticity effects on C_∞ as anticipated by theory [1, 158, 162, 240], although the experimental results for PP are widely scattered [241]. Interestingly, recent SANS results for atactic PP melts [39] yield chain dimensions identical to those for isotactic PP, also obtained by SANS [249]. These data are supported by rheological studies on atactic and isotactic PP melts, where identical low shear viscosities are obtained for both materials at a given $\bar{M}_w$ [250]. These findings, combined with the lack of agreement of the various RIS models for PP, strongly suggest the need for additional investigations in this area.

6 Relationships Between Chain Flexibility and Bulk Properties

The potential for a quantitative interdependence of unperturbed chain dimensions and bulk viscoelastic properties was first recognized by Fox and Allen [251] and later extended upon by Berry and Fox [252]. These authors found a correlation between $\langle R^2 \rangle_0 /M$ and M_c (the molecular weight at which the break (upturn) in the log η_o − log M plot, where η_o is melt viscosity, occurred). Later, Graessley and Edwards [2] were able to demonstrate via a scaling approach that the plateau modulus G_N^o, associated with the terminal spectrum, was related to the Kuhn step length, the chain contour length and the number of chains per unit volume. A packing approach utilized by Lin [4] showed that the number of entanglement strands per cube of the tube diameter is a constant; an assumption that holds only if topological constraints are important. That approach has also been adopted by Kavassalis and Noolandi [5].

An alternative approach has been utilized by Aharoni [3] and Wu [253]. In essence, their model is based on the concept that the length of an entanglement strand, or the entanglement molecular weight, scales with the characteristic ratio C_∞ as follows:

$$M_e \sim C_\infty^2. \tag{32}$$

It is well-known that M_e (or M_c) will increase with temperature (melt or solution) for all flexible chains. However, the temperature dependence of $\langle R^2 \rangle_o$ ($d\ln \langle R^2 \rangle_o/dT$) can be negative, zero or positive. Thus, the assumption of a direct proportionality between the entanglement parameter and the unperturbed chain dimensions, as expressed by C_∞, is incorrect.

The following development involves melt density and $\langle R^2 \rangle_o/M$. It will be shown that these two parameters permit the calculation of G_N^o, M_e, and the tube diameter, d_t.

The equation that relates these three parameters is as follows

$$d_t^2 = [4/5\rho \, RT/G_N^o]\langle R^2 \rangle_o/M = M_e \langle R^2 \rangle_o/M \tag{33}$$

where both dynamic and static parameters are involved. The chain dimension $\langle R^2 \rangle_o$ can be cast as in the following:

$$\langle R^2 \rangle_o = C_\infty \overline{M_w} m_o^{-1} \ell^2 \tag{34}$$

or

$$\langle R_g^2 \rangle_o = C_\infty \overline{M_w} m_o^{-1} \ell^2/6 \tag{35}$$

and

$$\langle R_g^2 \rangle_o = C_\infty \overline{M_w} m_o^{-1} \ell_o^2 0.167. \tag{36}$$

A sphere volume for a single chain can be defined as

$$V_{sp} = A'\langle R_g^3 \rangle_o \tag{37}$$

where A' denotes a constant. Thus

$$\langle R_g^2 \rangle_o^{3/2} = C_\infty^{3/2} \overline{M}_w^{3/2} m_o^{-3/2} \ell^3 [6.8 \times 10^{-2}] \tag{38}$$

and

$$V_{sp} = A' C_\infty^{3/2} \overline{M}_w^{3/2} m_o^{-3/2} \ell^3. \tag{39}$$

N may be defined as the number of chains in V_{sp} for the case where $\overline{M}_w = M_e$

and N = 1. This yields

$$N = \frac{V_{sp}\rho}{\bar{M}_w} = A'C_\infty^{3/2}M_w^{1/2}m_o^{-3/2}\ell^3\rho \tag{40}$$

and

$$N/A' = C_\infty^{3/2}\bar{M}_w^{1/2}m_o^{-3/2}\ell^3\rho. \tag{41}$$

In the case of $\bar{M}_w = M_e$, $N/A' = 31.95$ with $A' = 3.13 \times 10^{-2}$ (std. deviation 3.10×10^{-3}).

A portion of the data set that was collected at 413 K and upon which the above value of A' is based, is given in Table 10.

With the foregoing in hand it is now possible to generate a quantitative expression for M_e in terms of ρ and $\langle R^2\rangle_o/M$ where N is equal to 1.

$$C_\infty = (A')^{-2/3}\ell^{-2}\rho^{-2/3}M_e^{1/3}m_o \tag{42}$$

$$C_\infty = 10.07\,\ell^{-2}\rho^{-2/3}M_e^{1/3}m_o. \tag{43}$$

Rearrangement then yields

$$M_e = (9.93\times10^{-2}\ell^2\rho^{2/3}m_o^{-1}C_\infty)^{-3} \tag{44}$$

$$M_e = (9.93\times10^{-2}\rho^{2/3}\langle R^2\rangle_o/M)^{-3} \tag{45}$$

Table 10. Molecular characteristics of model polymers at 413 K

Polymer	$\langle R^2\rangle_o/M$[a] (Å mol g^{-1})	ρ (g cm^{-3})	$\langle R^2\rangle_o/V$ (Å^2 mol cm^{-3})	$G_N^0\times10^{-6}$ (dynes cm^{-2})	$M_e\times10^{-3}$ (g mol^{-1})	d_t (Å)
PE	1.22	0.788	0.961	23.0	0.95	32.2
PEB-2	1.17	0.788	0.922	21.5	1.01	34.7/38.5[b]
PEB-4.6	1.15	0.788	0.906	19.0	1.22	35.3
PEO	0.817	0.996	0.814	12.0	2.28	39.3
1,4-PBD	0.876	0.826	0.724	12.0	1.90	44.1
alt-PEP	0.834	0.790	0.658	9.5	2.28	48.6/48.0[b]
PEB-24.6	0.799	0.797	0.637	6.7	3.27	50.2
1,4-PT	0.662	0.830	0.550	4.4	5.18	58.1/52.0[b]
alt-PBE	0.799	0.794	0.550	5.2	4.19	60.3
a-PP	0.670	0.791	0.530	4.7	4.65	66.0
PIB	0.570	0.849	0.484	3.2	7.74	69.0
PMMA	0.425	1.13	0.480	3.1	10.0	73.6
PS	0.434	0.969	0.421	2.0	13.9	78.1
PDMS	0.457	0.895	0.409	2.0	12.3	78.1
PEE	0.507	0.802	0.407	2.0	11.0	78.5
PVCH[c]	0.322	0.900	0.290	0.65	38.0	110.2

[a] Via small angle neutron scattering in the melt.
[b] Measured by neutron-spin-echo.
[c] Measurement at 435 K

$$M_e = 1.02 \times 10^3 \rho^{-2}(\langle R^2 \rangle_o/M)^{-3}. \tag{46}$$

The relationship between G_N^o and M_e is

$$M_e = \frac{4/5\rho RT}{G_N^o}. \tag{47}$$

Thus, the following expression for G_N^o (at 413 K) is obtainable

$$G_N^o = 2.69 \times 10^7 (\rho \langle R^2 \rangle_o/M)^3. \tag{48}$$

or

$$G_N^o = 2.69 \times 10^7 (\langle R^2 \rangle_o/V)^3 \tag{49}$$

where $M\rho^{-1}$ is V, the chain volume in the melt state.

Conversely, since $d_t^2 = (\langle R^2 \rangle/M)M_e$ then

$$d_t^2 = 1.02 \times 10^3 \rho^{-2}(\langle R^2 \rangle_o/M)^{-2} \tag{50}$$

and

$$d_t^2 = 1.02 \times 10^3(\langle R^2 \rangle_o/V)^{-2} \tag{51}$$

The preceding development has used parameters determined at 413 K. This is a necessary step since both $\langle R^2 \rangle_o/M$ and G_N^o can show temperature dependencies. Previous efforts have not considered this potential error. It is also germane to point out that Kavassalis and Noolandi [5] have predicted that $G_N^o \sim \rho^3$ and $M_e \sim \rho^{-2}$. Those exponents are present in this development.

Figure 6 shows most of the data contained in Table 10. The gradient in question is drawn to the predicted gradient 1/3. As can be seen, good agreement is found over a range of G_N^o of 2.3×10^7 (polyethylene)–6.5×10^5 dyne cm^{-2} for poly(vinyl cyclohexane). The latter material is hydrogenated polystyrene.

Polysulfone and polycarbonate, for example, are high plateau modulus materials, often referred to as engineering plastics. These are listed in Table 11, which, in the main, is based on a collection of data given by Roovers, Toporowski and Ethier [254]. As can be seen in Fig. 7, generally good agreement is found with the predicted behavior, which is based on the data given in Table 11. This demonstrates that the behavior reported is independent of polymer type.

The foregoing discussion shows that a direct connection exists between the various viscoelastic parameters, unperturbed chain dimensions, and ρ. Predictive capacities exist that show that the melt viscoelastic properties are determined by the joint interplay of the radius of gyration of a chain and its accompanying volume. Control of the plateau modulus, for example, can be exerted by manipulating $\langle R^2 \rangle_o/V$, e.g., an increase in this value results in an increase in G_N^o.

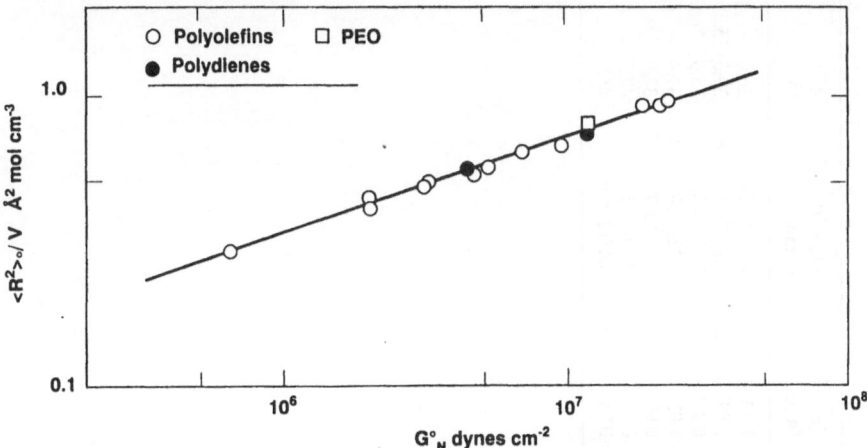

Fig. 6. Relationship between $\langle R^2 \rangle_0 / V$ and G_N^0 for various model polymers

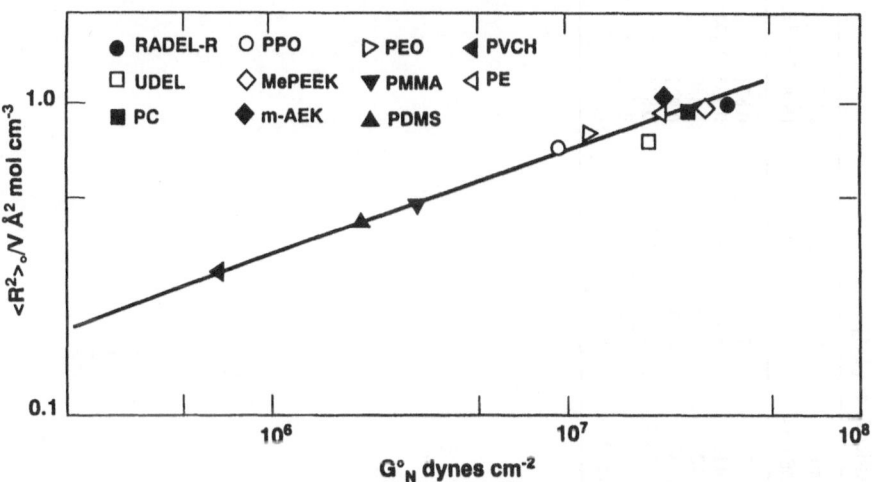

Fig. 7. Relationship between $\langle R^2 \rangle_0 / V$ and G_N^0 for varioius polymers including engineering plastics

7 Concluding Remarks

We have briefly reviewed methods for estimation of polymer unperturbed dimensions and the variation of unperturbed dimensions with temperature. The impact of various structural features on chain conformations has been elucidated. Special attention has been devoted to polymethacrylates, polydienes, and polyolefins. Stiff chain polymers have also been discussed briefly.

Table 11. Characteristics of engineering plastics from [254]

Polymer	Temp.[a] (K)	$G° \times 10^{-7}$ (dynes cm^{-2})	m_o (g mol^{-1})	l_o (Å)	ρ (g cm^{-3})	$\langle R^2 \rangle_o/M$ (Å mol g^{-1})	C_∞	$\langle R^2 \rangle_o/V$ (Å mol cm^{-3})	$M_e \times 10^3$ (g mol^{-1})	d_t (Å)
RADEL-R	555	3.7(2.8)	133	7.39	1.22	0.821	1.96	1.00	1.23(1.65)	36.8
M_ePEEK	523	3.1(2.0–2.8)	110	5.76	1.16	0.755–0.834	2.4–2.7	0.898–0.992	1.26(2.15–1.60)	39.3–35.6
PC	473	2.7(2.5)	127	7.00	1.14	0.850	2.20	0.969	1.33(1.45)	35.0
m-AEK	473	2.2(3.4)	85	4.94	1.20	0.901	3.14	1.08	1.72(1.09)	31.4
PET	200	(2.1)	38.4	2.87	1.08	0.835	3.89	0.902	(1.69)	37.6
UDEL	513	2.0(1.2)	110	5.76	1.16	0.645	2.14	0.747	2.00(3.45)	47.3
PPO	493	0.95(1.1)	120	5.40	1.0	0.741	3.05	0.741	3.45(2.95)	46.7

[a] Measurement temperature of G_N^o

We have attempted to shed light on the state of play for chain dimensions and their temperature dependence regarding their evaluation via theta solvent measurements, RIS theory, and SANS measurements in the melt. In some cases, e.g., alternating ehylene/propylene copolymers, theory and the various experimental measurements agree. In a great number of instances, however, discrepancies exist. With RIS models it seems clear that reduction of all possible local configurations to a very small subset of rotational isomers can lead to errors. Also, in many instances there are questions regarding the potential energies that are assigned to each of these. Thus, these methods should probably best be viewed as approximate.

With regard to differences in polymer behavior in solution versus the bulk state, several points must be made. Clearly, it is now well-established that the choice of theta solvent can affect chain dimensions to some extent [42–44, 46, 47]. Hence, only the chain in an amorphous melt of identical neighbors can be considered to be in the unperturbed state. Particularly striking are some of the differences noted in temperature coefficients measured by different techniques. Is it possible that the thermal expansion of a polymer molecule is fundamentally different in the bulk and in solution? Can specific solvent effects exist and vary in a systematic way within a series of chemically similar theta solvents? Does the different range of temperatures usually employed in bulk versus solution studies affect κ? Are chains in the bulk (during SANS and thermoelastic experiments) allowed adequate time to completely relax to equilibrium? All of these issues need further attention. Other topics perhaps worthy of consideration include the study of the impact of deuterium labelling on chain conformation (H^2 has lower vibrational energy than does H^1) and the potential temperature dependence of the Flory hydrodynamic parameter Φ_o.

Finally, we have shown that important bulk properties such as the plateau modulus directly reflect unperturbed chain dimensions. Enhanced understanding of the interrelationship between structure, conformation, and bulk properties will clearly lead to a more rational design of macromolecules to meet specific property requirements.

Acknowledgements. JWM thanks the National Science Foundation for support (Grant CTS-9107025) during the period in which this review was written. This review was initiated during 1992 when ZX was a Visiting Professor at UAB; this visit was also made possible by NSF support. We also thank Ray Chen for his help in putting together this manuscript and Prof. H. Fujita for his many insightful comments during the preparation of this atricle.

8 References

1. Flory PJ (1969) Statistical mechanics of chain molecules. Interscience, New York
2. Graessley WW, Edwards SF (1981) Polymer 22: 1389
3. Aharoni SM (1983) Macromolecules 16: 1722
4. Lin Y-H (1987) Macromolecules 20: 3080
5. Kavassalis T, Noolandi J (1988) Macromolecules 21: 2869

6. Iwata K, Edwards SF (1989) J Chem Phys 90: 4567
7. He T, Porter RS (1992) Makromol Chem Theory Simul 1: 119
8. Colby RH, Rubinstein M, Viovy JL (1992) Macromolecules 25: 996
9. Fetters LJ (1993) to be published
10. Flory PJ (1974) Macromolecules 7: 381
11. Kuhn W (1934) Kolloid-Z 68: 2; (1936) Kolloid-Z 76: 258; (1939) Kolloid-Z 87: 3
12. Guth E, Mark H (1934) Monatsh Chem 65: 93
13. Eyring H (1932) Phys Rev 39: 746
14. Flory PJ (1969) Statistical mechanics of chain molecules. Interscience, New York, p 38
15. Benoit H (1947) J Chem Phys 44: 18
16. Kuhn H (1947) J Chem Phys 15: 843
17. Taylor WJ (1948) J Chem Phys 16: 257
18. Volkenstein MV (1963) Configurational statistics of polymer chains. Wiley-Interscience, New York
19. Birshtein TM, Ptitsyn OB (1966) Conformation of macromolecules. Wiley-Interscience, Nev York
20. Kratky O, Porod G (1949) Recl Trav Chim 68: 1106
21. Porod G (1949) Monatsh. Chem 2: 251
22. Fujita H (1990) Polymer solutions. Elsevier, Amsterdam, p 140
23. See, for example: Fujita H (1990) Polymer solutions. Elsevier, Amsterdam, pp 11–12, 140–141
24. Flory PJ (1969) Statistical mechanics of chain molecules. Interscience, New York, p 111
25. Yamakawa H, Fujii M (1973) Macromolecules 6: 407
26. Yamakawa H, Fujii M (1974) Macromolecules 7: 128
27. Yamakawa H, Yoshizaki T (1980) Macromolecules 13: 633
28. For a review see: Yamakawa H (1984) Annu Rev Phys Chem 35: 23
29. Koyama H, Yoshizaki T, Einaga Y, Hayashi H, Yamakawa H (1991) Macromolecules 24: 932
30. Fujii Y, Tamai Y, Konishi T, Yamakawa H (1991) Macromolecules 24: 1608
31. Abe F, Einaga Y, Yamakawa H (1991) Macromolecules 24: 4423
32. Ballard DGH, Schelten J, Wignall GD (1973) Eur Polym J 9: 965
33. Cotton JP, Decker D, Benoit H, Farnoux B, Higgins J, Jannink G, Ober R, Picot C, des Cloizeaux J (1974) Macromolecules 7: 863
34. Wignall GD, Ballard DGH, Schelten J (1974) Eur Polym J 10: 861
35. See, for example: Richards RW (1978) In: Dawkins JV (ed) Developments in polymer characterization, vol. 1. Elsevier, Amsterdam
36. Butera R, Fetters LJ, Huang JS, Richter D, Pyckhout-Hintzen W, Zirkel A, Farago B, Ewen B (1991) Phys Rev Lett 66: 2088
37. Boothroyd AT, Rennie AR, Boothroyd CB (1991) Europhys Lett 7: 715
38. Zirkel A, Richter D, Pyckhout-Hintzen W, Fetters LJ (1992) Macromolecules 25: 954
39. Zirkel A, Urban V, Richter D, Fetters LJ, Huang JS, Kampmann R, Hadjichristidis N (1992) Macromolecules 25: 6148
40. Flory PJ (1953) Principles of polymer chemistry. Cornell Univ Press, Ithaca, pp 423–426
41. Flory PJ (1953) Principles of polymer chemistry. Cornell Univ Press, Ithaca, pp 601–602
42. Orofino TA (1966) J Chem Phys 45: 4310
43. Mays JW, Hadjichristidis N, Fetters LJ (1985) Macromolecules 18: 2231
44. Bohdanecky M, Berek D (1985) Makromol Chem Rapid Commun 6: 275
45. Konishi T, Yoshizaki T, Yamakawa H (1991) Macromolecules 24: 5614
46. Lifson S, Oppenheim J (1960) J Chem Phys 33: 109
47. Bahar I, Baysal BM, Erman B (1986) Macromolecules 19: 1703
48. Sanchez IC (1979) Macromolecules 12: 980
49. de Gennes PG (1979) Scaling concepts in polymer physics. Cornell Univ Press, Ithaca, sections IV. 3.2 and XI
50. Bruns W (1989) Macromolecules 22: 2829
51. Zimm B (1948) J Chem Phys 16: 1093
52. Brandrup J, Immergut EH (1975) Polymer handbook, 2nd ed, Wiley-Interscience, New York, p IV-310 See also: Bareiss RE (1983) Eur Polym J 19: 847
53. Baumann H (1965) J Polym Sci, Polym Lett 3: 1069
54. Yamakawa H (1971) Modern theory of polymer solutions. Harper and Row, New York, Chap 7
55. Flory PJ (1949) J Chem Phys 10: 51
56. Fox Jr TG, Flory PJ (1949) J Phys Colloid Chem 53: 197

57. Flory PJ (1953) Principles of polymer chemistry. Cornell Univ Press, Ithaca, Chap XIV
58. For a listing of values see Table III of reference 43. A value of 2.5×10^{21} for poly(α-methyl-styrene) in cyclohexane is derived from the combined data of Cowie JMG, Bywater S, Worsfold DJ (1967) Polymer 8: 105 and Lindner JS, Hadjichristidis N, Mays JW (1989) Polym Commun 30: 174. This is larger than the value for the same system reported in reference 43. Very recently $\Phi_o = 2.4 \times 10^{-21}$ has been measured for poly(α-methylstyrene) in cyclohexane (Li J, Mays JW, to be published)
59. Zimm BH (1980) Macromolecules 13: 592
60. Bareiss RE (1975) In: Brandrup T, Immergut EH (eds) Polymer handbook, 2nd edn, Wiley-Interscience, New York, section IV, See also: Bareiss RE (1986) Makromol Chem 187: 955
61. Kurata M, Stockmayer WH (1963) Fortschr Hochpolym-Forsch 3: 196
62 Jenkins R, Porter RS (1980) Adv Polym Sci 36: 1
63. Burchard W (1960) Makromol Chem 50: 20
64. Stockmayer WH, Fixman M (1963) J Polym Sci Part C 1: 137
65. Kurata M, Yamakawa H (1958) J Chem Phys 29: 311; Kurata M, Yamakawa H, Utiyama H (1959) Makromol Chem 34: 139
66. Stockmayer WH (1977) Br Polym J 9: 89
67. Tanaka G (1982) Macromolecules 15: 1028
68. Stickler M, Panke D, Wunderlich W (1987) Makromol Chem 188: 2651
69. Lewis ME, Nan S, Mays JW (1991) Macromolecules 24: 197
70. Lindner JS, Hadjichristidis N, Mays JW (1989) Polym Commun 30: 174
71. Fetters LJ, Hadjichristidis N, Lindner JS, Mays JW, Wilson WW (1991) Macromolecules 24: 3127
72. Norisuye T, Fujita H (1982) Polym J 14: 143
73. See for example: Reddy GV, Bohdanecky M (1987) Macromolecules 20: 1393 and references therein
74. Benoit H, Doty P (1953) J Phys Chem 57: 958
75. Helminiak TE, Benner CL, Gibbs WE (1967) Polym Preprints 8: 284
76. Tanner DW, Berry GC (1974) J Polym Sci, Polym Phys Ed 12: 941
77. Vitovskaya MG, Tsvetkov VN (1976) Eur Polym J 12: 251
78. Morcelet M, Loucheux C (1978) Biopolymers 17: 593
79. Motowoka M, Norisuye T, Teramoto A, Fujita H (1979) Polym J 11: 665
80. Yanaki T, Norisuye T, Fujita H (1980) Macromolecules 13: 1462
81. Tsvetkov VN, Andreeva LN (1981) Adv Polym Sci 39: 95
82. Fujita H (1990) Polymer solutions. Elsevier Amsterdam, pp 153–154
83. Bohdanecky M (1983) Macromolecules 16: 1483
84. Katime-Amashta IA, Sanchez G (1975) Eur Polym J 11: 223
85. Niezette J, Hadjichristidis N, Desreux V (1976) Makromol Chem 177: 2069
86. Tricot M (1986) Macromolecules 19: 1268
87. Kurata M, Tsunashima Y (1989) In: Brandrup J, Immergut EH (eds) Polymer Handbook, Wiley-Interscience, New York, VII-1-60
88. For earlier reviews on unperturbed dimensions of polymethacrylate see: Mays JW, Hadjichris-tidis N (1988) J Macromol Sci-Rev Macromol Chem Phys C28(3 & 4): 371; Hadjichristidis N, Xu Z, Mays JW (1991) Chimika Chronika, New Ser. 20: 39
89. Fox Jr TG (1962) Polymer 3: 111
90. Mays JW, Nan S, Wan Y, Li J, Hadjichristidis N (1991) Macromolecules 24: 4469
91. Chinai SN, Samuels RJ (1956) J Polym Sci 19: 463
92. Chinai SN, Valles RJ (1959) J Polym Sci 39: 363
93. Chinai SN (1957) J Polym Sci 25: 413
94. Xu Z, Hadjichristidis N, Fetters LJ (1984) Macromolecules 17: 2303
95. Chinai SN, Guzzi RA (1959) J Polym Sci 41: 475
96. Lee HT, Levi DW (1960) J Polym Sci 42: 449
97. Simionescu CI, Ioan S, Flondor A, Simionescu BC (1988) Makromol Chem 189: 2331
98. Karandinos A, Nan S, Mays JW, Hadjichristidis N (1991) Macromolecules 24: 2007
99. Didot FE, Chinai SN, Levi DW (1960) J Polym Sci 43: 557
100. Siakali-Kioulafa E, Hadjichristidis N, Mays JW (1989) Macromolecules 22: 2059
101. Hadjichristidis N, Devaleriola M, Desreux V (1972) Eur Polym J 8: 1193
102. Hadjichristidis N, Desreux V (1972) J Macromol Sci, Chem A6: 1227
103. Hadjichristidis N, Mays JW, Ferry W, Fetters LJ (1984) J Polym Sci, Polym Phys Ed 22: 1745
104. Tricot M, Bleus JP, Riga JP, Desreux V (1974) Makromol Chem 175: 913

105. Gargallo L (1975) Colloid Polym Sci 253: 288
106. Gargallo L, Niezette J, Desreux V (1977) Bull R Soc Liege 1-2: 82
107. Matsumoto A, Tanaka S, Otsu T (1992) Colloid Polym Sci 270: 17
108. The literature values for C_∞ of poly(benzyl methacrylate) vary from ca. 9–10.9. See, for example: Richards RW (1977) Polym 18: 114; Gargallo L, Munoz MI, Radic D (1983) Polym Bull 10: 264
109. Mays JW, Hadjichristidis N, Lindner JS (1990) J Polym Sci, Polym Phys Ed 28: 1881
110. Niezette J, Hadjichristidis N, Desreux V (1976) Makromol Chem 177: 2069
111. Alexopoulis JB, Hadjichristidis N, Vassiliadis A (1975) Polymer 16: 386
112. Alexopoulis J, Hadjichristidis N (1979) Makromol Chem 180: 455
113. Ojeda T, Radic D, Gargallo L (1980) Makromol Chem 181: 2237
114. Gargallo L, Hamidi N, Radic D (1990) Polymer 31: 924
115. Gargallo L, Hamidi N, Radic D (1991) Polymer International 24: 1
116. Mays JW, Ferry W, Hadjichristidis N, Fetters LJ (1985) Macromolecules 18: 2330
117. Niezette J, Hadjichristidis N (1977) Polymer 18: 200
118. Tricot M, Desreux V (1970) Makromol Chem 149: 185
119. Hadjichristidis N (1977) Makromol Chem 178: 1463
120. Becerra M, Radic D, Gargallo L (1978) Makromol Chem 179: 2241
121. Hadjichristidis N, Mays J, Vargo RD, Fetters LJ (1983) J Polym Sci, Polym Phys Ed 21: 189. The monomer used in this work was sold by Polysciences Incorporated as tetrahydropyranyl methacrylate. Subsequently to publication of this paper it was brought to our attention by WH Hertler of DuPont that this monomer was probably tetrahydropyranylmethyl methacrylate. We confirmed Dr. Hertler's suspicions by NMR and thank him for pointing out this discrepancy.
122. Gargallo L, Munoz MI, Radic D (1986) Polymer 27: 1416
123. Hadjichristidis N, Touloupis C, Fetters LJ (1981) Macromolecules 14: 128
124. Kokkiaris D, Hadjichristidis N (1977) Polymer 18: 639
125. Kokkiaris D, Touloupis C, Hadjichristidis N (1981) Polymer 22: 163
126. Xu Z, Han Z, Bu L, Pang Y, He Z, Zhang X (1988) J China Univ Sci Technol 18: 534
127. Xu Z, Han Z, Li S, Yang J, Su C (1988) J Functional Polym 1: 25
128. Stejskal J, Janca J, Kratochvil P (1976) Polym J 8: 549
129. Chawla RK, Stejskal J, Strakova D, Kratochvil P (1987) Croatica Chem Acta 60(1): 1
130. Dolezalova M, Petrus V, Tuzar Z, Bohdanecky M (1976) Eur Polym J 12: 701
131. Rudolph HD, Dreizler H, Maier W (1960) Naturforsch A15: 742
132. Pierce L, Hayashi M (1961) J Chem Phys 35: 479
133. Pateropoulou D, Siakali-Kioulafa E, Hadjichristidis N, Nan S, Mays JW (1994) Makromol Chem 195: 173
134. Eliel EL, Allinger NL, Angyal SJ, Morrison GA (1965) Conformational analysis. Wiley-Interscience, New York, pp 244–45
135. Katime I, Roig A (1975) Eur Polym J 11: 603
136. Kuwahara N, Ogino K, Kasai A, Ueno S, Kaneko M (1965) J Polym Sci A-3: 985
137. Mohite RB, Gundiah S, Kapur SL (1968) Makromol Chem 116: 280
138. Noguchi Y, Aoki A, Tanaka G, Yamakawa H (1970) J Chem Phys 52: 2651
139. Kuwahara N, Ogino K, Konuma M, Iida N, Kaneko M (1966) J Polym Sci, Part A-2 4: 173
140. Altares Jr T, Wyman DP, Allen VR (1964) J Polym Sci A-2: 4533
141. Cowie JMG, Bywater S (1965) Polymer 6: 197
142. Einaga Y, Miyaki Y, Fujita H (1979) J Polym Sci, Polym Phys Ed 17: 2103
143. Hadjichristidis N, Xu Z, Fetters LJ, Roovers J (1982) J Polym Sci, Polym Phys Ed 20: 743
144. Xu Z, Hadjichristidis N, Carella JM, Fetters LJ (1983) Macromolecules 16: 925
145. Mays JW, Hadjichristidis N, Fetters LJ (1984) Macromolecules 17: 2723
146. Xu Z, Mays JW, Chen X, Hadjichristidis N, Schilling F, Bair HE, Pearson DS, Fetters LJ (1985) Macromolecules 18: 2560
147. Mays JW, Fetters LJ (1989) Macromolecules 22: 921
148. Hattam P, Gauntlett S, Mays JW, Hadjichristidis N, Young RN, Fetters LJ (1991) Macromolecules 24: 6199
149. Abe A, Jernigan RL, Flory PJ (1966) J Am Chem Soc 88: 631
150. Ciferri A, Hoeve CAJ, Flory PJ (1961) J Am Chem Soc 83: 1015
151. Flory PJ, Ciferri A, Chiang R (1961) J Am Chem Soc 83: 1023
152. Chiang R (1965) J Phys Chem 64: 1645

153. Stacy CJ, Arnett RL (1965) J Phys Chem 64: 3109
154. Nakajima A, Hamada F, Hayashi S (1966) J Polym Sci, Part C 15: 285
155. Chiang R (1966) J Phys Chem 70: 2348
156. Stacy CJ, Arnett RL (1973) J Phys Chem 77: 1986
157. Lieser G, Fischer EW, Ibel K (1975) J Polym Sci, Polym Lett Ed 13: 39
158. Flory PJ, Mark JE, Abe A (1966) J Am Chem Soc 88: 639
159. Asakura T, Ando I, Nishioka A (1976) Makromol Chem 177: 1493
160. Mark JE (1972) J Chem Phys 57: 2541
161. Mattice WL (1986) Macromolecules 19: 2303
162. Wittwer H, Suter UW (1985) Macromolecules 18: 403
163. Mark JE (1966) J Am Chem Soc 88: 4354
164. Mark JE (1967) J Am Chem Soc 89: 6829
165. Abe Y, Flory PJ (1971) Macromolecules 4: 219
166. Abe Y, Flory PJ (1971) Macromolecules 4: 230
167. Morton M (1963) Anionic polymerization: principles and practice, Academic Press, New York
168. Wagner HL, Flory PJ (1952) J Am Chem Soc 74: 195
169. Bur AJ, Fetters LJ (1976) Chem Rev 76: 727
170. Ciferri A (1982) In: Ciferri A, Krigbaum WR, Meyer RB (eds) Polymer liquid crystals. Academic Press, New York, pp 66–67
171. Bianchi E, Ciferri A, Conio G, Cosani A, Terbojevich M (1985) Macromolecules 18: 646
172. Conio G, Bianchi E, Ciferri A, Tealdi A, Aden MA (1983) Macromolecules 16: 1264
173. Aden MA, Bianchi E, Ciferri A, Conio G, Tealdi A (1984) Macromolecules 17: 2010
174. Bianchi E, Ciferri A, Conio G, Lanzavecchia L, Terbojevich M (1986) 19: 630
175. Mays JW (1988) Macromolecules 21: 3179
176. Green MM, Gross RA, Crosby C, Schilling FC (1987) Macromolecules 20: 992
177. Murakami H, Norisuye T, Fujita H (1980) Macromolecules 13: 345
178. Conio G, Bianchi E, Ciferri A, Krighaum WR (1984) Macromolecules 17: 856
179. Ying Q, Chu B (1987) Macromolecules 20: 871
180. Krigbaum WR, Tanaka T, Brelsford G, Ciferri A (1991) Macromolecules 24: 4142
181. Ying Q, Chu B (1984) Makromol Chem, Rapid Commun 5: 785
182. Record Jr MT, Woodbury CP, Inman RB (1975) Biopolymers 14: 393
183. Godfrey JE, Eisenburg H (1976) Biophys Chem 5: 301
184. Sato T, Norisuye T, Fujita H (1984) Macromolecules 17: 2696
185. Coviello T, Kajiwara K, Burchard W, Dentini M, Crescenzi V (1986) Macromolecules 19: 2826
186. Utiyama H, Sakato K, Ikehara K, Setsuiye T, Kurata M (1973) Biopolymers 12: 53
187. Saito T, Iso N, Mizuno H, Onda N, Yamato H, Odashima H (1982) Biopolymers 21: 715
188. Yanaki T, Norisuye T, Fujita H (1980) Macromolecules 13: 1462
189. Kashiwagi Y, Norisuye T, Fujita H (1981) Macromolecules 14: 1220
190. Ciferri A, Krigbaum WR, Meyer RB (eds) (1982) Polymer liquid crystals. Academic Press, New York
191. Flory PJ (1982) In: Ciferri A, Krigbaum WR, Meyer RB (eds) Polymer liquid crystals. Academic Press, New York, Chap 4
192. Flory PJ (1956) Proc R Soc London, Ser A 234: 73
193. Flory PJ (1984) Adv Polym Sci 59: 1
194. Laivins GV, Gray DG (1985) Macromolecules 18: 1746
195. Laivins GV, Gray DG (1985) Macromolecules 18: 1753
196. Mays JW, Hadjichristidis N (1991) In: Barth HG, Mays JW (eds) Modern methods of polymer characterization. Wiley-Interscience, New York
197. Abe A (1970) Polym J 1: 232
198. Yoon DY, Sundararajan PR, Flory PJ (1975) Macromolecules 8: 776
199. Mays JW, Nan S, Whitfield D (1991) Macromolecules 24: 315
200. Quadrat Q, Bohdanecky M, Mrkvickova L (1981) Makromol Chem 182: 445
201. O'Reilly JM, Teegarden DM, Wignall GD (1985) Macromolecules 18: 2747
202. Dettenmeier DM, Maconnachie A, Higgins JS, Dausel HH, Nguyen TQ (1986) Macro-molecules 19: 773
203. Ito H, Russell TP, Wignall GD (1987) Macromolecules 20: 2213
204. Flory PJ, Fox Jr TG (1951) J Am Chem Soc 73: 1904, 1909, 1915
205. Mark JE, Flory PJ (1964) J Am Chem Soc 86: 138
206. Mark JE, Flory PJ (1965) J Am Chem Soc 87: 1423

207. Mark JE, Thomas GB (1966) J Phys Chem 70: 3588
208. Mays JW, Hadjichristidis N, Graessley WW, Fetters LJ (1986) J Polym Sci, Polym Phys Ed 24: 2553
209. Fujita H (1993) personal communication
210. Fetters LJ, Mays JW (1993) unpublished observations
211. Mark JE (1973) Rubber Chem Technol 46: 593
212. Mark JE (1976) J Polym Sci, Macromol Revs 11: 135
213. McCrum NG (1984) Polym Commun 25: 213
214. McCrum NG (1986) Polymer 27: 47
215. Smith Jr KJ, Mark JE (1988) Polymer 29: 292
216. Bohdanecky M (1968) Coll Czech Chem Comm 33: 4397
217. Ciferri A (1961) Trans Faraday Soc 57: 853
218. Bluestone S, Mark JE, Flory PJ (1974) Macromolecules 7: 325
219. Kotera A, Onda N, Saito T (1974) Rep Prog Polym Phys Japan 17: 25
220. Tobolsky AV, Carlson D, Indictor N (1961) J Polym Sci 54: 175
221. Lath D, Bohdanecky M (1977) J Polym Sci, Polym Phys Ed 15: 555
222. Fox Jr TG, Flory PJ (1951) J Am Chem Soc 73: 1909
223. Bazuaye A, Huglin MB (1979) Polymer 20: 44
224. Kotera A, Saito T, Yamaguchi N, Yoshizaki K, Yanagisawa Y, Tsuchiya H (1971) Rep Prog Polym Phys Japan 24: 31
225. Abdel-Azim A-AA, Huglin MB (1983) Polymer 24: 1429
226. Kawahara N, Sacki S, Konno S, Kaneko M (1974) Polymer 15: 66
227. Abe M, Fujita H (1965) J Phys Chem 69: 3263
228. Schulz GV, Baumann H (1963) Makromol Chem 60: 120
229. Dusek K (1966) Collect Czech Chem Commun 31: 1893
230. Dusek K (1967) Collect Czech Chem Commun 32: 2264
231. Orofino TA, Mickey JW (1963) J Chem Phys 38: 2512
232. Orofino TA, Ciferri A (1964) J Phys Chem 68: 3136
233. Boothroyd AT (1993) personal communication
234. Fetters LJ to be published
235. Roovers J, Toporowski PM (1990) Rubber Chem Technol 63: 734
236. Abe A (1970) Polym J 1: 232
237. Boyd RH, Breitling SM (1972) Macromolecules 5: 279
238. Biskup U, Cantow HJ (1972) Macromolecules 5: 546
239. Allegra G, Calligasis M, Randaccio L, Moraglio G (1973) Macromolecules 6: 397
240. Suter UW, Flory PJ (1975) Macromolecules 8: 765
241. Jenkins R, Porter RS (1980) Adv Polym Sci 36: 1
242. Jenkins R, Porter RS (1982) Polymer 23: 105
243. Sakurada I, Nakajima A, Yoshizaki D, Nakamae (1962) Kooloid-ZZ Polym 186: 41
244. Vacatello M, Flory PJ (1986) Macromolecules 19: 405
245. Sundararajan PR (1986) Macromolecules 19: 415
246. Kirste RG, Kruse WA, Ibel K (1975) Polymer 16: 120
247. Boothroyd AT, Rennie AR, Wignall GD (1993) to be published
248. Fetters LJ, Krishnamorti R, Graessley WW, Lohse D, Colby RH (1993) to be published
249. Ballard DGH, Cheshire P, Longman GW, Schelten J (1978) Polymer 19: 379
250. Pearson DS, Younghouse LB, Fetters LJ, Mays JW (1978) Macromolecules 21: 478
251. Fox TG, Allen VR (1964) J Chem Phys 41: 344
252. Berry GC, Fox TG (1968) Adv Polym Sci 5: 261
253. Wu S (1992) Polym Eng Sci 32: 823
254. Roovers J, Toporowski PM, Ethier R (1990) High Performance Polymers 2: 165

Editor: Prof. H. Fujita
Received December 1993

Polyelectrolytes in Solution

S. Förster
Max Plauch Institut für Kolloid- und Greuzflächenforschung, D-14513 Teltow-Seehof, FRG

M. Schmidt
Universität Bayreuth, Makromolekulare Chemie II, D-95440 Bayreuth, FRG

Dedicated to Prof. Walther Burchard on the occasion of his 65th Birthday

1 Introduction

Despite the increasing theoretical and experimental effort, particularly during the last ten years, the solution properties of polyelectrolytes are not well understood. Investigations of various polyelectrolytes by different experimental methods very often lead to controversial conclusions which are sometimes subject to the bias of individual scientists. This situation is favoured by the fact that experiments on polyelectrolytes almost always turned out to be extremely difficult and the usually enormous scatter of experimental data effectively prohibited, for a long time, a unique picture on the structure and dynamical behaviour of polyelectrolyte solutions.

During the last 40 years many books and reviews have been written on polyelectrolytes in solution [1–7], almost all of which are worth reading in order to understand the historical development of how to ask the right question and, quite often, to obtain the wrong answer or vice versa. The most recent book by Schmitz [8] summarizes the present knowledge on polyions in solution in a very fundamental manner and partial coincidence with the present article is neither intended, nor accidental, but unavoidable.

The present review mainly, but not exclusively, deals with linear flexible polyions and focuses on single chain properties such as the "electrostatic persistence length" of intermolecular interactions of a mainly electrostatic nature, on static and dynamic properties at the dilute-semidilute cross-over regime and, only briefly, on the occurrence of "unidentified" polyelectrolyte structures, sometimes referred to as "extraordinary phase", "cluster", "association" or, according to its most striking phenomenology, "slow mode".

As a guideline through the whole review, we will now develop a simple "contour map" of various regimes occurring in polyelectrolyte solutions in dependence on polyion and salt concentration. Here we shall omit all complications and only briefly describe the most generally observed phenomena which, to our knowledge, are unambiguously proven.

One of the most perplexing (and not yet understood) properties of polyelectrolyte dynamics is the fact that, at a certain ratio λ^* of polyion-concentration c_p^m (in mol monomer or mol charges, abbreviated "monomol/l") to added salt concentration c_s^m (mol/l), a slow mode is observed in dynamic light scattering with a concomitant drastic increase in scattering intensity.

The value of this ratio λ^* is only known approximately, as literature values vary from $1 \leqslant \lambda^* \leqslant 5$ for monovalent counterions. There is no question, however, that such a "slow mode" is observed for essentially all investigated polyelectrolytes and the phenomenon was attributed to the formation of an "extraordinary phase" (EO-phase). It is worth mentioning that there is some dispute as to whether the spectacular name is justified, i.e., whether this phase is a phase in the thermodynamic sense with the usually observed discontinuities for phase-transitions or not. We shall not participate in this speculative and partly semantic discussion but simply adopt the term "extraordinary phase", because it was introduced as such and is currently used in polyelectrolyte literature. Together with the first appearance of the EO-phase, the ordinary phase ("ordinary phase" obviously being used as a synonym for "(fairly) well understood")[1] showed a cross-over from the diffusion of single polyions to the much faster, strongly coupled diffusion of polyions and gegen- or coions, details of which will be discussed below. The important point is that polyelectrolyte solutions do show a universal behaviour which is governed by the ratio λ^* as defined above (see Fig. 1). For excess salt ($\lambda^* < 1$) the properties of single polyions are at least principally observable, whereas for an excess of polyion charges ($\lambda^* > 1$) the strong coupling of polyions and coions together with the formation of large, unidentified objects (extraordinary phase) effectively prohibits a quantitative interpretation of most experimental quantities.

The behaviour of polyelectrolytes would not be as puzzling as it is if the facts described above were the only obstacles for experimental investigations. For all values of λ^*, the properties of the polyions are additionally governed by the absolute value of added salt concentration. At high salt content and low polyion concentration, i.e., $\lambda^* < 1$, the polyions indeed behave "normally". This is the well-known condition for characterizing polyelectrolyte chains without too

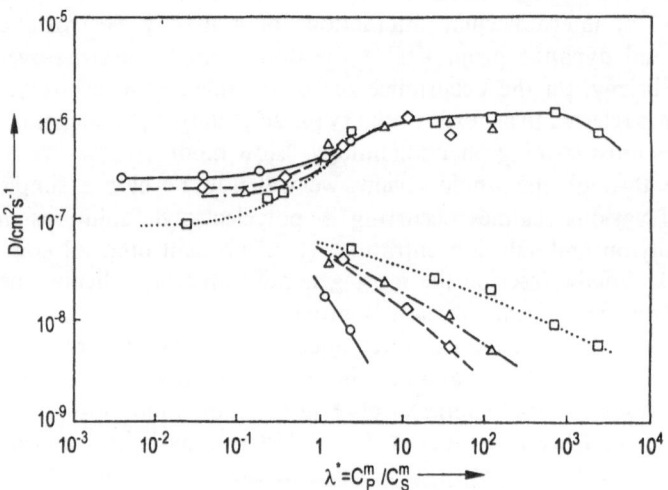

Fig 1. Diffusion coefficient of QPVP ($M_w = 200\,000$ g/mole) as a function of $\lambda = c_p^m/c_s^m$ for different added salt concentrations. $\square$ = "salt-free" $\approx 10^{-5}$ m, $\triangle = 10^{-3}$ m, $\diamond = 10^{-2}$ m, $\bigcirc = 0.1$ m. Data taken from [9]

---

[1]Until recently, the nature of the ordinary phase was also controversely discussed as the strong increase of the diffusion coefficient with increasing polyion concentration at constant salt concentration was interpreted in terms of scaling laws for semi-dilute polyelectrolyte concentrations. As it is now proven that the onset of the cross-over region does not depend on molar mass, the coupled mode picture unambiguously describes the ordinary phase best. See Sect. 3.2.3

many difficulties, although strong hydrophobic effects may lead to aggregation (especially for weakly charged chains) and obscure experimental results.

At the limit of very low "added" salt concentration[2], the strong electrostatic forces lead to the formation of surprisingly stable intermolecular structures which are most easily detected experimentally by scattering methods as inter-molecular interference (structure peaks). These intermolecular structure peaks are observed at very low added salt concentrations with an even lower polyion concentration ($\lambda^* < 1$) and for high polyion concentration with low added salt ($\lambda^* > 1$). It is this intermolecular interaction which obscures experimental results on the conformation of single polyions at low ionic strength, even in the comparatively simple regime $\lambda^* \ll 1$. And it is this intermolecular interaction which makes the already complex behaviour of polyelectrolytes for $\lambda^* > 1$ even less transparent.

In contrast to the rather well-defined "regimes" in terms of the parameter λ^*, the significance of the intermolecular structure on measurable quantities as a function of added salt is uncertain. At low salt concentration intermolecular interaction is easily detected as a structure (Bragg-) peak in the angular dependent scattering intensity of light, neutrons or X-rays. The addition of salt makes the maximum in the scattering intensity disappear. The problem is that the absence of the Bragg-peak in a scattering experiment does not necessarily prove intermolecular interaction to be completely ineffective. In fact, there is some evidence that intermolecular interactions affects experimental quantities in very dilute polyion solution at salt concentrations as high as 10^{-3} to 10^{-2} mol/l[3].

So far, the discussion has been concerned with the dynamical behaviour and structure of electrostatically interacting particles irrespective of the particle shape. However, for rigid particles, such as spheres or rods, the particle shape is known and can be more or less easily incorporated into theoretical model calculations, whereas the shape of flexible polyelectrolytes is changing with ionic strength. In addition to the above mentioned problems, such a conformational change imposes another complication on the understanding of flexible poly-electrolytes. The established picture of the conformation of flexible poly-electrolytes comprises a conformational transition from an almost ideal flexible coil at high ionic strength (where the electrostatic interaction is effectively screened) to a rod-like structure at low ionic strength. This textbook knowledge seemed to be at least qualitatively supported, if not proven, by a large number of theoretical and experimental investigations during the last 40–50 years. As will be discussed below, new experiments and calculations are most likely to disprove the rod-like nature of flexible polyions at low ionic strength. Instead, an

---

[2]"Added" salt also includes low molar mass ionic impurities which often cannot be excluded experimentally.
[3]Of course, the above arguments are not valid for infinitely dilute polyion concentrations. They do hold true, however, for the lowest concentration range of $c > 5 \cdot 10^{-4}$ g/l accessible by light scattering, viscosity and Kerr effect experiments.

expanded coil-like structure is favoured, which might also be quite well described by a flexible worm-like chain conformation.

In the following we will present details of the complex phenomena in polyelectrolyte solutions described above and hope to convince the reader that the basic understanding of polyelectrolytes is still far from being complete and currently subject to partial revision.

2 Intermolecular Interaction of Polyelectrolytes

As early as 1941 Bernal and Fankuchen [10] observed peaks in the scattering curve of concentrated virus solutions. Later, such peaks were also reported for concentrated solutions of bovine serum albumin (BSA), hemoglobin, and DNA [11]. It soon became clear that this phenomenon originates from intermolecularly oriented structures in the solution caused by the electrostatic interaction between the polyions. In recent years this general behaviour of polyelectrolytes has been extensively investigated. Nevertheless, many properties still remain unexplained and we are still far from accounting quantitatively for many experimental results. Even the qualitative interpretation of experiments is still a matter of considerable discussion and controversy.

As is already evident from the historical sketch, scattering experiments yield the most direct information on the structure of polyelectrolyte solutions. The scattered intensity $\mathfrak{J}(q)$ is related to the normalized scattering intensity $I(q)$ by the number of particles N in the scattering volume and by the squared contrast factor b

$$\mathfrak{J}(q) = Nb^2 I(q) \tag{2.1}$$

where q represents the usual scattering vector depending on the wavelength in the medium, λ, and on the scattering angle θ as $q = (4\pi/\lambda) \sin(\theta/2)$.

The normalized scattering intensity is a complex function of both the intermolecular structure factor S(q), which describes the distribution on the polyelectrolyte molecules in solution, and the intramolecular particle scattering factor P(q) which is the Fourier transform of the segment distribution (probability density) within one macromolecule. Only for rigid spheres the simple relation

$$I(q) = P(q) \cdot S(q) \tag{2.2}$$

holds true, but may also be applied to (form-) anisotropic particles if the mean distance between the particles is much larger than the particle dimension.

From $I(q)$ other physical properties can be calculated, e.g., $I(q = 0)$ is directly related to the osmotic compressibility, $d\pi/dc$, of the solution. The apparent diffusion coefficient, as measured by inelastic scattering techniques, is for many systems inversely proportional to the scattered intensity $I(q)$. Therefore it is also possible to extract information on the structure of polyelectrolyte solutions from diffusion measurements. For example, in the interesting case where a peak is observed in $I(q)$, the diffusion coefficient exhibits a minimum (see Sect. 3.1 for details). Also the shear viscosity is related to the distribution of polyions although this relation is much more complicated.

2.1 Theory

In this section we discuss various origins of intermolecular structures of the solute particles, i.e., their spatial distribution reflected in $S(q)$ only. This distribution may be completely random (disordered: $S(q) = 1$) or may originate from a well-defined crystalline lattice (ordered). The distribution of polyions varies between these two extreme cases, of which a pictorial description is provided by the pair-correlation function $g(r)$, expressing the probability of finding two particles separated by a distance r.

For a crystal, where particles are located at their lattice points, $g(r)$ consists of a series of sharp peaks at values r corresponding to the distances between the lattice points. For less perfect crystals (lattice defects, paracrystals) these peaks broaden, especially the higher order peaks. With increasing perturbation the long-range order is eventually completely lost and a liquid-like structure is obtained. The first and usually largest peak corresponds to a shell of nearest neighbours, and further peaks indicate further shells which are increasingly less correlated to the central particle. In the limit $r \rightarrow \infty$, $g(r)$ approaches unity, the normalized mean density of the system. Less correlated particles may merely exhibit a correlation hole characterized by the monotonic increase of $g(r)$ (see Fig. 2c). The inflection point of $g(r)$ corresponds to a mean distance below which other particles are expelled. For particles with a finite size, but otherwise non-interacting, $g(r)$ exhibits a discontinuity at the hard-core radius of the particle. For non-interacting point particles, $g(r) = 1$ (ideal gas).

Thus, depending on the strength of the electrostatic interaction, polyelectrolytes may form intermolecular structures caused by pure hard-core or by crystalline-like lattice potentials. Accordingly, the concepts describing the structure of polyelectrolyte solutions may be divided[4] into lattice models and polymer models with the application of liquid state theories.

Lattice theories apply to the limit of strong electrostatic interactions where the polyions form lattice-like structures. Polymer theories combine concepts developed for neutral polymers together with results from liquid-state theory. The latter focus on weak interactions, the rationale being that, in the case of vanishing electrostatic interaction, polyelectrolytes should behave like neutral polymers. So far there is no theory that covers the whole range from weak to

---

[4]Such a classification may appear artificial from the view of theoretical physics as will be discussed in detail, later (see Sect. 2.1.3)

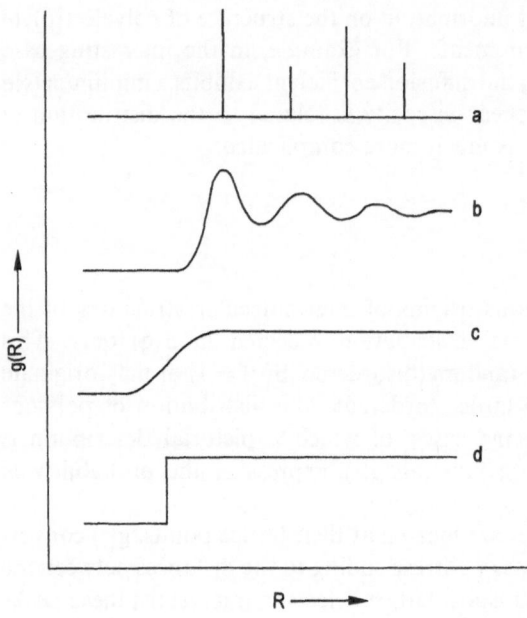

g(R)

R ⟶

Fig. 2. Schematic sketch of common density profiles for (*a*) crystalline, (*b* and *c*) liquid-like and (*d*) hard core structure

strong interactions. Here, Monte Carlo simulations may develop into a powerful tool in the future, particularly with the recent advent of sophisticated computers and algorithms.

2.1.1 Lattice Structures

When the polyions form a crystalline structure, sharp reflexes in the scattering curve are observed. From the first reflection at a scattering vector q_{max} the average nearest neighbour distance $\bar{d}$ is derived by applying Bragg's law

$$\bar{d} = \frac{2\pi}{q_{max}}. \tag{2.3}$$

Higher order reflections identify the type of lattice.

2.1.1.1 Spheres

It is known that highly charged latex particles crystallize into bcc or fcc lattices with interparticle spacings of many times the particle diameter. The ordering is, however, not perfect due to the Brownian motion of the latex particles, the finite size of the crystalline region, lattice vibrations, lattice defects, and lattice distortions [12]. As a result, the scattering curve exhibits no sharp reflexes as for crystals, but a rather broad first maximum with a few smaller higher order peaks. Some concepts have been developed to describe the relation between the degree of ordering and the shape of the scattering curve.

Lattice vibrations are accounted for by a Debye–Waller factor [13]. Usually, an increase in the temperature leads to large amplitudes of vibration which cause peak broadening and a decrease in peak intensity. Lattice distortions are described by paracrystal theories [13, 14]. Assuming a Gaussian distribution of particles around their equilibrium position, quantitative expression are derived for the scattering curves. Increasing lattice distortions also lead to peak broadening and strongly reduce peak intensities of higher order reflections. The peak positions are not affected except that, for large distortions, higher order reflections eventually disappear and only one broad main peak remains.

For polyelectrolyte solutions the relation between the mean interparticle distance $\bar{d}$ and the concentration is of great interest. Assuming that polyions fill the space homogeneously due to their strong mutual repulsion, the simple relation

$$\bar{d} = f_s \, c_p^{-1/3} \qquad (2.4)$$

holds, where c_p is the number concentration of the polyion and the factor f_s is determined by the lattice type (bcc, fcc, etc.) and follows simple geometrical considerations.

2.1.1.2 Rods

It was long believed that flexible polyions exhibit a rod-like conformation at low ionic strength. Due to the large anisotropy of rods, a different kind of packing is expected at concentrations higher than the overlap concentration $c_p^* \approx L^{-3}$, where L is the length of the rod. A simple cell model, which goes back to Lifson and Katchalsky in the 1950s [15] assumes the rods to adopt a parallel arrangement which leads to the following relation for the concentration dependence of $\bar{d}$:

$$\bar{d} = f_r \, c_p^{-1/2}, \quad c_p \gg c_p^* \qquad (2.5)$$

where again f_r depends on the lattice geometry (hexagonal, square etc.). The latter model appeared particularly attractive because the exponent $-1/2$ is frequently observed for the solution of flexible polyions at low ionic strength.

2.1.1.3 Flexible Polyions

De Gennes et al. [16] proposed a detailed model of various concentrations regimes of flexible polyelectrolytes at low ionic strength. At very low concentrations (dilute disordered regime) different polyion chains do not overlap and are very far from each other. At a slightly higher concentration the chains still do not overlap but the electrostatic interaction between the polyions becomes larger than the thermal energy. Here it is expected that the polyions build up a three-dimensional periodic lattice (dilute ordered regime). At still higher concentrations ($c_p > c_p^*$, semi-dilute ordered regime) where different chains overlap each other, de Gennes discussed several arrangements: (a) a hexagonal lattice of rigid rods, (b) a cubic lattice of rigid rods, and (c) an isotropic phase of entangled and thus partially flexible chains. It was suggested that case (c) represents the most probable structure. In the isotropic model each chain consists of connected, randomly oriented segments of length 1. Within one segment, electrostatic effects dominate and induce an elongated segmental conformation, whereas between different segments the interactions are completely screened. The average nearest neighbour distance for this model is derived as

$$\bar{d} = f_c\, c_p^{-1/2}, \quad c_p \gg c_p^* \tag{2.6}$$

with $f_c = \sqrt{2} \cdot f_r$. Thus, an exponent of $-1/2$ does not necessarily indicate an ordered array of rigid rods, but might also originate from an isotropic network of entangled chains.

Addition of salt yields the transitions from the ordered dilute to the disordered dilute regime and from the semi-dilute ordered to the disordered transient network.

On the basis of his well-known theory on the chain flexibility of intrinsically flexible polyions as a function of the ionic strength, Odijk [17] recalculated the transition concentrations and postulated one additional transition from the semi-dilute ordered regime to a disordered transient network with increased polyion concentration, even if no salt is added. An expression, derived from the lattice theory, for the corresponding "melting" transition is given. Odijk's paper on concentration regimes in polyelectrolyte solutions was the first quantitative approach to the difficult field of salt-free polyelectrolyte solutions based on purely microscopic properties such as contour length, charge density, and intrinsic persistence length of the polyions. It initiated a considerable number of experimental studies in order to test their quantitative predictions. Unfortunately, as polyelectrolyte solutions represent a fairly mean state of matter, hardly any of the theoretical predictions for the transition concentrations could be experimentally verified to better than a few orders of magnitude, except for some apparent accidents.

Kaji et al. [18] constructed a phase diagram for salt-free polyelectrolyte solutions as a function of polyelectrolyte concentration and molecular weight by means of five different concentration regimes. The semi-dilute regime is divided into a lattice, transition and an isotropic phase and the dilute regime into an ordered and disordered phase. The critical concentrations for the phase boundaries were calculated according to the theories developed by de Gennes and Odijk, utilizing the theoretical, electrostatic persistence length derived by Le Bret [19].

2.1.2 Simple Electrolytes

Lattice theories assume a strong electrostatic repulsion between polyions and yield information on the peak positions and intensities as a function of concentration and degree of ordering. They do not, however, provide quantitative information on weakly interacting particles. Moreover, it is desirable to calculate the structure factor S(q) for detailed microscopic models, e.g., ions of a certain diameter and charge at a given concentration.

The theory of simple electrolytes provides a scheme for calculating S(q) for a given interaction potential U(r) where the interaction strength varies via a screening parameter κ. From the structure factor S(q) important properties may be derived, e.g., the charge-charge structure factor which is related to the spatial distribution of the ionic charges in the solution, the osmotic compressibility and the shear viscosity. The diffusion coefficient which is connected to S(q) will be treated in Sect. 3.

2.1.2.1 Structure Factor

The theories on simple electrolytes most commonly applied to the study of polyelectrolytes are the Debye–Hückel (DH)[5] theory, the Mean Spherical Approximation (MSA)[5], and the Hypernetted Chain (HNC)[5] theory. Underlying models are systems of hard spheres with charge z and diameter d. In the simplest model, the restricted primitive model, the diameter d of anions and cations are identical and the charge is fixed at ± 1. There are, however, generalizations to an arbitrary number of different ions with different diameters and charges. Alternatively, the One Component Plasma (OCP)[5] consists of discrete point charges in the uniformly smeared, structureless background of opposite sign. The OCP was originally applied to the study of liquid metals, where the discrete mobile species corresponds to the metal ions and the background represents the valence electrons. The aim of these theories is to calculate the pair-correlation function g(f) for model systems of known interaction potential U(r). The experimentally observable structure factor S(q) is essentially the Fourier transform of this pair-correlation function.

The DH-potential has the form of a screened Coulomb potential

$$U(r) = \frac{ze}{4\pi\varepsilon\varepsilon_0 r} \exp(-\kappa \cdot r) \tag{2.7}$$

κ^{-1} is called Debye length or "screening" and represents the distance beyond which the electrostatic potential around one central ion is effectively screened by the locally induced charge distribution. The Debye length is connected to the concentration of all ionic species i in the solution by

$$\kappa^2 = \frac{N_a e^2}{\varepsilon\varepsilon_0 kT} \sum_i z_i^2 c_i. \tag{2.8}$$

---

[5]For a more profound introduction to these theories the reader is referred to the book of Schmitz [8]

In the limit $U \ll kT$, $g(r)$ simplifies to $1 - U(r)/kT$ and $S(q)$ is calculated to

$$S_g(q) = \frac{q^2}{q^2 + \kappa^2} \tag{2.9}$$

when the index g denotes $S(q)$ being derived from $g(r)$ instead of from $c(r)$ introduced below. For mathematical convenience the total correlation between particles expressed by $g(r)$ may be separated into a direct correlation in terms of the direct correlation function $c(r)$, and an indirect correlation which propagates via other particles. The relation between $g(r)$ and $c(r)$ is given by the Ornstein–Zernike (OZ) equation:

$$g(r) - 1 = c(r) + N \int c(|r - r'|)(g(r') - 1)dr' \tag{2.10}$$

and the structure factor in terms of $c(r)$ is given by

$$S_c(q) = \frac{1}{1 - Nc(q)} \tag{2.11}$$

with $c(q)$ the Fourier transform of $c(r)$.

The reason for introducing $c(r)$ is that it is shorter ranged than $g(r)$ and has simpler structure. It is therefore more convenient to make approximations to the form of $c(r)$ than to $g(r)$. A common approximation to the direct correlation function for weakly interacting particles is $c(r) \approx U(r)/kT$. The OZ-equation requires a closure, i.e., a relation that further specifies $g(r)$ and $c(r)$. Different closures are provided by the MSA [20, 21] and the HNC [22]. The main appeal of the MSA in the calculation of $g(r)$ and $S(q)$ is that, for the restricted primitive model both quantities are obtained analytically in a closed form. MSA results show good agreement with MC simulations and experiments for concentrated electrolyte or charged latex solutions, but are less reliable for strongly interacting systems where, when applied to the restricted primitive model, the HNC represents a considerable improvement over the MSA. Numerical solutions of the HNC equation are in excellent agreement with MC-simulations. Generalization of the MSA to multicomponent systems with an arbitrary number of components, different diameters and charges have also been given in closed analytical form [23–25].

In the following, we will briefly discuss the influence of ionic strength, polyion concentration, and particle charge on the ionic structure, i.e., $S(q)$ and $g(r)$, on the basis of results obtained by MSA and HNC. With increasing particle charge $S(0)$ initially decreases corresponding to a decrease in the isothermal compressibility and $S(q)$ becomes a monotonically increasing function q. At even higher charge densities $S(0)$ continues to decrease, but one broad main peak appears in $S(q)$ and eventually higher order peaks develop. The corresponding main maximum in $g(r)$ at $r = \bar{d}$ defines a shell of nearest neighbours and further secondary peaks originate from further shells. With increasing polyion concentration the structure becomes more pronounced and the peak in $S(q)$ is shifted to higher q-values.

The addition of salt screens the interactions and gradually destroys the structure. At constant polyion concentration the position of the main maximum is shifted to larger scattering vectors as a function of ionic strength [26], because the weakening of the electrostatic repulsion allows the particles to penetrate into previously excluded interparticle space thereby decreasing the distance $\bar{d}$. The resulting rearrangement in the first shell of neighbouring particles is reflected in a change of the lattice type expressed by the value of f_s [27] (Eq. 2.4). For example at low ionic strength the number of nearest neighbours corresponds to a fcc structure, whereas with increasing ionic strength it approaches the value of a simple cubic structure. At high salt concentration the pair distribution function will eventually change towards a simple step function describing the pure hard-core interaction in simple liquids.

2.1.2.2 Mixtures of Electrolytes

Often we have to consider mixtures of electrolytes such as a polyelectrolyte solution with added salt. The generalization of the above equations to multi-component systems is, in principle straightforward. From a given pair potential $U_{\alpha\beta}(r)$ for components α and β, the correlation functions $g_{\alpha\beta}(r)$ or $c_{\alpha\beta}(r)$ are calculated, from which the partial structure factors $S_{g,\alpha\beta}(q)$ or $S_{c,\alpha\beta}(q)$ may be derived. The scattered intensity is a linear combination of all partial structure factors:

$$\Im(q) = \sum_{\alpha\beta} b_\alpha b_\beta I_{\alpha\beta}(q) \tag{2.12}$$

$$I_{\alpha\beta}(q) = P_\alpha(q) S_{\alpha\beta}(q) \tag{2.13}$$

where b_α and $P_\alpha(q)$ are the contrast factor and form factor of particle α, respectively. Often, especially for polyelectrolyte solutions, it is customary to treat solutions of different electrolytes as effective one-component systems. For example, if the structure of the main component A (polyelectrolyte) is experimentally accessible, the structure of component B (salt) is of minor importance only, and merely considered to screen electrostatic interactions of A. To implement this simplification into the calculation scheme $U(r) \rightarrow (g(r), c(r)) \rightarrow (S_g(q), S_c(q))$, the potential $U_A(r)$ (Eq. 2.7) is calculated for component A with the charge z_A of component A and a Debye length κ_B^{-1} due to the screening provided by component B. Of course this differentiation into contributions from A and B is artificial. It could be argued that component A also contributes to the screening length κ^{-1} which it usually does via the contribution of the gegenions to the screening length. Sometimes it is helpful to split κ into contributions from all ionic components:

$$\kappa^2 = \sum_\alpha \kappa_\alpha^2 \tag{2.14}$$

$$\kappa_\alpha^2 = \frac{e^2 z_\alpha^2 c_\alpha}{\varepsilon\varepsilon_0 kT} \tag{2.15}$$

For the often encountered case that polyelectrolyte and salt dissociate into the same type of counterions, we have three contributions, κ_p due to the polyions, κ_c due to co-ions, and κ_g due to the counterions (gegen-ions). The question now arises as to which of these contributions should enter the total screening length κ^{-1}. The discussion is simplified if the salt concentration is much higher than the polyelectrolyte concentration. Then the screening of the salt ions dominates the total screening and $\kappa^2 \approx \kappa_s^2$, where $\kappa_s^2 = \kappa_c^2 + \kappa_g^2$. The subscript s indicates the contribution due to the salt ions. If the salt concentration is comparable with the polyelectrolyte concentration, the contribution due to the counterions dissociated by the polyelectrolyte are often taken into account as well.

The effective one-component approach works particularly well if the polyions are much larger in size and charge compared to the co-ions and a distinction between both types of ions is obvious.

2.1.2.3 Coupling Constants

The thermodynamic state of an ionic system is conveniently characterized by the reduced density $\rho^* = d^3/\bar{d}^3$, with d the ion size and $\bar{d}$ the mean distance between different ions and a reduced Coulomb parameter $\beta^* = l_B/\bar{d}$ [27], where l_B is the Bjerrum length given by

$$l_B = \frac{e^2}{4\pi\varepsilon\varepsilon_0 kT}. \tag{2.16}$$

β^* simply represents the ratio of electrostatic and thermal energy at a distance $\bar{d}$ between two charges. Thus, a single coupling constant may be introduced for simple electrolytes interacting via a Debye–Hückel potential:

$$\xi^* = \kappa^2\bar{d}^2 = 4\pi\rho^*\beta^*. \tag{2.17}$$

Both MSA and HNC reduce to the DH-result in the weak coupling limit $\xi^* \ll 1$ because the DH-limiting law is valid when ion-size effects become negligible, i.e., $d \ll \bar{d}$. In the strong coupling limit, g(r) develops oscillations characteristic of systems with short-range order. Such oscillations are a result of competition between hard-core packing and local charge neutrality.

The above relations have been introduced for the restricted primitive model. There is no straightforward generalization for multi-component systems. A solution containing polyelectrolyte and salt is often treated as an effective one-component system of polyions, where the counterions and salt ions merely screen the electrostatic interactions between polyions. Accordingly, the coupling between polyions may be expressed by the ratio of coupling constants $\xi_p^*/(1 + \xi^*)$ with ξ_p^* calculated from the charge and concentration of polyions and ξ_s^* from the charge and concentration of counterions and salt ions. The coupling at a length scale of the scattering vector q^{-1} is given by [27]

$$\chi^*(q) = \frac{\kappa_p^2}{q^2 + \kappa_s^2}. \tag{2.18}$$

At the thermodynamic limit $q \to 0$, $\chi^*(0)$ reduces to the "Donnan" second virial coefficient. For multi-component systems similar coupling constants are derived. For example, in Sect. 3.2 the coupling constant λ^* already mentioned in the introduction is introduced which yields the degree of coupled diffusion between polyions and counterions, the only difference being that the index s is replaced by the index c corresponding to co-ions.

2.1.2.4 Charge–Charge Structure Factor

An interesting property of ionic systems is reflected in the charge–charge structure factor $S_{zz}(q)$ which is expressed by the linear combination of partial structure factor [28]

$$S_{zz}(q) = \sum_{\alpha\beta} z_\alpha z_\beta S_{\alpha\beta}(q) \tag{2.19}$$

where z_x are the charges of different particles α and β. The importance of the charge–charge structure factor arises from its direct relation to the dielectric constant ε which in turn relates the ionic structure to dielectric properties of the system.

In the DH-approximation, the charge–charge structure factor is given at the limit $q \to 0$:

$$S_{zz}(q) = \frac{q^2}{\kappa^2}. \tag{2.20}$$

This equation may be used as an operational definition of κ^{-1} as a screening length obtained from the slope of $S_{zz}(q)$ at low q. For simple electrolytes, a DH-type q-dependence of $S_{zz}(q)$ similar to Eq. (2.9) has been derived. This corresponds to a correlation hole around a given charge with depletion of like charges and an enhancement of opposite charges. In the low q limit, as expected, Eq. (2.9) reduces to Eq. (2.20). MD-simulations on molten salts shows that charge ordering results in a very sharp main peak in the charge-charge structure factor. The generalized q-dependent dielectric constant $\varepsilon(q)$ is introduced as [29]

$$\frac{1}{\varepsilon(q)} = 1 - \frac{\kappa^2}{q^2} S_{zz}(q) \tag{2.21}$$

which is usually defined through the density of dipoles ρ as $\varepsilon(q) = 1 - \rho_{ind}(q)/\rho(q)$, where ρ_{ind} is induced by the external field. The density of dipoles is directly related to the charge–charge structure factor $S_{zz}(q)$ which describes the spatial distribution of charges. In the limit $q \to 0$ the assumption of "perfect screening" requires $\varepsilon(0) \to \infty$.

Since the charges are embedded in a solvent, the problem of the dielectric response of the solvent and respective solute–solvent interactions is encountered. In a first approximation the solvent is simply treated as a medium with dielectric constant ε that alters the electrostatic interaction through Eqs. (2.7) and (2.8). Polar solvent molecules may exhibit their own dielectric structure, thus giving rise to a contribution that rescales the total dielectric constant as $\varepsilon(q)/\varepsilon_{LM}(q)$.

2.1.2.5 Osmotic Pressure

The osmotic compressibility is related to the scattered intensity by

$$\frac{kT}{I(0)} = \left(\frac{\partial \Pi}{\partial c}\right). \tag{2.22}$$

Π is usually developed into a virial expansion where the second virial coefficient depends on the distribution and potential of the particles:

$$\frac{\Pi}{cRT} = \frac{1}{M} + A_2 c + \cdots \tag{2.23}$$

The second virial coefficient A_2 may be calculated directly from $g(r)$. Since the potential $U(r)$ is divided into a hard-core repulsion and an electrostatic interaction, the second virial coefficient is composed of the excluded volume v and an electrostatic part $(A_2 c = \kappa_p^2/\kappa_s^2)$ simply given by $\chi^*(0)$.

2.1.2.6 Viscosity

Statistical mechanics also relate the shear viscosity of simple electrolytes to the structure factor $S(q)$. Hess and Klein [30] derived the excess shear viscosity η for weak and strong interactions, the latter of which was obtained from a mode-mode coupling approach with the result

$$\eta = \frac{kT}{2} \frac{c_p^2}{(2\pi)^3} \int\int S(q,t)^2 \left(\frac{q}{S(q)^2}\right)^2 \left(\frac{\partial h(q)^2}{\partial q}\right)^2 dq dt. \tag{2.24}$$

In Eq. (2.24) the total correlation function is defined as $h(q) = [S(q) - 1]/N$ and $S(q,t)$ is the dynamic structure factor which will be discussed in Sect. 3. As the weak coupling limit, Eq. (2.24) simplifies for an effective one component system to [30]

$$\eta = A f_0 \frac{\kappa_p^4}{\kappa_s^3} \tag{2.25}$$

with A a numerical constant and f_0 the friction factor.

2.1.3 Polyelectrolytes

The theory of simple electrolytes provides expressions for the intermolecular structure factor $S(q)$ and allows the calculation of the charge-charge structure factor, the osmotic pressure and shear viscosity. Polyions differ enormously in size and shape from simple electrolytes. Even for simple spherical polyions, additional forces have to be considered, such as van-der-Waals interactions, hydrodynamic interactions and effective surface potentials which complicate the theoretical calculations considerably. Still, compared to rod-like and flexible polyions, the behaviour of spherical polyions is well understood. The static and dynamic behaviour of non-spherical and flexible polyelectrolytes is additionally governed by a significant increase of the internal degrees of freedom which introduce cross-correlations between inter- and intramolecular properties (leading, e.g., to the breakdown of Eq. (2.2)) and by the different characteristics of the concentration regimes below and above the overlap concentration, c_p^*.

An important complication arises here because the intermolecular structure factor as introduced in Eq. (2.13) has now become a function of the form factor $P(q)$, i.e., the distribution of polyions depends on their mutual orientation and their shape and vice versa. It is only in the case of spherical polyions that $S(q)$ and $P(q)$ are separable by the use of center-of-mass coordinates. For rod-like polyions the mutual orientation and the spatial distribution are correlated, and for flexible polyions the chain conformation and the spatial distribution of chains depend on each other. Assuming weak interactions, several approximations were introduced to separate form- and structure factor. However, for strong, long-range electrostatic interactions intra- and intermolecular correlations cannot yet be properly separated [28]. This is an important limitation to all current theories except for Monte Carlo simulations.

2.1.3.1 Spheres

Solutions of charged latex spheres have been extensively studied and as a result an essential quantitative understanding of the equilibrium properties has emerged. For spheres the form factor $P(q)$ is already known and the scattered intensity is simply given by $I(q) = P(q) S(q)$. The charge–charge structure factor for these systems is easily obtained under the zero-average-contrast condition (zac), $\sum_i b_i c_i = 0$, where simply $I_{zac}(q) = P(q) S_{zz}(q)$. In this case not all partial structure factors as indicated in Eq. (2.19) have to be measured because cross-correlations do not exist.

Large spherical polyions are usually treated as an effective one-component system where the interaction between the polyions is given by a hard sphere potential plus a repulsive screened Coulomb potential (DLVO model) [31]. The screening of the polyion interactions is entirely due to the charges and concentrations of counterions and salt ions. As a result, the polyions interact via an effective charge z_{eff} or an effective surface potential. The value of z_{eff} depends on how the correlations between the polyions themselves and between polyions and counterions are theoretically formulated. All models discussed so far lead to an effective interaction in terms of screening arguments. A more detailed theory is required to consider the small ions in the system explicitly. Different approaches taking into account small ions [32–34] yield concrete values for the effective charge or surface potential. In practice, z_{eff} is often determined by fitting the height of the main maximum in S(q) to the MSA or rescaled MSA (RMSA) theory, because z_{eff} is not easily available by experimental techniques with no underlying theoretical model. Solutions of the Poisson–Boltzmann (PB) equation gave good agreement between fitted and calculated values of the effective charge.

In addition to hard-core and electrostatic interactions, van-der-Waals dispersion forces also have to be included, especially if the polyions are only weakly charged. For dynamic properties at higher concentrations, hydrodynamic interactions become important.

2.1.3.2 Rods

In contrast to spherical ions, a proper evaluation of the static structure factor for rod-like polyions by means of pair distribution functions is not available so far. As mentioned earlier, the difficulty arises from the coupling of the form factor and structure factor, i.e., the orientation of the rods depends on the position and orientation of the neighbouring rods. The distribution function for rod-like particles may be written as $g(r, u_1, u_2)$ which includes both the rod-to-rod distance r and the mutual orientation of two rods given by the vectors u_1 and u_2 [35]. From this distribution function the averaged distribution function $\bar{g}(r)$, the analog to g(r) in the case of spherical ions, and the structure factor S(q) are principally accessible.

For neutral rods the hard-core interaction leads to orientational correlations at higher concentrations. As predicted by Onsager and Flory [36, 37], at a critical concentration a solution of rods phase separates into an isotropic and an anisotropic nematic phase. The latter is characterized by a long-range orientational and a liquid-like positional order. Charged rods may be simply viewed as rods with a larger effective diameter of the order of κ^{-1}, but the orientation of the rods is complicated by the fact that electrostatic interactions tend to disorient the rods because the electrostatic repulsion is strongest for a parallel array and weakest for a perpendicular orientation.

A perturbation approach leads to an analytical expression for the pair-distribution function of weakly interacting short rods and is in good agreement with MC-simulations [38]. A more general treatment was recently proposed by Maeda [39] for charged rods in the isotropic regime.

Given the structure factor S(q), the scattering intensity for dilute solutions with negligible orientational correlation may still be calculated directly by I(q) = P(q)S(q) with the known form factor for rods. This procedure might also be applied to more concentrated solutions, e.g., above the overlap concentration, but could yield erroneous results as the degree of orientational correlation is not known.

For rod-like polyions the effective charge z_{eff} has been calculated from a computed distribution of counterions around the polyion. e.g., on the basis of the Poisson–Boltzmann equation employing a cell model [40, 41].

2.1.3.3 Flexible Polyions

Flexible polyions exhibit many internal degrees of freedom and the treatment of inter/intra-molecular coupling is quite hopeless for current theoretical approaches. Therefore, approximate treatments assume a given conformation for the polyions which is not altered by intermolecular interactions and utilizes the simplified expression for the scattered intensity $I(q) = P(q) S(q)$. Ignoring cross correlations between $P(q)$ and $S(q)$, the ionic strength dependence of both quantities may be calculated separately, i.e., the form factor from Fixmann's or Odijk's treatment of the chain conformation (see Sect. 4) and $S(q)$ from simple electrolyte theories possibly modified by single contact approximations. Also, multi-component theories have been explicitly formulated for the structure factor which become quite complicated as a set of OZ-equations has to be solved for charged polymer segments, co-ions and counterions.

2.1.3.4 Structure Factor

The simplest approach to calculate the structure factor for flexible polyions utilizes $S_g(q)$ as defined in Eq. 2.9 [42]. Interestingly enough, the calculated scattering intensity $I(q)$, which is approximately evaluated by the product of $P(q)$ and $S(q)$, exhibits a single broad peak as is observed in many experiments on polyelectrolyte solutions. (Fig. 3). The structure factor $S_g(q)$ resembles the so-called correlation hole first introduced by de Gennes in polymer melts [43]. In this model each polymer is surrounded by a correlation hole of radius ξ, from which other polymers are strongly expelled. For polyions the length ξ is identified with the Debye length κ^{-1}. Mathematically, the peak originates from the product of a monotonously decreasing, $P(q)$, and a monotonously increasing, $S(q)$, function of q. Assuming a rod-like conformation of the polyions and keeping to the high q-regime ($P(q) \propto q^{-1}$) leads to a $q_{max} \propto c_p^{1/2}$ -law as observed in many experiments.

Similar but more detailed calculations have been performed by Benmouna et al. [44] utilizing the form factors of spheres and rods and introducing various effective potentials to account for the Coulomb repulsion. All calculations lead to a peak in $I(q)$ as a consequence of the correlation hole as described previously. Koyama [45–47] calculated the scattering curve assuming a Gaussian distribution of rod-like segments and derived qualitatively the same results as Hayter et al. [42]. The structure factor for the correlation hole was approximated by the function $S_k(q) = 1 - h \exp[-\xi^2 q^2/4]$, with h being an adjustable interaction parameter. The results show quantitative agreement with the experimentally

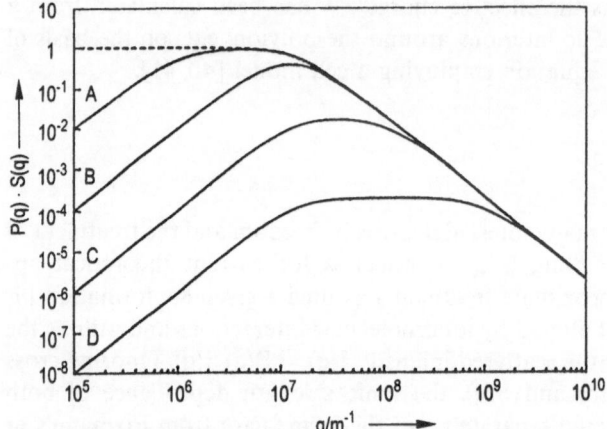

Fig. 3. Calculated scattering curves for a correlation hole utilizing Eq. (2.9) for the structure factor with $\chi^{-1} = 10^6$ Å (A), 10^7 Å (B), 10^8 Å (C) and 10^9 Å (D). The form factor was calculated by the Debye function for random coils with $R_g = 1000$ Å (from [121])

determined q_{max}-values in the dilute and semi-dilute regime by somewhat arbitrarily adjusting the interaction parameter to $h = 1$. The calculated c_p-dependence of q_{max} changes from $c_p^{1/2}$ in the semi-dilute regime to $c_p^{1/3}$ in the dilute regime with a cross-over concentration proportional to M^2 typical for the overlap concentration c_p^* of rigid rods.

Since 1986, theories have been developed which employ the concept of direct correlation functions. The total correlation is factorized into a contribution from intermolecular correlations, $c_{eff}(q)$, and intramolecular correlations, $P(q)$, the form factor of the polyion. Benmouna et al. [48, 49] presented a generalization of the single contact approximation due to the Zimm [50] to express the direct correlation function as

$$c_{RPA}(q) = c_{eff}(q) P(q) \qquad (2.26)$$

with similar assumptions as in the Random Phase Approximation (RPA) or the Gaussian Approximation in the Edwards Hamiltonian formalism [51]. As a result, these theories are expected to be valid for systems with small density fluctuations, i.e., for concentrated or dilute solutions where the single contact approximation between polymer segments and polymer coils, respectively, represent a good approximation. For neutral polymers, c_{eff} corresponds to the excluded volume parameter v and becomes identical to c(r) for weakly interacting particles, i.e., $c_{eff}(q) \approx U(q)/kT$. Generally, such approximations yield correction hole structure factors, i.e., S(q) is an increasing function of q and exhibits no peak. Using the DH-potential for U(q), the scattered intensity is derived as

$$I_{RPA} = \frac{P(q)}{1 + \chi^*(q)P(q)}. \qquad (2.27)$$

$\chi^*(q)$ is the q-dependent polyelectrolyte coupling constant discussed earlier in this section (Eq. 2.18).

Utilizing the unusual series expansion of the form factor $P(q)^{-1} = 1 + \frac{1}{2}R_g^2 q^2$, the position of the peak is calculated as [52]

$$q_{max} = \left(\frac{\sqrt{2\kappa_p}}{R_g} - \kappa_s^2 \right)^{1/2} \tag{2.28}$$

which yields the experimentally testable predictions $q_{max} \propto c_p^{1/4} z_p^{1/2}$. The intensity at peak maximum is

$$I(q_{max}) = \left(1 + \sqrt{2}R_g \kappa_p - \frac{1}{2}R_g^2 \kappa_s^2 \right)^{-1} \tag{2.29}$$

leading to the power relation $I(q_{max}) \propto c_p^{-1/2} z_p^{-1}$.

The theories discussed so far treat the polyelectrolyte solution as an "effective one component" problem, i.e., the co-ions and counterions are only included by screening the interactions and allowing for total electroneutrality. Grimson et al. [53] introduced a multi-component generalization of the earlier models which explicitly and consistently treats the finite size and non-uniform distribution of the co-ions and counterions. The partial structure factors of the various ions are calculated from a set of OZ-equations which complicates calculations. In order to decouple form factor and structure factor it is assumed that the monomers of one chain interact with the monomers of a different chain as if they were located at the center-of-mass of each chain. In the currently most sophisticated model, Genz et al. [54, 55] provide a detailed outline of the assumptions necessary to decouple the set of equations and thereby decoupling intra- and intermolecular correlations. The concrete calculations neglect the influence of direct interaction between different polymers on the chain conformation and the form factor is evaluated from Odijk's theory (see Sect. 4). Such multi-component theories principally allow one to study separately the influence of each ionic component on the total scattered intensity as a function of polyelectrolyte concentration, salt concentration, charge density, and molecular weight.

The theoretical predictions for some important quantities are summarized in Table 1.

Recently, weakly charged polyelectrolytes have attracted considerable interest. With decreasing electrostatic interactions the excluded volume v becomes more important and appears as an additional interaction term in the expression for the scattered intensity:

$$I_{RPA} = \frac{P(q)}{1 + [v + \chi^*(q)]P(q)}. \tag{2.30}$$

The excluded volume may adopt negative values for poor solvents, i.e., below the θ-temperature. As most polyelectrolytes consist of an intrinsically hydrophobic backbone, which is mainly kept in solution by the dissociating ionic groups in the side-chain, such a negative excluded volume is likely to occur. With negative v-values the denominator in Eq. (2.30) might become zero leading

Table 1. Exponents for $q_{max} \propto c_p^\beta$, $I(q_{max}) \propto c_p^\gamma$, $q_{max} \propto z_p^\delta$ and $q_{max} \propto c_s^\varepsilon$ according to various theoretical predictions described in the text

Theory	P(q)	S(q)	β	γ	δ	ε
Hayter	rod	$S_g(q)$	1/2	$-1/2$	1/2	
Koyama	rod	$S_k(q)$	$c_p \gg c_p^*$: 1/2	$-1/2$		
			$c_p \ll c_p^*$: 1/3			
RPA	coil	$S_{RPA}(q)$	1/4	$-1/2$	1/2	≈ 0
Genz	worm-like		1/3			

to a diverging scattering intensity at a certain scattering vector q_{sp} which is usually interpreted as spinodal decomposition. If $q_{sp} = 0$, then phase separation occurs macroscopically as observed in polymer blends or solutions. $q_{sp} \neq 0$ leads to microphase separation as encountered in block copolymer melts.

Theoretical studies of weakly charged polyelectrolytes postulate a number of interesting phenomena such as the existence of several concentration regimes with characteristic conformational properties [56, 57], chain collapse [58], microphase separation with lamellar, cylindrical or spherical morphologies [59] as observed in block copolymer melts [60], or micellar systems [61]. A qualitative physical argument for the occurrence of microphase separation is that macroscopic phase separation would result in too severe a loss of counterion translational entropy. Peaks in the scattered intensity of weakly charged polyelectrolytes are thus not to be interpreted solely as correlation holes or ordered structures, but also in terms of microphase separation. In the first case, individual polyelectrolyte chains preferably assume a certain mean particle distance, whereas, in the latter, polyelectrolyte microphases are formed, which are separated by the continuous aqueous phase on a length scale of q_{sp}^{-1}. The latter explanation, however, encounters the problem that experimentally observed q_{max}-values correspond to distances of the order of, or even smaller than, the dimensions of a single polyion resulting in phase separation on a submolecular scale.

For mixtures of two different polyelectrolytes of like or opposite sign [62], no considerable change in the scattering intensity is predicted on the basis of Eq. (2.30), since the sign of the charges enters the interaction terms as z^2, which does not alter the results qualitatively. Application of Eq. (2.30) to polyion mixtures of unlike charges appears to be unnecessary because experiments reveal the difficulty from mixtures of positively and negatively charged polyelectrolytes forming so-called polyelectrolyte complexes or simplexes, which eventually macroscopically phase separate. The topic concerning mixtures of polyelectrolytes and neutral polymers has also been considered theoretically [62]. The addition of neutral polymers is expected to screen the excluded volume parameter v. Thus, in the case of weakly charged polyelectrolytes, the scattered intensity depends critically on the detailed balance between excluded volume v and electrostatic interaction $\chi^*(q)$. No decisive experiments to test these predictions have yet been performed.

Finally, we would like to comment on a controversy as to whether the peak in the scattered intensity originates from a correlation hole or from ordered structures. The $c_p^{1/2}$-dependence in the semi-dilute regime is in both theories due to a stretched conformation of polyions on the length scale of the mesh-size of the transient network of polyion chains. If the intermolecular distance $\bar{d}$ in the lattice equals the size of the correlation hole ξ then the correspondence is exact. The difference in both approaches is the expected q-dependence of the structure factor $S(q)$. In the case of the correlation hole, $S(q)$ is a steadily increasing function of q, whereas for an ordered polyelectrolyte solution $S(q)$ would exhibit peaks. Since the structure factor may vary gradually from a correlation hole-type to a liquid or lattice-type function depending on the interaction strength, a strict determination, as often formulated in publications on this topic, might not be entirely justified.

2.1.3.5 Charge–Charge Structure Factor

For flexible polyions, the charge-charge structure factor is not as easily obtained under zero-average contrast condition as for spherical polyions, because the partial form factors and structure factors are correlated. In general the charge-charge structure factor is derived from the partial structure factors as indicated by Eq. (2.19). For systems without added salt the distribution of counterions around a polyion has to be known in order to calculate partial structure factors. With the assumption that flexible polyions at low ionic strength are locally rod-like ("local stiffness approximation", see Sect. 4) the cell model as discussed for rod-like polyions was utilized to calculate effective charges and counterion distributions. [63]

2.1.3.6 Osmotic Pressure

The expression for the second virial coefficient of polyelectrolytes, $A_2 c = \kappa_p^2/\kappa_s^2$, is derived from the Donnan potential [64]. It originates from a non-uniform spatial distribution of the counterions in order to fulfill the local electroneutrality condition. As a consequence the concentration of the counterions is enhanced in the vicinity of the polyion.

Above the overlap concentration, i.e., in the semi-dilute regime, the relation between osmotic pressure and concentration has been successfully calculated by scaling arguments of the form

$$\Pi \propto c_p^\alpha, \quad c_p \gg c_p^*. \tag{2.31}$$

For neutral polymer solutions $\alpha = 9/4$, while for polyelectrolyte solutions (treated as an effective one-component system) Odijk calculated the exponent α

to be 9/8. Bloomfield and Wang [65, 66] employed the RNG theory to evaluate the osmotic pressure in the dilute and semi-dilute regime, the results of which were in essential agreement with those of Odijk.

2.1.3.7 Viscosity

The relevant parameter is the reduced viscosity η_r, which is defined by

$$\eta_r = \frac{\eta - \eta_0}{\eta_0 c_p} \tag{2.32}$$

with η and η_0 the viscosity of the solution and of the solvent, respectively. For neutral polymers and for polyelectrolytes at high ionic strength, the reduced viscosity is expanded in a virial expansion of the concentration,

$$\eta_r = [\eta](1 + k_H[\eta]c_p + o(c^2)\dots) \tag{2.33}$$

with $[\eta]$ the "Staudinger Index" and k_H the always positive Huggins coefficient which is not yet theoretically quantified for polyions in solution.

Contrary to Eq. (2.33) the reduced viscosity of polyelectrolyte solutions is observed to decrease strongly with increasing concentration at low ionic strength. In the late 1940s this experimental fact inspired Fuoss and Strauss [67–70] to propose their famous empirical linearization of the reduced viscosity

$$\eta_r^{-1} = A + Bc_p^{1/2} \tag{2.34}$$

with A and B some physically meaningless constants. For almost 50 years Eq. (2.34) remained the central relation for the evaluation of the viscosity data of charged polymers at low ionic strength, because by means of Eq. (2.34) essentially all experimental viscosity data could be linearized and extrapolated to zero concentration in order to obtain the intrinsic viscosity $[\eta] = A$. Whereas the determination of $[\eta]$ by the Fuoss-Strauss relation is now known to be one of the capital errors in polyelectrolyte history, Eq. (2.34) is easily derived from Eq. (2.25) by Hess and Klein at the limit $c_p \gg c_s$ (corresponding to $\chi^*(0) \gg 1$), the universal validity of which explains the "success" of the Fuoss-Strauss formula when applied to many different experimental data. Equation (2.25) also reproduces the well-known result of isoionic dilution, i.e., $\eta_r \propto c_p$ when $\kappa_s = \text{const.}$

In 1987 Witten and Pincus [71] presented a theory for the viscosity of polyelectrolyte solutions which was derived for concentrations near the overlap concentration. Again, at the limit $c_p \gg c_s$ the Fuoss law $\eta_r \propto c_p^{-1/2}$ was obtained. Later Rabin [72] derived a similar relation (for the viscosity of polyelectrolyte solutions) on the basis of the theory by Hess and Klein. With some bold simplifications Rabin arrived at

$$\eta_r = A' f_0 l_B^2 \frac{c_p}{\kappa_S^3} z_p^2 \tag{2.35}$$

where f_0 is the hydrodynamic friction coefficient of the polyion. Equation (2.35) is in formal analogy to Eq. (2.25), the difference being that a term z_p^2 has disappeared. This equation also reduces correctly to the Fuoss-law and the isoionic dilution limit. One feature of Eq. (2.25) and Eq. (2.35) is that both yield a maximum in the reduced viscosity as a function of concentration if a constant background ionic strength of the solvent is accounted for. If the friction coefficient is proportional to the molar mass (i.e., in the free draining limit) the position of the maximum c_{pmax} is proportional to c_s, and becomes independent of molecular weight, whereas the height of the maximum is proportional to M and is predicted to decrease with increasing salt concentration.

Borsali et al. [73] started with the general Eq. (2.24) derived by Hess and Klein and evaluated the shear viscosity by replacing the dynamic structure factor by the mean field expression $S(q, t) = \exp[-Dq^2 t / S(q)]$ and assuming Gaussian chain statistics for the calculation of the form factor. Numerical calculations of the resulting integral show the peak position to vary as $c_{p,max} \propto c_s^1 z^{-2} M^{-1}$, the peak height to decrease with increasing salt concentration and to increase with increasing charge and molecular weight. Hydrodynamic interactions do not alter the behaviour of the viscosity qualitatively, but reduce its value dramatically by almost two orders of magnitude. In a subsequent paper, Borsali and Rinaudo [74] generalized this theory to two-component systems, i.e., mixtures of polyelectrolytes.

Another interesting approach to the viscosity problem has been presented by Nishida et al. [75], who calculated the viscosity for interacting spheres (point

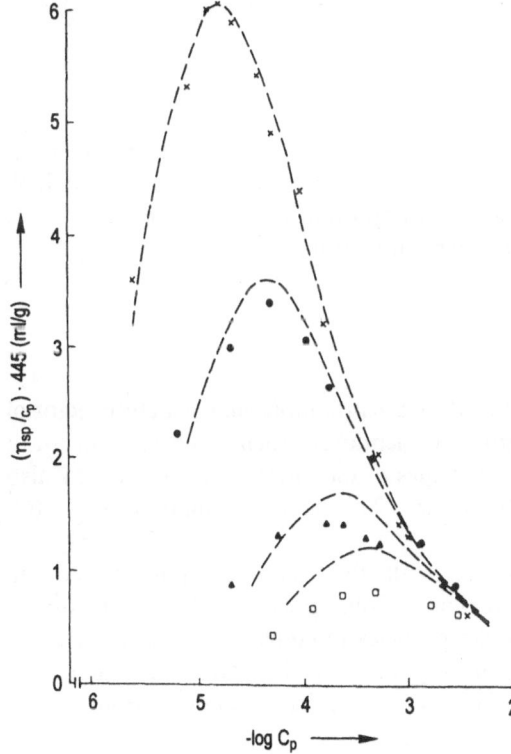

Fig. 4. The reduced viscosity η_r $= \eta_{sp}/c$ for a NaPSS ($M_w = 16\,000$) as a function of the polyion concentration c_p^m for different added salt concentrations: $x = 4 \times 10^{-6}$ m, $\bullet$ $= 1.1 \times 10^{-5}$ m, $\blacktriangle = 6 \times 10^{-5}$ m and $\square = 1.04 \times 10^{-4}$ m. The *dotted curves* represent the theoretical predictions by Eq. (2.35). Data taken from [72]

particles) according to an old theory of Rice and Kirkwood [76]. The inter-action between the particles is described by a Debye–Hückel potential which exhibits a maximum as a function of polyion concentration. The maximum in the interaction energy is reflected in the maximum of the viscosity. As will be later discussed in more detail (see Sect. 4.3.1), the calculation may be utilized to separate formally the influence of the interaction and of the conformational change of the polyion on experimental viscosity data.

2.2 Monte Carlo Simulations

Ionic systems, such as simple electrolyte solutions or molten salts, were success-fully characterized by MC-simulations which sometimes also provide the only "experimental" data available in order to test the validity of various approxima-tions employed in the respective theories when "real" experiments are difficult or impossible to perform. With the advent of powerful computers and algorithms, MC-simulations are also performed successfully on polyelectrolyte solutions. Important contributions of MC-simulations to the understanding of charged systems are particularly expected for polyions with internal degrees of freedom, such as rod-like and flexible polyions, where the coupling between intra- and intermolecular correlations imposes great problems and restrictions to analy-tical theories.

2.2.1 Spheres

For dilute solutions of spherical ions, MC-simulations confirm the calculations from MSA and HNC theories [26, 76–79] but, as expected, deviate at high charge densities. Because this topic is not important for the goal of the present review we shall not further discuss such simulations.

2.2.2 Rods

In the vicinity of the overlap concentration the intermolecular structure factor of rods depends increasingly on the mutual orientation. Such effects are difficult to handle analytically, because subtle changes of the interaction energy may also favour the formation of nematic phases. This is the domain where MC-simulations become very helpful.

Early MC-simulations studied successfully the cross-over regime from dilute to semi-dilute solutions of charged rigid rod-like polyions [80, 81]. In highly diluted solution ($c_p \ll c_p^*$) the rod-like particles resemble pointlike objects and the solution properties are similar to those of charged spheres. With increasing concentration the distribution function $g(r, u_1, u_2)$ increasingly depends on the orientation of the axes u_1 and u_2 of the rods and on the distance vector r, because the orientation starts to contribute significantly to the interaction energy between the rods. As a result, the structure factor becomes less pronounced and broader compared to spherical particle systems which is intuitively understood as follows. When the mean distance of the rods becomes comparable to the rod length, one given segment of a rod interferes with all segments of the more or less randomly oriented neighbouring rods, i.e., a certain "Bragg distance" is much less unique. The system of charged rods remains globally isotropic showing no sign of nematic ordering up to the highest investigated concentrations of 2.3 c_p^*, well in the regime with $c_p^{1/2}$-scaling. This is a result of the already mentioned tendency of strongly interacting rods to acquire a perpendicular rather than a longitudinal orientation. As a consequence, the scaling relation $d \sim c_p^{1/2}$ does not necessarily require a parallel arrangement of rods.

The addition of salt leads to the screening of interactions thereby decreasing the peak height and shifting q_{max} to larger q-values until finally the maximum vanishes completely. The shift in q_{max} is qualitatively the same as for spherical particles, i.e., a rearrangement of particles in the nearest neighbour shell.

2.2.3 Flexible Polyions

Many MC-simulations have been performed to study the conformation of single polyions (see Sect. 4.2). So far only one group has focused on many chain systems, thereby additionally taking the counterions explicitly into account.

Stevens and Kremer [82–84] calculated the osmotic pressure of a system of up to 16 chains of 64 beads. Their simulation showed two scaling regimes. In the high density regime they found a scaling exponent of 9/4, characteristic of semi-dilute neutral polymer solutions. For densities below the overlap concentration, the data are consistent with a 9/8 value as predicted by Odijk's scaling theory. The simulation extended to lower densities where the noninteracting limit $\Pi = kTc_p(1 + 1/N)$ is reached.

One striking aspect of the osmotic pressure data is the chain length dependence of the cross-over density. The cross-over density corresponds approximately to the condition $\xi^* = 1$. At higher concentrations, Coulomb interactions are completely screened and the osmotic pressure is that of neutral polymers. As seen in Fig. 5a and b, the simulation data compare well with the experimental results.

The structure factor exhibits a peak whose position varies with concentration. The calculated q_{max} exhibits two scaling regimes as shown in Fig. 6. At low concentrations the peak position scales as $c_p^{-1/3}$, whereas at high concentrations the peak position scales as $c_p^{-1/2}$. The cross-over density corresponds approximately to the overlap concentration c_p^*. These results are in qualitative agreement with Eqs. (2.5) to (2.6).

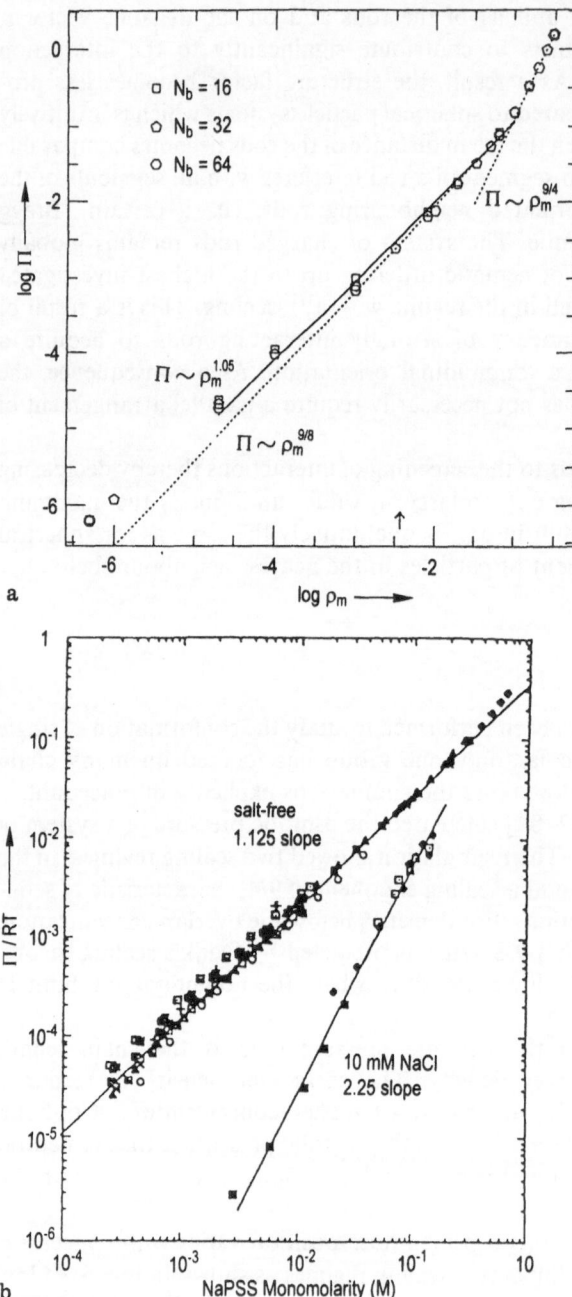

Fig. 5. a Double logarithmic plot of the osmotic pressure π vs the monomer density ρ_m derived by molecular dynamics simulations. *The dotted curves* represent the theoretical exponents 9/8 and 9/4, respectively, and *the full curves* are the best fit to the data (from [83]). **b** Experimental values for the osmotic pressure π as a function of the polyion concentration c_p^m for various molar masses. (Taken from [66])

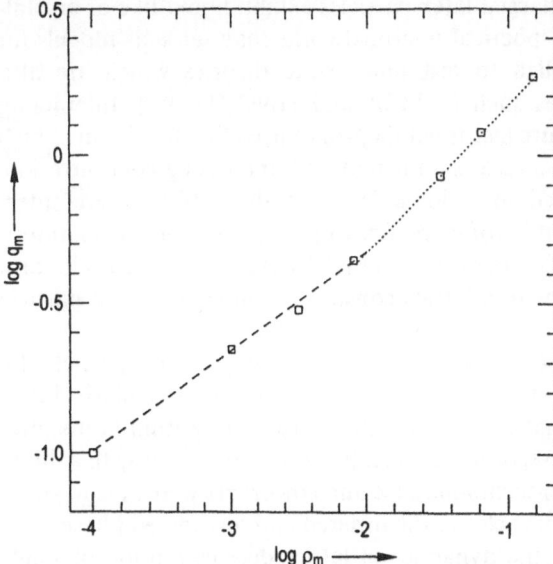

Fig. 6. Position of the peak on the structure factor, q_m, as a function of the monomer density ρ_m. At low densities $q_m \sim \rho_m^{1/3}$ (*dashed line*) and at high densities $q_m \sim \rho_m^{1/2}$ (*dotted line*). Taken from [82]

2.3 Experiments

Scattering experiments provide the most direct insight into interacting particles in solution and will be predominantly discussed in the following. Polyelectrolyte solutions resemble multicomponent systems and accordingly, the total scattered intensity I(q) contains contributions from all components, which make it difficult to extract the quantity of interest. Fortunately, contributions from different components vary considerably due to their difference in contrast factors.

If one of the contrast factors, say b_α, is much larger than all the other contrast factors, then $I_{\alpha\alpha}$ dominates the total scattered intensity. This is anticipated by deuterium labelling in SANS studies or by heavy atom labelling or scattering at the absorption edge in SAXS experiments. In light scattering the polyions usually dominate the scattered intensity as long as the scattered intensity is proportional to the molar mass which is no longer true for infinitely low polyion concentration at infinitely small ionic strength. In SAXS studies one has to be even more careful, since small ions such as Na, Br, etc. scatter more effectively than the hydrocarbon polyions.

2.3.1 Spheres

Charged spheres have been thoroughly studied experimentally and correspond to a model system in many respects [85, 86]. Depending on the charge and concentration, solutions of charged latex particles exhibit liquid-like or crystal-line-like order. Lattices and spherical micronetworks may serve as models for simple liquids and are applied to test liquid-state theories which are also employed for polyelectrolytes such as MSA and HNC [86–96]. Interacting microspheres exhibit a structure factor with a pronounced first maximum which shifts to higher q-values and increases in height with increasing concentration. Liquid-like order is observed at volume fractions $\Phi < 0.001$, where inter-molecular interactions are still strong enough to produce strong correlations.

If $\Phi > 0.001$, formation of colloidal crystals is observed [94, 96–101] in form of regular bcc or fcc lattices with a lattice constant as large as several particle diameters.

Ise and coworkers [102–104] studied solutions of charged latex particles by SAXS and ultramicroscopy. Their micrographs showed a particularly inter-esting phenomenon, the formation of a two-state structure. Within hours after the sample preparation, a dense, ordered phase is formed coexistent with a dilute disordered phase. Using video-techniques, Ise and coworkers were able to study the Brownian motion of the particles in the ordered and disordered phase, thus obtaining information about the dynamics of lattice defects, motion of single particles around their lattice position, and the dynamics of particles at the interphase between the ordered and disordered regime.

The driving force for the formation of the ordered phase is still unclear and has been a matter of much controversy. It has been claimed by Ise and Sogami [105, 106] that the densely ordered phase originates from attractive interactions between polyions, a point of view that has been heavily criticized by others [107]. For a detailed discussion of this controversy see [8].

Comparing micrographs and SAXS-patterns obtained from identical sam-ples, Ise has correlated the degree of ordering (order parameter) and the shape of the scattering curves, which exhibit rather broad scattering peaks. The increase in temperature leads to larger amplitudes of the vibrations causing a decreasing peak intensity and peak broadening which has been qualitatively observed for poly(L-lysine) (PLL). A description of the degree of ordering, i.e., lattice distor-tions, is provided by the paracrystal theory. A fit to the scattering curve for lysozyme, a globular protein, yielded a second moment of distribution of polyions around the lattice points which indicated quite large distortions of the polyion lattice [13].

The viscosity of spherical polyions has also been studied by Ise et al. [108, 109]. A peak of the reduced viscosity was observed as a function of concentra-tion similar to flexible polyions. The significance of this observation will be discussed in Sect. 2.3.3.

2.3.2 Rods

The Tobacco Mosaic Virus (TMV) and fd viruses were the first model systems for rod-like structures investigated by SLS and SANS [110–113]. Whereas solutions of the TMV are easily studied below the overlap concentration c_p^*, fd viruses are mainly investigated in the semidilute regime, as the overlap concentration is extremely low because of the very large contour length.

Under low salt conditions, the scattered intensity $I(q)$ exhibits pronounced maxima, the intensity of which increases and its position shifts to higher q values with increasing concentration. The structure factor $S(q)$ was calculated from the total scattered intensity $I(q)$ via the approximate relation $S(q) = I(q)/P(q)$ utilizing the form factor $P(q)$ measured at low concentrations, i.e., without interactions. This appeared to be justified by the observation that the solutions were optically isotropic, i.e., exhibit no orientational order. A study of the magnetic field birefringence in the isotropic phase [114] shows local angular correlations to occur only at high concentration. The evaluated $S(q)$ progressively develops a peak with increasing concentration. The interparticle distance $\bar{d}$ as calculated from the position of the main peak scaled as $c_p^{-1/3}$ below and as $c_p^{-1/2}$ above the overlap concentration. Also the prefactors agreed qualitatively with the predictions given in Sect. 2.1.1. The cross-over from the dilute to the semidilute regime is clearly seen in Fig. 7.

A solution of the TMV was also studied under variation of the ionic strength (denoted as series 1–3 in Fig. 7), which was simply assumed to be proportional to the conductivity of the solutions. The addition of salt leads to a decrease in the peak height and to a shift of the peak position to larger q-values. This is in agreement with studies on spherical polyions where rearrangements in the nearest neighbour shell upon weakening interactions are the cause of the shift in the peak position (see Sect. 2.1.2).

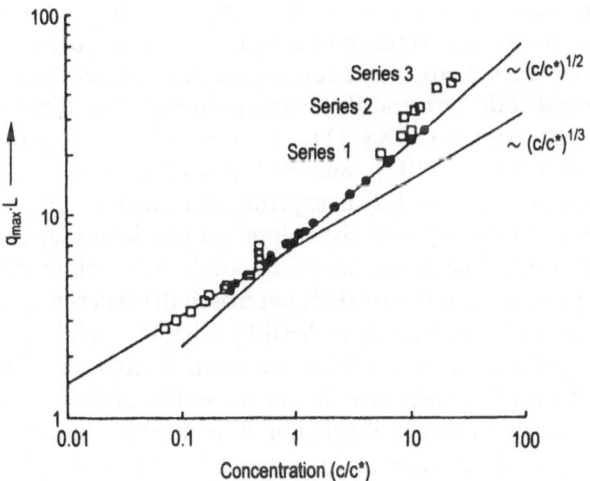

Fig. 7. Product $q_{max}L$ (L the rod length) as a function of the reduced polyion concentration c_p/c_p^* with c_p^* the overlap concentration for different rod-like polyions (●: fd-virus: □: TMV series 1: $c_s^m \approx 10^{-3}$ m, series 2: $c_s^m \approx 4.7 \times 10^{-3}$ m, series 3: $c_s^m \approx 1.7 \times 10^{-2}$ m). *The solid lines represent the scaling behavior with the exponents 1/3 and 1/2, respectively (taken from [113])*

2.3.3 Flexible Polyions

The determination of S(q) from the measured intensity I(q) is not unambiguously possible because the chain extension and the respective form factor of an intrinsically flexible polyion are not a priori known. Experimentally, a peak in the scattering intensity is often observed at low ionic strength, the physical origin of which is not easily related to one of the theoretical models discussed in Sect. 2.1.3, i.e., correlation hole, lattice structures or microphase separation. Nevertheless, the peak position and height have been studied as a function of c_p, c_s, and the polyion charge z and the corresponding mean interparticle distances $\bar{d}$ have been extracted by formal application of the Bragg equation. Also, isotopic labelling techniques were utilized to determine the partial structure factors and the charge–charge structure factor, respectively. The osmotic pressure and the viscosity of polyelectrolytes have been studied in detail over a wide range of concentrations and will be discussed below in detail.

2.3.3.1 Mean Interparticle Distance

In Fig. 8a the concentration dependence of the average intermolecular distance $\bar{d}$ is plotted as a function of monomer molar concentration c_p^m for poly-sodiumstyrenesulphonate (NaPSS) (SLS: [115, 116], SANS: [118, 119]) and poly-2-vinyl-pyridiniumbenzylbromide (QPVP) (SLS: [9, 120, 121], SANS: [9]).

A power law behaviour $\bar{d} \sim c_p^\alpha$ is observed in each of two concentration regimes which are characterized by the exponent $-1/2$ and $-1/3$ respectively, the latter of which is not unambiguously detected in the light scattering studies at very low polyion-concentration. The cross-over from $-1/3$ to $-1/2$ is clearly observed at approximately the overlap concentration c_p^* of the polyions[6] in Fig. 8b which shows the expected cross-over concentration to become larger with smaller molecular weight of the intrinsically flexible polyelectrolyte. Similar results are also obtained with NaPSS (SAXS: [117, 122], SANS: [123]), poly-sodiumacrylic acid (NaPAA) (SAXS: [109]), and PLL (SAXS: [110]). In the semi-dilute concentration regime ($\alpha = -1/2$) all experimental data fall into one common line, which coincides perfectly with the theoretical prediction for the isotropic network model (Eq. 2.6). The dotted line corresponds to the cell-model of fully stretched rod-like polyions and clearly does not match the experimental data at high concentrations. This was already realized by Nierlich et al. [118], who pointed out that the polyions must exhibit some chain flexibility in this regime and postulated a kind of parallel ordering of worm-like chains. Subsequently, magnetic birefringence studies by Weill [126] have demonstrated that no strong orientational correlations, such as in a parallel array or rod-like

---

[6]It should be stressed again that the change of the exponent α constitutes the only vague experimental hint for the existence of an overlap concentration in solutions of intrinsically flexible polyions at low ionic strength

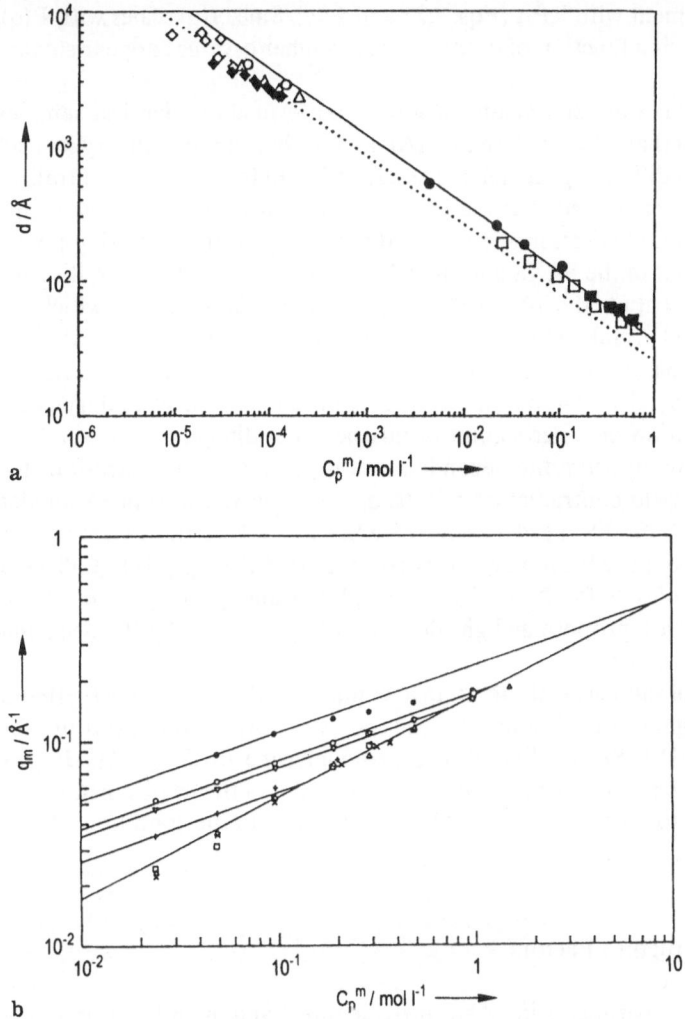

Fig. 8. a Double logarithmic plot of the "Bragg" distance d (derived by $d = 2\lambda/q_m$) as a function of the polyion concentration c_p^m for different polyions as reported in the literature (taken from [121]). **b** q-vector of the maximum scattering intensity, q_m, as a function of the polyion concentration for different molar masses (NaPPS). ●: $M_w = 1.8 \times 10^3$ g/mol; ○: 4.6×10^3 g/mol; ▽: 8.0×10^3 g/mol; + : 1.8×10^4 g/mol; △: 10^5 g/mol; ×: 2.2×10^5 g/mol; □: 1.2×10^6 g/mol. (Taken from [18])

polyions, exist in the semidilute regime. Thus, neither rod-like nor possibly extended flexible polyions form nematic phases in highly concentrated solutions without added salt.

When the charge density is lowered by variation of the pH-value for BSA (SAXS: [13]), NaPSS (SAXS: [124, 127]), and PLL (SAXS: [125], SANS: [128, 129]), q_{max} is shifted to lower q-values and the maximum intensity, $I(q_{max})$, decreases with increasing polyelectrolyte concentration or charge density in qualitative agreement with RPA (Eqs. 2.28 and 2.29). Since $I(q)$ scales with $P(q)$, which is a decreasing function of q, the peaks at higher q-values appear smaller in height.

SAXS and SANS measurements have been performed on identical samples of polysodiummethacrylic acid (NaPMA) and poly-lithiummethacrylic acid (LiPMA) yielding different peak heights as a function of the degree of neutralization [130]. It was suggested that the observed differences might be due to the formation of a dense hydration shell around the polyions, the scattering power of which is different in the SAXS and the SANS experiments. However, it seems more natural to attribute the increase to the presence of Na or Li ions, which are introduced upon charging of the polyions because of their rather large scattering length for neutrons. In any case, the expected trend of a decreasing scattering intensity caused by electrostatic interactions is obscured by a secondary effect and an interpretation of experimental results becomes difficult.

Experiments monitoring the dependence of q_{max} on salt concentrations for flexible polyions yield contradicting results. q_{max} is observed to shift to smaller values for NaPAA (SAXS: [124]), PLL (SAXS: [131]), NaPSS (SAXS: [122]), and t-RNA (SAXS: [132]). However, it is also reported that q_{max} is not affected by the addition of salt for NaPSS (SANS: [118]) and proteoglycane (SLS: [133]). RPA theories propose a slight decrease of q_{max} caused by the decaying form factor at smaller q.

The amount of salt necessary to remove completely the peak in the scattered intensity is roughly proportional to the polyelectrolyte concentration for NaPSS (SLS: [115], SANS: [118]) and proteoglycane (SLS: [133]). It was conjectured that the Debye length must be of the order of the nearest neighbour distance in order to form a structure. This corresponds to the condition $\xi^* = 1$ (see Sect. 2.1.2).

2.3.3.2 Partial Structure Factors

Isotopic labelling experiments in SANS provide direct and unambiguous access to the structure factor $S(q)$. A first experiment based on this technique was performed by Williams et al. [134] and subsequently by Nierlich et al. [119, 131] on deuterated NaPSS, and Plestil et al. [135] on NaPMA. The solvent consists of a mixture of H_2O and D_2O, the composition of which is chosen to reduce the contrast of the non-deuterated polyion to zero. The form factor and the structure factor are then obtained by extrapolating the results from a series of measurements with increasing amounts of deuterated polyions at constant total polyion concentration to a vanishing concentration of deuterated and non-deuterated polymer, respectively. The experiments demonstrate that the structure factor $S(q)$ exhibits a peak. The peak position shifts to higher q-values with increasing polyelectrolyte concentration. Also, in some SAXS [136] and light scattering experiments [116], attempts were made to obtain information about $S(q)$ by assuming a particular form factor for the polyions.

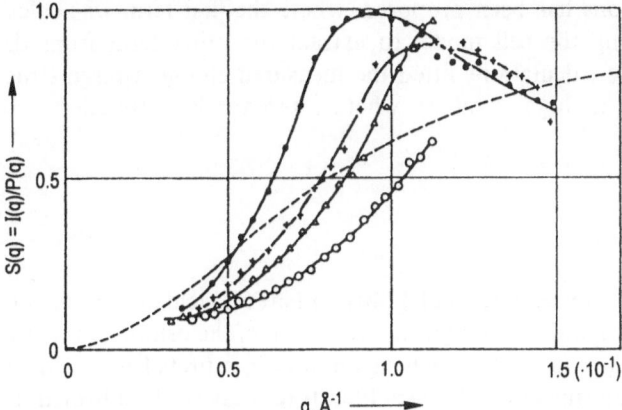

Fig. 9. Experimentally determined structure factor $S(q) = I(q)/P(q)$ for NaPSS at different polyion concentrations. $c_p^m = 0.211$ mol/l (●), 0.338 mol/l (+), 0.422 mol/l (△) and 0.591 mol/l (○). *The dashed line* represents the theoretical curve according to Eq. (2.9). (Taken from [119])

All labelling and "normal" scattering studies encounter serious difficulties at low q-values because an anomalous steep increase of the scattered intensity is observed for polyelectrolyte solutions. The significance of this intense small angle scattering has not yet been fully acknowledged and is mostly ignored in SAXS or SANS literature. The small-angle scattering has, however, its dynamic correspondence in a slow-mode as observed in inelastic light scattering experiments. Interpretations of this phenomenon will be discussed (Sect. 3.3).

With isotopic labelling experiments, not only the polyion-polyion structure factor, but also other partial structure factors such as counterion-counterion structure factors, have been determined [137, 138]. First experiments provide evidence that the counterion–counterion structure factor also exhibits a peak at nearly the same position as the peak in the polyion-polyion structure factor. This is interpreted to originate from condensation of counterions onto the polyions. A similar observation was also made using SAXS by studying polypotassiumstyrenesulphonate (KPSS) and NaPSS [139], where the scattering curve is mostly determined by the distribution of counterions because of their high contrast factor.

2.3.3.3 Charge–Charge Structure Factor

From the partial structure factors the charge–charge structure factor $S_{zz}(q)$ is calculated according to Eq. (2.19). Experiments show that $S_{zz}(q)$ is an order of magnitude greater in polyelectrolytes than in simple electrolytes. This difference is attributed to counterion condensation. A separation of the charge–charge structure factor into a term corresponding to condensed counterions and uncondensed counterions has been proposed, where the first term was calculated theoretically using the cell model to extract the other term from the experiment [63]. These calculations fitted the measured charge–charge structure factor quite well at high q-values, where interparticle correlations are negligible.

2.3.3.4 Osmotic Pressure

Detailed osmotic studies by Mandel et al. [140] on NaPSS yield an exponent of 9/8 at low concentration and 9/4 at high concentration of the osmotic pressure-concentration power law ($\pi \sim c_p^\alpha$) which again was interpreted as a dilute-semidilute concentration transition. A recent literature study [66] confirmed the experimental scaling exponents but clearly demonstrated that the cross-over concentration does not depend on the molar mass of the polyions.

Comparison to the MC-simulations and theory indicates that the osmotic pressure is most likely governed by intermolecular interactions and not by changes in the polyion conformation. Intermolecular interactions at low concentration correspond to an excluded volume of κ^{-1} at higher concentration compared to the hard-core diameter d. The cross-over concentration, as stated earlier, corresponds to $\kappa^{-1} = \bar{d}$ or, equivalently, to $\xi^* = 1$.

2.3.3.5 Viscosity

As early as 1932 Trommsdorf [141] and Staudinger [142] observed the viscosity of PMA in pure water to increase dramatically with decreasing polyion concentration. The same observations were made later by Kern [143]. This is in contrast to the viscosity behaviour usually found for neutral polymers. Later, Fuoss and Strauss developed their famous empirical extrapolation and explained the "polyelectrolyte effect" by a coil-to-rod transition because, upon dilution, the ionic strength decreases, eventually leading to a fully stretched polyion as anticipated at the time.

Later, some publications postulated the reduced viscosity to decrease again at very high dilution [144–147], thus questioning the validity of the Fuoss-Strauss extrapolation procedure. However, due to the enormous experimental difficulties (influence of dust particles, atmospheric carbon dioxide altering the ionic strength, adsorption problems) these results were mostly considered as unreliable. In recent years measurements with highly sensitive and sophisticated viscometers [148–158] provided overwhelming evidence for the presence of a maximum and proved the Fuoss-Strauss extrapolation to be one of the "big errors" in polyelectrolyte history.

The maximum in the reduced viscosity has also been observed for aqueous suspensions of charged spheres [108, 109] which cannot undergo a conformational transition, as already discussed in Sect. 2.3.1. These experiments strongly indicate that the conformation of the macroion and intramolecular effects are not the main factors in determining the dynamics of polyelectrolyte solutions, but rather intermolecular, electrostatic interaction as theoretically formulated in Sect. 2.1.3.

The features proposed by Eq. (2.35) seem to be supported by experiments, some of which were, however, not extrapolated to zero shear. More careful investigations showed the reduced viscosity to exhibit a pronounced shear thinning [159–162] which is particularly pronounced at the maximum of the viscosity [149, 150, 154–156].

The linear relation between the reduced viscosity and concentration upon isoionic dilution has been experimentally verified [163–167]. For high salt concentration where $\chi^* \ll 1$, κ is fairly constant upon dilution of the polyelectrolyte component, similar to the case of isoionic dilution. Thus there is no maximum and the data may be easily extrapolated to zero concentration in order to determine the intrinsic viscosity.

2.3.3.6 Dynamic Kerr Effect

In the dynamic Kerr effect experiment an electric field is applied to the solution which orientates the dissolved molecules exhibiting an induced or permanent dipole [168]. As a consequence the solution becomes birefringed, monitored by the light intensity passing through crossed polarizers. When the electric field is turned off, the decay of the birefringence is monitored and yields the rotational diffusion coefficient in the simplest case (for details see [168]). Unfortunately, the high electric fields usually applied (some thousand volts per centimeter) require low conductivity in the solution in order to avoid Joule heating and to maintain, with time, a constant electric field. This usually limits the applicability of this method to "salt-free" polyelectrolyte solutions for $c_p < 1\%$. Dynamic Kerr effect measurements were also applied to determine the shape of flexible polyions at low ionic strength (see Sect. 4.3.4) all of which ignore the role of intermolecular interaction between the polyions.

Only recently a dynamic Kerr effect study on QPVP and NaPSS [169] investigated the effect of intermolecular interaction on the field-free decay of the birefringence at very low polyion concentration without added salt, with the remarkable result that a maximum of the decay time is observed in each of the two relaxation processes at approximately the same concentrations where the viscosity displays a maximum and the scattering techniques show a structure peak (Fig. 10). Whereas the longer relaxation process is identified as the overall rotation of the polyion, the fast relaxation has become a matter of extensive speculation.

For stiff molecules such as DNA, which also exhibit two relaxation times in the field-free decay of the birefringence [170, 171], the fast relaxation is thought to originate from hydrodynamic effects which bend the rod-like molecules to arcs or even to horseshoe structures [172–174] when the molecules migrate

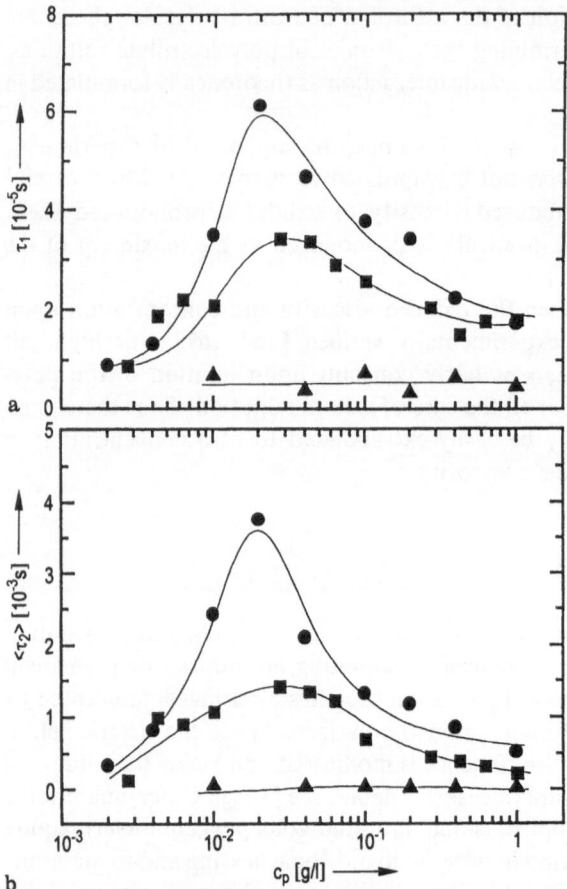

Fig. 10. a, b Fast and slow (respectively) relaxation times as a function of polyion concentration c_p in "salt-free" solution for different molar masses. ● $M_w = 2.4 \times 10^6$ g/mol, ■ $M_w = 9.75 \times 10^5$ g/mol, ▲ $M_w = 2.4 \times 10^5$ g/mol. (Taken from [169])

within a short pulse of the electric field. When the field is turned off the structure relaxes quickly to its unperturbed shape before the overall rotation eventually leads to the isotropic state.

If such a hydrodynamic deformation is postulated for intrinsically flexible polyions, the actual chain stiffness remaining more or less uncertain at such low ionic strength (see Sect. 4.3), the question remains to be answered how the intermolecular interaction can possibly affect the relaxation of the hydrodynamic distortion and cause longer decay times.

It has also been attempted to interpret the fast decay of the birefringence by the relaxation of partly coiled segments at one end of the polyion chain [175, 176], which could possibly originate from a locally increased ionic strength due to migration of the gegenions along the contour to one end of the chain in an electric field. When the field is turned off the gegenions quickly rearrange to the equilibrium distribution and the coiled segments relax to their equilibrium conformation before the rotational diffusion becomes effective. Again, it is not easily understood why intermolecular interaction should slow down the local relaxation of the chain.

At higher, probably semi-dilute, polyion concentration, sign-inversion of the decaying birefringence is observed, particularly if one long single electric pulse is applied to the polyelectrolyte solution [176, 177], the origin of which is also not clear. Ordering of polyion clusters perpendicular to the orientation of single polyions in the electric field has been proposed [177] as well as perpendicular orientation by hydrodynamic forces. Also, the experimental conditions become increasingly less defined at higher polyion concentrations because of the high conductivity of the solution.

In conclusion the dynamic Kerr effect measurements on polyions are far from being understood and more sophisticated experimental data, such as light scattering data of the polyions in an electric field, are needed.

3 The "Ordinary–Extraordinary" Concentration Regimes

First investigations in the 1970s, mostly on biopolymers such as DNA, demonstrated that the dynamic behaviour of polyelectrolyte solutions compared to neutral polymer solutions is much more spectacular because two separate relaxations were often observed. Both relaxations turned out to originate from diffusion processes and it was the theoretically still unexplained slow diffusion which stimulated the fantasies of scientists to create terms such as "ordinary–extraordinary transition", "jeu de molecules somnolentes", or more practical, "splitting of relaxation times" as often found in the literature. During the past twenty years numerous speculations on the nature of these relaxations arose considering effects such as coil-to-rod transitions, two-state structures, cage-effects, aggregation, coupled diffusion, non-draining–draining transitions, or polyion specific effects (e.g. for DNA). So far it has not been possible to discuss all experimental results within the framework of one particular theory.

Despite the uncertainties discussed above, considerable knowledge on the dynamical behaviour of charged particles has been gained from inelastic scattering experiments. In the following we will briefly describe the relevant theoretical background and experimental results. For details we refer the reader to the review by Schurr and Schmitz [178] or to the textbook by Schmitz [179].

3.1 Inelastic Scattering

The basic quantity measured by inelastic scattering techniques, such as dynamic light scattering or neutron spin echo experiments, is the mutual diffusion coefficient as introduced within the framework of irreversible thermodynamics

$$D_m = \frac{M}{N_A f}\left(\frac{\partial \Pi}{\partial c}\right) \tag{3.1}$$

with M the molar mass, N_A the Avogadro constant, f the friction coefficient, and $\partial \Pi/\partial c$ the osmotic compressibility, the concentration dependence of which is usually approximated by a virial expansion in the dilute regime (Eq. 2.23) or by scaling laws in the semi-dilute regime (Eq. 2.31). The mutual diffusion coefficient, D_m, sometimes also referred to as cooperative diffusion coefficient, generally differs from the tracer diffusion coefficient, $D_{tr} = kT/f$, which is measured by techniques probing the mean square displacement of particles with time. Although identical at infinite dilution both diffusion coefficients show a different behaviour as a function of concentration, as will become evident at the end of this section. In the following we will mostly discuss the mutual coefficient and therefore drop indices for the sake of simplicity.

Inelastic scattering techniques monitor correlations of concentration and/or density fluctuations. In dilute and semi-dilute concentrations, concentration fluctuations are related to the mutual diffusion coefficients by fluctuation-dissipation theories, whereas density fluctuations are generally neglected.

The experimentally accessible quantity is the intensity line correlation function $g_2(t)$ which is connected to the theoretically simpler correlation function of the electric field, $g_1(t)$ via the Siegert relation

$$g_1(t) = \left(\frac{g_2(t) - A}{A}\right)^{1/2} \tag{3.2}$$

with A an experimentally determined base time. Usually $g_1(t)$ is expressed as the ratio of the dynamic and the static structure factor

$$g_1(t) = I(q, t)/I(q). \tag{3.3}$$

In the following we confine the discussion to the first cumulant of the correlation function which is easily available by experiment and theory even though the time decay of $g_1(t)$ is quite complicated:

$$\Gamma \equiv [d \ln g_1(t)/dt]_{t \to 0} \tag{3.4}$$

Frequently, an apparent, q-dependent diffusion coefficient is introduced as

$$D_{app}(q) = \Gamma/q^2 \tag{3.5}$$

which yields the mutual diffusion coefficient in the zero q limit:

$$D = \lim_{q \to 0} D_{app}(q). \tag{3.6}$$

Besides polydispersity effects, which are neglected in the following discussion, internal modes of motion for flexible polymers, rotational diffusion for rigid rods, hydrodynamic interaction and long range intermolecular potentials may cause the partly pronounced q-dependence of $D_{app}(q)$.

For a detailed discussion of experimental observations given later, we briefly summarize the characteristic behaviour of $D_{app}(q)$ in terms of the intermolecular structure factor $S(q)$, the intramolecular form factor $P(q)$ and of the hydrodynamic interaction $H(q)$ for the important polyion structures: sphere, coil and rod. In Sections 3.1.2–3.1.4 we assume for simplicity that $I(q) \cong S(q)$. $P(q)$ which is, however, exact for rigid spheres, only. (See Section 2.)

3.1.1 Rigid Spheres

For rigid spheres only the intermolecular structure factor $S(q)$ causes the q-dependence D_{app} if hydrodynamic interactions are neglected, which only become important at very high concentrations [180]:

$$D_{app}(q) = \frac{D}{S(q)}. \tag{3.7}$$

For non-interacting spheres $D_{app} = D$ independent of q. The product of the two experimentally accessible quantities $I(q) \cdot D_{app}(q)$ exhibits the q-dependence of the particle form factor

$$I(q)\, D_{app}(q) = D \cdot P(q). \tag{3.8}$$

3.1.2 Free Draining Coils

The particle scattering factor additionally affects the apparent diffusion for coil-like structures which may be physically viewed as the effect of the internal modes of motion. Accordingly [179]

$$D_{app}(q) = \frac{D}{P(q)\,S(q)} \tag{3.9}$$

and

$$D_{app}(q) \cdot I(q) = D + f(q). \tag{3.10}$$

For non-interacting coils, i.e., $S(q) = 1$, Eq. (3.10) reduces to the result first obtained by Büldt [181]:

$$D_{app}(q) = \frac{D}{P(q)}. \tag{3.11}$$

3.1.3 Non-Draining Coils

The hydrodynamic interaction $H(q)$ explicitly enters the expression for the apparent diffusion coefficient as [179]

$$D_{app}(q) = \frac{D \cdot H(q)}{P(q)\,S(q)} \tag{3.12}$$

The product

$$I(q) \cdot D_{app}(q) = D \cdot H(q) \tag{3.13}$$

monitors the hydrodynamic effect on the internal modes of motion ("Zimm-modes") which at low q may be expanded as [182]

$$H(q) = 1 - C_{HI} \cdot q^2 + \cdots . \tag{3.14}$$

For $S(q) = 1$ and expanding the form factor at low q, Eq. (3.14) acquires the well-known form

$$D_{app}(q) = D \cdot (1 + C \cdot R_g^2 \cdot q^2 + \dots) \tag{3.15}$$

with the numerical constant $C = \dfrac{13}{75}$ for non-preaveraged hydrodynamics.

3.1.4 Rigid Rods

Equations (3.12)–(3.15) do formally apply for rigid rods too, when the hydrodynamic interaction term $H(q)$ is replaced by a term $C_\theta(q)$ originating from the rotational relaxation of rigid rods which differs numerically only from the function $H(q)$ given above. Eventually the numerical coefficient C in Eq. (3.15) becomes $C = 1/30$ and $C = 0.1$ with and without consideration of rotation-translation coupling, respectively [183]

We will now discuss theories and experiments on the "ordinary" and "extraordinary" diffusion separately, since there is currently no common basis for interpretation for both types of diffusion. Also, the discussion of the "ordinary" diffusion is much less speculative.

3.2 "Ordinary" Diffusion

The existing theoretical models may be characterized by the number of dynamic components (polyions, co-ions, counterions) being considered explicitly. One-component theories most closely resemble the theories developed for neutral polymer solutions. A rationale for using this approach is that, in the limit of vanishing electrostatic interactions, polyelectrolytes behave as neutral polymers. On the other hand, multi-component theories are based on models developed for low molecular weight electrolytes. This approach is most powerful for strong electrostatic interactions, where "polymer" effects are less important.

3.2.1 Polymer Theories

"Polymer" theories are effective one-component theories which only consider the dynamics of the polyions. Co-ions and counterions merely screen electrostatic interactions. Such theoretical treatments fail to explain the complicated dynamics of polyelectrolyte solutions when the dynamical behaviour of polyions and co- or counterions are strongly coupled. However, they are valid in the important weak coupling limit, where diffusion of individual polyions is observed and information on the conformation of single polyions may be obtained from diffusion coefficients.

In these theories the dynamics of the polyions is assumed to be very similar to the dynamics of neutral polymers. For neutral polymers in good solvents, different concentration regimes with different static and dynamical properties are predicted [43]. Below the overlap concentration the diffusion of individual chains is measured, which obeys a simple Stokes-Einstein relation

$$D = \frac{kT}{6\pi\eta_0 R_h}, \, c_p \ll c_p^* \tag{3.16}$$

where R_h is the hydrodynamic radius of the polyion controlled by its size and shape and η_0 the solvent viscosity. Above the overlap concentration, intermolecular entanglements produce cooperative modes analogous to those of a permanent network. The resulting cooperative diffusion is no longer correlated to the hydrodynamic radius R_h, but to a characteristic correlation length ξ, representing the dynamically effective length scale between successive entanglements along a chain. According to the "blob" theory developed by de Gennes, each chain is pictured as a succession of uncorrelated "blobs" with an average dimension ξ. The dynamics of this physical or temporary network is then attributed to the diffusion of individual "blobs". The Stokes-Einstein law is still formally applied to yield the dynamic correlation length ξ:

$$D = \frac{kT}{6\pi\eta_0\chi}, \quad c_p \gg c_p^*. \tag{3.17}$$

The factor 6π has no physical significance and has only been introduced to maintain the similarity between Eqs. (3.16) and (3.17). The cooperative diffusion coefficient increases with concentration, because the blob size ξ decreases as $c_p^{-3/4}$.

According to Odijk [17], similar relations should also apply for polyelectrolyte solutions in both the dilute and the semidilute regime, provided that the electrostatic repulsion between the like charges on the chain is taken into account, which leads to an increase in the average chain dimension. The effect has been accounted for within the framework of the two-parameter theory, i.e., with an "electrostatic" persistence length and an "electrostatic" excluded volume. From both quantities the radius of gyration, R_g, and the hydrodynamic radius, R_h, have been calculated by employing the worm-like chain model and excluded volume theories on various levels of sophistication (see Sect. 4). By means of scaling arguments Odijk derived relations for the correlation length ξ, which should scale as $c_p^{-1/2}$ for saltfree solutions and as $c_p^{-3/4}$ for solutions with excess added salt. Since the addition of salt screens electrostatic interactions at very high salt concentrations, polyelectrolytes should behave like neutral polymers. It is important to note that the overlap concentration, c_p^*, separating the dilute from the semidilute regime depends on the added salt concentration via the polyion dimension.

Scaling theory does not give the prefactors ("amplitudes") of the power laws. For neutral polymers in solution the Renormalization Group Theory (RNG) has been applied to describe the region from dilute to semi-dilute behaviour with a prediction of the prefactors [184, 185]. As a result, the reduced diffusion coefficient $D(c_p)/D(c_p = 0)$ should be a universal function of the reduced concentration $A_2 c_p M_w$, where A_2 is the second virial coefficient. For polyelectrolytes the excluded volume parameter z enters this universal relation which depends on the solvent quality. The RNG developed by Wang and Bloomfield [186] for polyions has been applied to explain the concentration dependence of the polyion diffusion coefficient.

The calculations above apply to the weak coupling limit, i.e., only intramolecular electrostatic interactions for the calculation of the hydrodynamic radius or the correlation length, the latter of which scales as $c_p^{-1/2}$ as a consequence of the local stiffness of the polyions.

3.2.2 Coupled Diffusion-Multi-Component Theories

In contrast to polymer theories, multi-component theories explicitly consider the dynamics of polyions, gegenions and coions and provide a detailed understanding of dynamic coupling phenomena. The theoretical formulation of the multi-component, coupled diffusion problem is quite complex and the mathematical treatment requires gross simplification. Current calculations treat all components as point particles exhibiting different mobilities and charges which obviously represent a rather poor approximation for the polyion structure.

Assuming a mean force field around each particle rather than treating local force field fluctuations (mean field theory), the multi-component approach is expected to yield meaningful results in the low q-limit, i.e., on large length scales, where local fluctuations are averaged to a mean value.

Despite such serious approximations and restrictions, multi-component theories provide essential features of polyelectrolyte dynamics, such as coupled diffusion, plasmon modes and Nernst-Hartley diffusion.

Even in the simplest version the quantities in the relaxation Eq. (3.2) become matrices with a rank equal to the number of components i leading to a set of differential equations, the solutions of which are sums of exponentials $\Sigma_i a_i e^{-t/\tau_i}$ with characteristic amplitudes a_i and relaxation times τ_i. Analogous to Sect. 3.1, corresponding "apparent" diffusion coefficients $D_{app,i} = (Nq^2 \tau_i)^{-1}$ result.

The first theoretical calculation performed by Stephen [187] on a basis of the DH-potential leads to the important result that the effective diffusion coefficient of polyelectrolytes is enhanced by the forces exerted by the ionic atmosphere. A rigorous, fundamental formulation of the theory was developed by Lee and Schurr [188] and later by Berne and Pecora [189], Tivant et al. [190] and Schmitz [179]. Accordingly, in addition to the concentration gradient as given by Fick's diffusion law, the local electric field due to the neighbouring ions also contributes to the particle flux of ions. In the resulting Smoluchowski-type diffusion equation, the local field ∇U experienced by an ion is replaced by its thermal average $\nabla \bar{U}$, which is given by the Poisson-Boltzmann equation.

In the following we briefly discuss the relaxation equation for the three-component system composed of polyions (1), gegenions (2) and coions (3) given as

$$\frac{\partial}{\partial t} \underline{S}(q, t) = - \underline{\Omega}(q) \underline{S}(q, t) \tag{3.18}$$

$$\Omega_{ij} = q^2 D_i \left(\delta_{ij} + \frac{z_i}{z_j} \lambda_j \right) \tag{3.19}$$

$$\lambda_i = \frac{\kappa_i^2}{q^2}. \tag{3.20}$$

The detailed calculation requires the concrete formulation of the structure factor $S(q)$. For the following results $S(q)$ is identified to $S_g(q)$ defined in Eq. (2.9), i.e., a correlation hole structure factor. The off-diagonal elements λ_i couple this set of diffusion equations. In the limit $\lambda_i \to 0$ the diffusion equations decouple and reduce to Fick's second law of diffusion for each ionic species. Interestingly, the coupling constant λ had already been introduced by Hermans in 1949 [191] when he derived from simple arguments that fluctuations of ions become coupled if $\kappa^2 V^{2/3} \gg 1$, where V is the experimentally probed volume. In scattering experiments fluctuations are probed in a volume $V = q^{-3}$, so that Herman's coupling condition results in $\kappa^2/q^2 \gg 1$ which corresponds to the definition of λ in Eq. (3.20).

It is usually assumed that the dominant contribution to the dynamic structure factor comes from the polyion–polyion dynamic structure factor $S_{11}(q, t)$ which depends in a crucial way on the ratio

$$\lambda^* = \frac{\lambda_p}{1 + \lambda_c} = \frac{\kappa_p^2}{q^2 + \kappa_c^2} \tag{3.21}$$

where the indices p and c indicate polyions and co-ions. λ^* defines the strength by which the motion of polyions is coupled to the motion of counterions. The identity of λ^* and χ^* should be noted although λ^* has been introduced explicitly for multi-component systems, whereas χ^* is formulated for an effective one-component system.

Table 2 summarizes the fundamental features of the dynamic behaviour of ionic multicomponent systems. Focusing on the low-q regime, where the theory is expected to be applicable and where most experiments are performed, the coupling constant is given by $\lambda^* = \kappa_p^2/\kappa_c^2$. The "apparent" diffusion coefficients are derived from the respective relaxation times given in Table 2. It turns out that the amplitudes of the relaxations τ_2 and τ_3 always remain negligible. Thus, in the following, the relaxation τ_1 will be mainly discussed.

Table 2. Relaxation times τ_i of $S_{11}(q, t)$ calculated from Eq. (3.19)

Regime	λ^*	τ_1	τ_2	τ_3
I	$\ll 1$	$(q^2 D_1)^{-1}$	+	+
II	≈ 1	$[(q^2 + \kappa_p^2)D_1]^{-1}$	+	+
III	$\gg 1$	$(q^2 D_{NH})^{-1}$	+	τ_{p1}

+ not important for the present discussion as it contains coupled dynamics of
co- and counterions, only

For $\lambda^* \ll 1$ (weak coupling, regime I), the apparent diffusion coefficient is identical to the diffusion coefficients D_1 of the polyions. In the intermediate coupling, regime II, the diffusion coefficient is an increasing function of κ_p, i.e., it increases with increasing polyion concentration. The corresponding diffusion coefficient could also be derived from Eq. (3.1) inserting the Donnan-potential for the osmotic compressibility (Eq. 2.22). The origin for the enhanced mobility of the polyions is the coupling to the motion of the small counterions as a consequence of local electroneutrality. Alternatively, the Donnan potential as a direct consequence of the electroneutrality condition may be viewed as the additional driving force for the diffusion of polyions.

In the coupled diffusion, regime III, the interpretation of the diffusion processes in molecular terms is more complicated and certain linear combinations of the D_i such as interdiffusion modes or cooperative diffusion coefficients are used to describe the underlying physical processes. As can be seen from Table 2, two relaxation times are proportional to q^2, indicating diffusive behaviour. The third relaxation (τ_3) is q-independent and is often referred to as the "Debye mode", "plasmon mode" or "interdiffusion mode" (although not diffusive at all). Its relaxation time is, however, related to the polyion and coion diffusion according to

$$\tau_{p1}^{-1} = \kappa_p^2 D_p + \kappa_c^2 D_c \tag{3.22}$$

and describes the motion of the counterions relative to the polyions in order to meet the local electroneutrality condition. Its amplitude is negligibly small because of the large energy penalty for the resulting charge–charge fluctuations. The relaxation τ_2 corresponds to the simple diffusion of co-ions. The dominating relaxation τ_1 represents a diffusive process sometimes termed "cooperative", "Nernst-Hartley" or ambipolar diffusion [192] and corresponds to the strongly coupled diffusion of polyions and counterions according to

$$D_{NH} = \frac{(z_p + z_c)D_p D_c}{z_p D_p + z_c D_c}, \tag{3.23}$$

$D_{NH} > D_p$ because the counterions accelerate the motion of the polyions.

The simultaneous occurrence of two different relaxations corresponding to the plasmon and Nernst-Hartley mode is not unique in polyelectrolyte solutions, but rather a general feature of ionic systems such as ionic crystals, plasmas, metals, and molten salts. In both ionic crystals and plasmas, optical and acoustic modes originate from charge fluctuations generating a local electric field which acts as a restoring force on the local charge separation. The acoustic mode responds in a diffusive manner in liquids (large damping) and as a propagating mode in ionic crystals and plasmas (undamped) in analogy to the NH-diffusion: ions of opposite sign move "in phase" and the resulting frequency or relaxation time vanishes in the limit $q \rightarrow 0$. The optical mode resembles the Debye mode in as much as ions of opposite sign move "out of phase" and the resulting frequency or relaxation time approaches finite values in the limit $q \rightarrow 0$. This plasma frequency is often called "Plasmon mode" in analogy to the "Debye mode"in ionic solutions. In a way, polyelectrolyte solutions represent a unique system where the "restoring force" in terms of λ^* is easily varied over a wide range of values by simply changing the ionic strength of the solution. Moreover, the large scattering of the polyions, as compared to the other components, allows the study of the dynamics of the charge fluctuations in a relatively straightforward manner. Unfortunately, only the NH-diffusion is experimentally accessible, because the amplitude of the plasmon mode is negligible.

Recent developments have attempted to include more realistic electrostatic potentials, polymer form factors, hydrodynamic interactions [193] and activity coefficients [194, 195] and also focused on weakly charged polyelectrolytes [52, 196]. However, each of these improvements made the mathematics hardly tractable and, except for a few limiting cases, no analytical solution to the problem is provided.

A comparison of the transport coefficients $[\eta]$ and D reveals some striking similarities. At a certain ratio $\lambda^* \propto c_p^m/c_s^m$, a transition in the diffusion behaviour is observed, where the reduced viscosity exhibits a pronounced maximum. Both relations have been derived from simple point charge models without any considerations of conformational effects of the polyions. This indicates that, under most conditions, electrostatic interactions between different ionic components dominate over "polymer-specific" effects.

3.2.3 Experiments

It turned out that the dynamical behaviour of polyelectrolyte solutions is even more spectacular then theoretically anticipated. In the early 1970s mostly biopolymers such as DNA were studied and often two separate relaxations were observed which were then attributed to internal relaxations [197–202]. During the past twenty years numerous studies on synthetic polyelectrolytes (NaPSS, NaPMA, NaPAA, QPVP), proteins (BSA, PLL), polynucleotides (DNA, RNA) and charged polysaccharides (heparin, chondroitin-6-sulfate, proteoglycan hyalonurate) have been performed. The dynamical behaviour of all these polymers exhibits common features which are attributed to the ionic character of the polyelectrolytes. So far, most studies have focused on the dependence of the apparent diffusion coefficient on polyelectrolyte concentration, salt concentration, and molecular weight. Fewer studies investigated the influence of charge density and scattering vector. It has taken more than a decade of sometimes seemingly contradicting results for a uniform picture of polyelectrolyte diffusion to emerge. It is as though each study contributed a piece to a puzzle which is now almost completed and visualized by most experts in this field. The discussion below mainly addresses the question of how to interpret this picture in order to trace the physical origin of the displayed characteristics.

Most studies agree that at high salt and low polyion concentration (regime I) one relaxation time is observed representing the translational diffusion of single polyions (see Fig. 1). At low salt and high polyion concentration (regime III), a much faster, concentration independent diffusion is detected. A gradual increase of the diffusion coefficient from the value in regime I to the value in regime III is monitored in the transition regime II, located at approximately equal concentrations of polyions and salt $\lambda^* = c_p^m/c_s^m \approx 1$. The behaviour described so far is termed "ordinary" diffusion and it is believed to match the theoretical predictions discussed in the previous section. However, together with the increasing diffusion in the transition regime II, a slow diffusion process gradually appears which eventually becomes the dominant relaxation in regime III. The presence of the slow mode is also reflected in a drastic increase in scattering intensity at low q, the angular dependence of which results in an apparent radius of gyration much larger than a single chain radius. This is the "extraordinary" diffusion which will be discussed in more detail later (Sect. 3.3).

Not all experimental studies have identified all of the characteristic features in each of the concentration regimes as summarized above. Many experiments covered too small a range of concentrations, other focused on the fast mode only, thereby missing the slow mode or, conversely, focused on the slow and missed the fast mode. The latter may be excused by the great difference in amplitude and relaxation time of both modes which are difficult to detect with "single-tau" correlators utilizing a constant sample time. With the advent of modern "multi-tau" correlators covering many decades in sample time, these difficulties have been overcome.

It should be mentioned, however, that the clear picture developed above might also be obscured by the influence of the intermolecular structure factor on the apparent diffusion coefficient as outlined in Sect. 3.2 which becomes particularly important at low ionic strength. In fact, under very low added salt conditions, the transition regime II becomes most complicated because all three phenomena show up simultaneously: fast diffusion, slow diffusion and intermolecular structure.

Separation of all the three effects turns out to be a most delicate task and could not yet be unambiguously performed. For more details see the end of this section.

The transition from the concentration regime I to the concentration regime III has been experimentally monitored either by variation of salt concentration at constant polyelectrolyte concentration, or by variation of polyelectrolyte concentration at constant salt concentration. A third way to arrive at this transition is to vary the charge density by adding strong bases (e.g. NaOH) or acids (e.g. HCl) at constant total ionic strength in order to suppress or enhance dissociation of the ionic groups along the polyion chain.

Numerous studies have been done on the c_p-dependence of the diffusion coefficient D at constant salt concentration. The presence of two diffusion processes in partly hydrolyzed polyacrylamide (PAM/PAA) solution was already reported by Jamieson and Presley in 1973 [203]. They also noticed the slow diffusion D_s decreasing with increasing concentration. One year later, Raj and Flygare [204] observed the increase of D in the transition regime for BSA solutions. In a series of papers by Mathiez et al. [205–208] on the dynamics of poly(adenilic acid) the fast and slow diffusions were measured simultaneously as a function of concentration. In 1981 Grüner et al. [209] found the characteristic behaviour of the fast diffusion D_f in each of the three concentration regimes for PSS, but missed the slow mode. Later it was shown [9, 121] that the transition regime starts at a certain polyion concentration which is independent of molecular weight and proportional to the salt concentration, a fact, which may already be concluded from the data of Raj and Flygare. More pieces to this puzzle were collected by further experiments of chondroitin-6-sulfate [190, 210], NaPSS [188, 210–212], QPVP [9], hyalonurate [213], proteoglycan [113], and heparin [210]. The onset of the transition regime II at $\lambda^* \approx 1$ was observed and realized for BSA [204], QPVP [9], heparin [210], and chondroitin-6-sulfate [210] whereas Koene et al. [214–216] preferred to interpret their data of NaPSS in the transition regime II in terms of a dilute–semidilute crossover transition and derived the corresponding scaling laws in the "semidilute" regime.

The increase of D as a function of salt concentration c_s at constant polyelectrolyte concentration is already reported in one of the earliest studies on polyelectrolyte solutions by Doherty and Benedek on BSA [217]. The appearance of the slow diffusion process upon variation of c_s at constant c_p was first detected by Schurr et al. on PLL solutions [218]. Drifford and Dalbiez [219] demonstrated that the slow and fast diffusion processes occur simultaneously. Moreover, the increase of D with decreasing salt at constant polyelectrolyte concentration has been observed for BSA [204, 220], poly(L-lysine) [221–223], DNA [133]. heparin [190], NaPSS [210, 219], QPVP [9], hyalonurate [213], proteoglycane [133], chondroitin-6-sulfate [210], and crosslinked NaPAA [224] (Fig. 11). The onset of the transition regime II at $\lambda^* \approx 1$ has been found for poly(L-lysine) [218, 221], hyalonurate [213], proteoglycan [133] and NaPSS [219].

The transition from regime I to regime III has also been monitored by changing the charge density of the weak polyacid NaPMA or of the weak polybase poly(L-lysine) [225].

The overwhelming majority of experiments unambiguously demonstrate that the coupling of polyions and counterions is the basic mechanism which determines the dynamics of polyelectrolytes as a function of polyion concentration and added salt. The results cannot be explained by one-component theories, i.e., a dilute/semi-dilute cross-over, although the experimental scatter of some of

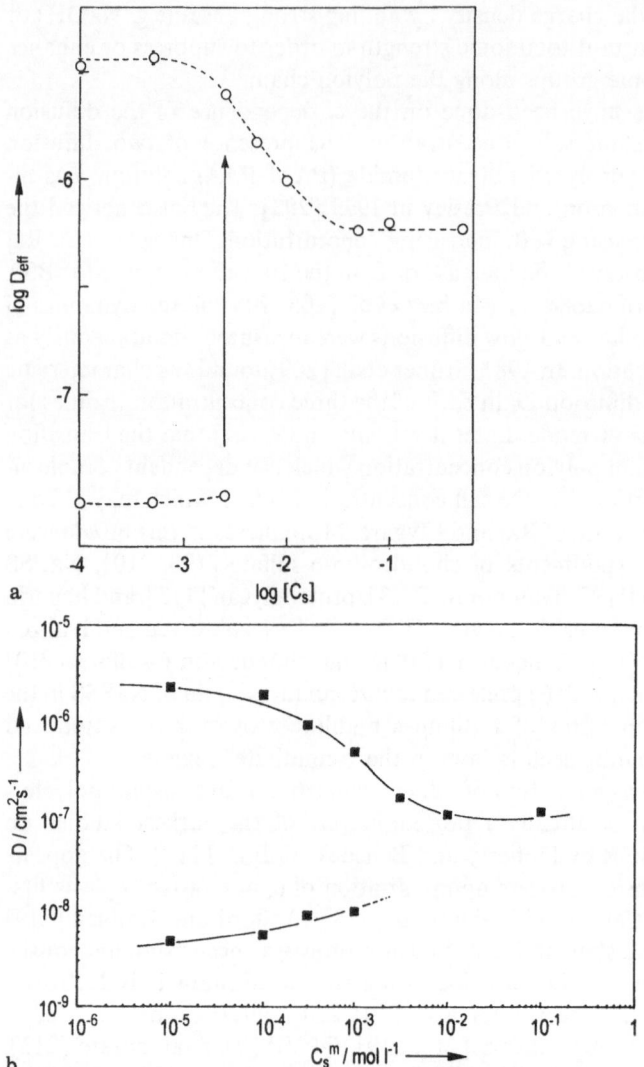

Fig. 11. a, b Diffusion coefficients D as a function of c_s^m at constant c_p for: **a** NaPSS (M_w = 10^5 g/mol), c_p = 5 g/l (data taken from [219]; **b** QPVP (M_w = 8.2×10^5 g/mol, 75% quaternized), c_p = 0.33 g/l (data taken from [9])

the data is large enough to allow interpretation by scaling concepts. It is experimentally observed that the cross-over transition is often well below the overlap concentration, that it is molecular weight independent and that there is a plateau value of D at higher concentrations (Fig. 12). These observations contradict an interpretation as a dilute–semidilute cross-over in agreement with the osmotic pressure experiment and the molecular dynamics simulations outlined in Sect. 2.2.3.

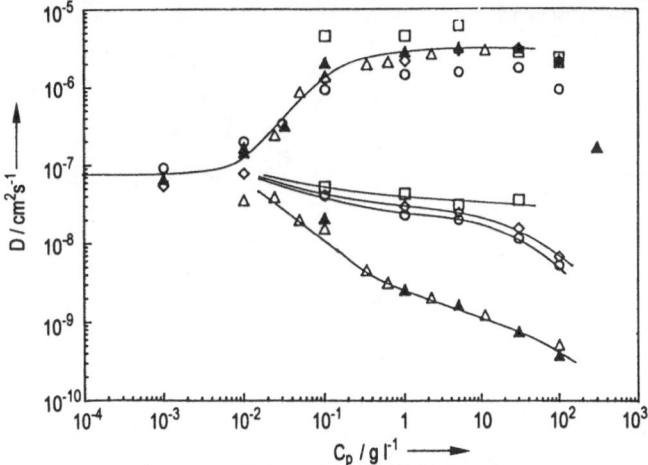

Fig. 12. Diffusion coefficient D as a function of polyion concentration c_p in "salt free" solution for different contour lengths L and degrees of quaternization Q. $\triangle$: 8800 Å, Q = 40%; $\blacktriangle$: 8800 Å, Q = 75%; $\bigcirc$: 1900 Å, Q = 60%; $\diamondsuit$: 1330 Å, Q = 65%; $\square$: 270 Å, Q = 100%. (Taken from [121])

Despite this striking evidence, even nowadays scaling concepts are applied to interpret the increasing diffusion coefficient in the transition regime II [226]. This is achieved by performing inconclusive experiments as will be shown by the detailed examination of the most recent example where two polydisperse samples of polyethylenimine are investigated at different added salt concentrations. In Fig. 13a the originally published data are plotted as a function of the reduced concentration c_p/c_p^* with the overlap concentration calculated from Odijk's theory neglecting the polydispersity. An inspection of the superimposing curves leads the unbiased reader to believe that interpretation of the data in terms of scaling concepts is appropriate. Figure 13b shows the same data, now replotted against $\lambda^* = c_p^m/c_s^m$. An even better superposition of the different curves is obtained which demonstrates that the performed experiments are not conclusive for the controversial discussion above, due to the limited molar mass range investigated. Simple measurements of the radius of gyration (rather than calculating it under questionable assumptions) would most likely provide clear evidence for the one or the other interpretation. Unfortunately, the authors confess that the radii of gyration were too small to be measured, although a "state of the art" light scattering goniometer (a similar instrument as used in [9]) was utilized and the R_g-values should definitely exceed 200 Å.

According to Sect. 3.1, the structure factor S(q) affects the q-dependence of the apparent diffusion coefficient. The first study of the q-dependence of the diffusion coefficient was performed using inelastic neutron scattering [20]. The results demonstrated that the diffusion coefficient exhibits a minimum at a q-value where the scattered intensity shows a maximum in qualitative agreement with theory. Dynamic light scattering experiments on PSS in the concentration

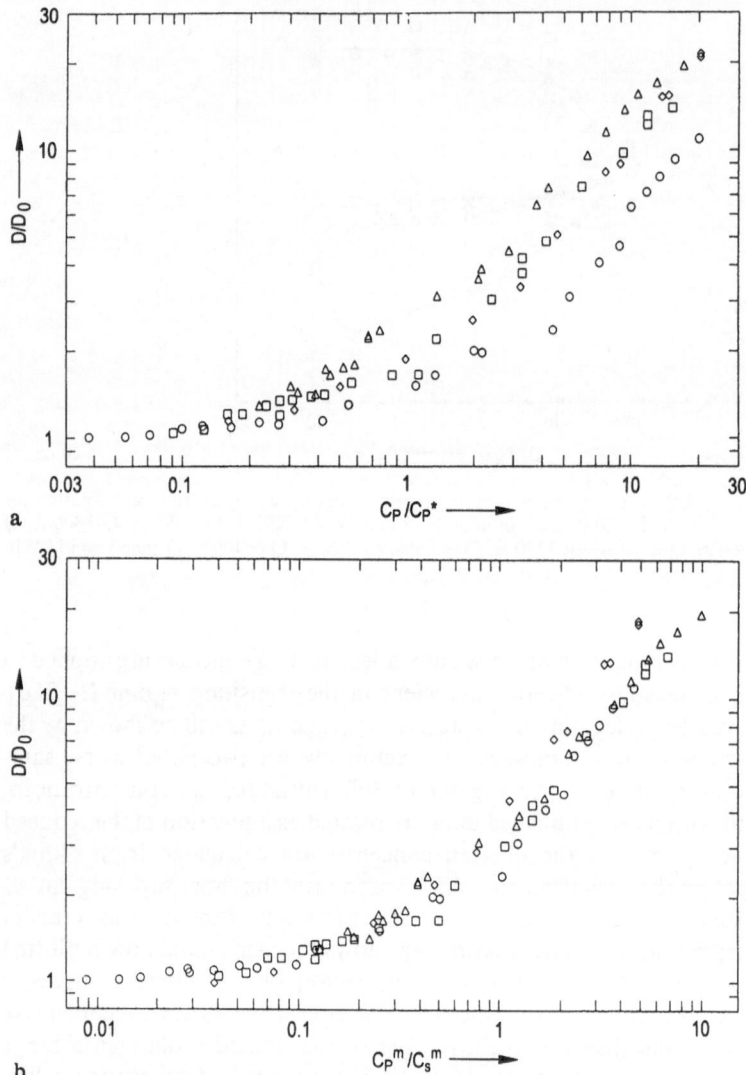

Fig. 13. a, b Reduced diffusion coefficient $D/D_{cp=o}$: **a** as a function of c_p/c_p with c_p^* the overlap concentration calculated by the OSF theory (see Sect. 4.1); **b** a function of c_p^m/c_s^m (b) for different molar masses and different added salt concentrations. ○: $M_w = 4.15 \times 10^4$ g/mol, $c_s^m = 0.1$ m, □: $M_w = 4.15 \times 10^4$ g/mol, $c_s^m = 0.025$ m, △: $M_w = 4.15 \times 10^4$ g/mol, $c_s^m = 0.01$ m, ◇: $M_w = 8.82 \times 10^4$ g/mol, $c_s^m = 0.025$ m. (Data taken from [226])

regime I by Drifford and Dalbiez [115] and by Li and Reed [133] lead to similar results (Fig. 14).

According to Eqs. (3.13) and (3.14) for nondraining coils and rods the product $D_{app}(q)I(q)$ should exhibit a slightly negative slope which should increase with chain stiffness. Experiments [121, 123, 137] using INS and DLS do

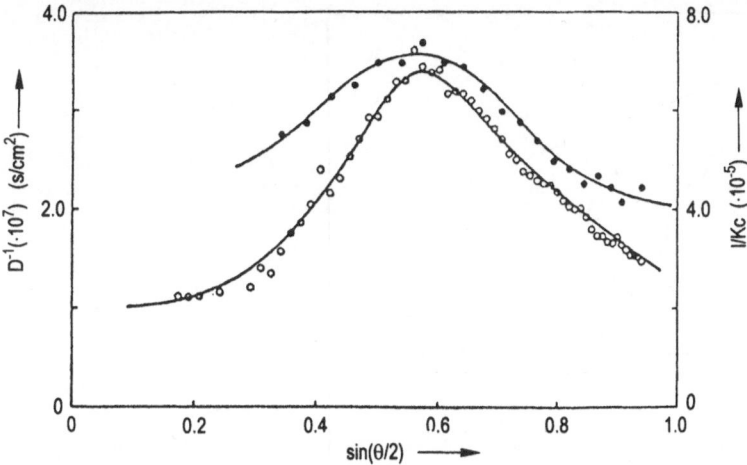

Fig. 14. Reciprocal diffusion coefficient D^{-1} (q) (*left scale*) and the reduced scattering intensity (*right scale*) as a function of sin $\theta/2$ with θ the scattering angle for the proteoglycan monomer ($M_w \approx 1.6 \times 10^6$ g/mol) at $c_p = 0.1$ g/l in "salt-free" solution. (Taken from [133])

not give a clear picture; in some cases the product $D(q)I(q)$ is q-independent, whereas in others it seems to decrease with increasing scattering vector as expected. Since the q-dependence may be as complicated as a concentration dependence, there is a great need for experiments which cover more than one decade in the scattering vector.

For QPVP some few experimental data have been collected on the effect of the structure factor $S(q)$ on the fast and slow diffusion in the transition regime II. From several, mostly non-reproducible experiments, one common feature is not believed to be purely of random nature: this is that only the fast diffusion coefficient exhibits a similar, if not identical q-dependence as the inverse structure factor $S(q)^{-1}$ (i.e., a pronounced minimum or a negative slope, see Fig. 15a). The resulting mobility $DH(q)$ displays no q-dependence within experimental error. Contrary to the fast diffusion the simultaneously measured slow diffusion coefficient appears essentially unaffected by $S(q)$ as it shows the normally observed slightly positive slope when plotted vs q^2 (Fig. 15b).

It is a matter of fact that multi-component theories which take into account the dynamics of polyions, co-ions, and counterions propose the existence of exactly the three concentration regimes I–III, all of which are experimentally found. The predicted and observed transition regime II occurs at $\lambda^* \approx 1$, independent of molecular weight. However, a major drawback of the multi-component theory is the fact that a quantitative comparison of the measured diffusion coefficient in regime III with Eq. (3.23) yields an effective polyion charge $z_{eff} \approx 1$. It is frequently observed that the apparent charge of the polyion, when calculated from dynamic light scattering experiments, is orders of magnitude too small compared to the charge expected from the number of charged

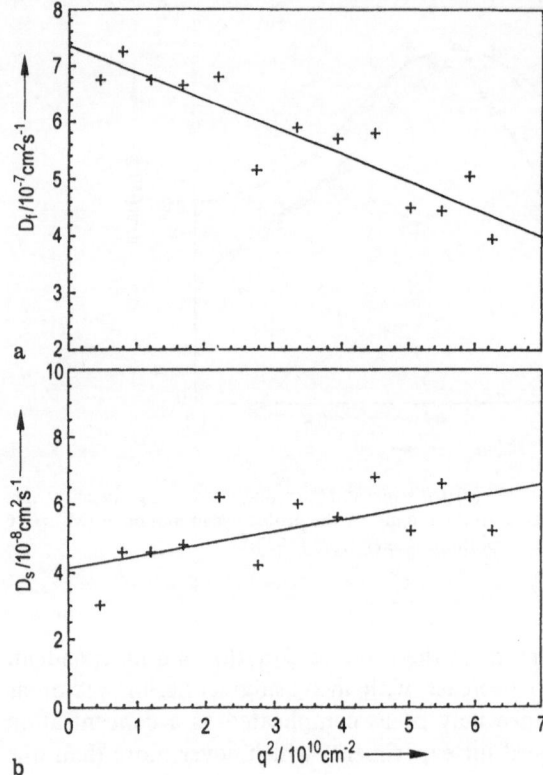

a

b

Fig. 15. a The apparent fast diffusion D_f as a function of q^2 for
QPVP ($M_w = 8.26 \times 10^5$ g/mol, c_p
$= 0.1$ g/l) in "salt-free" solution. **b**
the simultaneously measured slow
diffusion coefficient D_s. (Data taken
from [121])

units on the polyion [133, 210, 213, 219, 222]. Together with the independence
from molecular weight it seems that not the whole polyion, but rather single
charged segments of the polyion, couple to the motion of the counterions. This is
similar to the observation that the peak position in the relative viscosity is
independent of molecular weight (see Sect. 2.3.3). Both facts are not at all
understood.

3.3 Extraordinary Diffusion

The presence of a "slow diffusion" seems to be characteristic for most polyelectrolyte solutions. The slow mode was first detected by Schurr et al. [218] for
poly(L-lysine) upon variation of salt concentration. A sudden drop of the
diffusion coefficient was observed as the salt concentration was decreased below
some critical value. Since then, this transition is called "ordinary–extraordinary
transition", the fast diffusion being referred to as "ordinary" and the slow
diffusion as "extraordinary". Drifford and Dalbiez [219] later gave an empirical
expression which describes the relation between this critical salt concentration

and the polyelectrolyte concentration. For monovalent co- and counterions a critical ratio $c_p^m/c_s^m = 4$ which is postulated closely resembles the experimental value of $\lambda^* \approx 1$

The slow mode is only observed in regimes II and III, i.e, the coupled diffusion regimes. The related diffusion coefficient decreases with increasing polyelectrolyte concentration and molecular weight [9, 211, 212]. The presence of a slow mode is always accompanied by a drastic increase in the scattering intensity at low q. The enhanced scattering has been observed by light scattering, SAXS and SANS. From the diffusion coefficient and from the q-dependence of the scattering curve, hydrodynamic radii and radii of gyration are derived which are much larger than those of single chains, even if they are fully stretched. Recently, more attention has been paid to the enhanced scattering at low q [227] which, interestingly enough, is also observed for bulk ionomers [228–231].

There have always been controversies on whether the slow modes or the increased small angle scattering were due to impurities, such as dust particles or aggregates. The increased small angle scattering in SAXS and SANS experiments has been routinely considered to be caused by impurities. Reed et al. [210, 232] found, for a variety of polyelectrolyte systems, that the slow mode is removed by filtering the solution through small pore size filters, i.e., smaller than the dimension of clusters or aggregates. In most of the solutions no or negligible loss of polyion concentration was detected by UV-absorption. Once the slow mode was removed, it did not reappear within a period of two weeks. A trivial explanation for the Reed experiments is not easily found. A subsequent study by Sedlak [233] showed that the amplitude and diffusion coefficient of the slow mode were indeed affected if small pore size filters were used. It was however, not possible to remove the slow mode completely. For a more detailed discussion see [8].

Interpretations of the slow diffusion process have been very speculative so far. Since it is not predicted by mean-field theories, it is natural to suspect large concentration fluctuations to cause this phenomenon. These fluctuations become increasingly dominant with increasing concentration, c_p. Since no slow diffusive process could be observed by tracer diffusion techniques [223, 234], it is not certain that these fluctuations originate from any kind of mass transport. On the other hand, tracer techniques might not be sensitive enough to detect a small amount of large clusters. The same argument may also apply to dynamic Kerr-effect measurements which could not detect a change of the field free decay of the birefringence when passing the critical concentration at λ^* [169].

These fluctuations have been named "random inhomogeneities" [9], where the term "random" indicates that these inhomogeneities are not ordered, i.e., influenced by S(q). For the fluctuating concentration, regions of high concentration are frequently called "temporal aggregates" or "clusters" and models have been proposed for the formation of these structures on a molecular level. They consider ionic concentration fluctuations that lead to long range correlations [235,236]. Different models propose possible distributions of polyions and counterions which could lead to correlations. A good summary of these models has been given in a book by Schmitz [179]. There are basically three models: (1) the equilibrium distribution of counterions should have a maximum *between* the polyions thus giving rise to attractive interactions and formation of clusters, (2) the electroviscous effect giving rise to attractive interactions and formation of clusters, (2) the electroviscous effect giving rise to asymmetric ion cloud, and (3) van-der-Waals type dispersive ionic fluctuations. The major problem for all these explanations is to prove theoretically the formation of asymmetrically distributed counterions between different polyions or around a single polyion.

4 The Conformation of Flexible Polyions as a Function of the Ionic Strength

The conformation of flexible polyions particularly at low ionic strength constitutes a historical problem in polyelectrolytes which has not yet been convincingly solved, neither theoretically nor experimentally.

4.1 Theory

Theoretically, the key problem is to calculate the intramolecular electrostatic potential which is formed by the distribution of gegenions surrounding the polyion charges. For rigid polyion structures it is at least numerically possible to solve the Poisson–Boltzmann equation of the distribution of gegenions and calculate the desired electrostatic potential. The conformation of flexible polyions, however, does change under the influence of this potential leading to a modified conformation which in turn will modify the electrostatic field (i.e. distribution of gegenions) and so on. Up to now, two approximations of the polyion conformation have been proposed in order to simplify the problem described above. As these theories have been reviewed a number of times, they will only be briefly described in the following.

4.1.1 The Gaussian Coil Approximation

This is the old Kuhn–Katchalsky approach [1, 237–242], which overcomes the above-mentioned difficulties by assuming that the segment distribution of an ideal flexible coil remains Gaussian upon expansion by electrostatic repulsion of the like charges along the chain. The basic approach is to balance the electrostatic expansion by the entropic restoring force, utilizing the Kuhn model of random flight statistics. As a certain number of monomer units (charges) are represented by one statistical Kuhn segment of length l, several charges have to be placed on this Kuhn segment which, for simplicity, were concentrated on the flexible joints. A Coulomb or a screened Debye–Hückel potential was used to calculate the electrostatic interaction. As a result, the calculated expansion of the polyelectrolyte chain was much higher than deduced from experiments even at relatively high ionic strength. At low ionic strength a rod-like conformation was readily approached, thus creating an increasing inconsistency with the underlying theoretical model.

However, the obvious tendency to form rod-like conformations at low ionic strength was in remarkable agreement with the "Fuoss-Strauss" extrapolated intrinsic viscosities[7]. Thus, the picture of the rod-like conformation of flexible polyions at low ionic strength has been manifested for some decades since.

4.1.2 Numerical Calculation Utilizing a Worm-Like Chain Probability Density

The Kuhn–Katchalsky approach was recently modified [243, 244] by modelling the segment distribution function more realistically by a worm-like chain probability density according to Koyama [245], which describes the spatial distance R of two segments of a chain separated by a contour distance s:

$$4\pi R^2 W(R, s) = \frac{R}{2AB\sqrt{\pi}} \left\{ \exp\left(\frac{-(R - B)^2}{4A^2}\right) - \exp\left(\frac{-(R + B)^2}{4A^2}\right) \right\} \quad (4.1)$$

with $A^2 = \langle R^2 \rangle (1 - y)/6$ and $B^2 = y\langle R^4 \rangle$ and y a parameter depending on the second and fourth moments of the Kratky-Porod worm-like chain, $\langle R^2 \rangle$ and $\langle R^4 \rangle$, $2y^2 = 5 - 3\langle R^4 \rangle / \langle R^2 \rangle^2$. The charges were distributed in equal distances along the contour of the continuously bent cylinder. Again, the electrostatic repulsion F_{DH} is compared with the elastic restoring energy, F_{el}, as a function of the total persistence length l_t of the chain, composed of an electrostatic l_{pe}, and an intrinsic part l_p which are derived as follows.

Electrostatic free energy

$$F_{DH}(L, l_p, l_{pe}) = \sum_{s=1}^{q-1} (q - s)\phi_{DH}(s) \quad (4.2)$$

with q the number of charges on the chain and with Debye–Hückel interaction of a pair of charges on the polyion

$$\phi_{DH}(s) = l_B KT \int_0^\infty 4\pi R^2 \frac{\exp(-\kappa R)}{R} W(R, s) dR$$

$$= l_B KT \frac{\exp(\kappa^2 A^2)}{2B} \left\{ \exp(-\kappa B)\mathrm{erfc}\left(\kappa A - \frac{B}{2A}\right) \right.$$

$$\left. - \exp(\kappa B)\mathrm{erfc}\left(\kappa A + \frac{B}{2A}\right) \right\}. \quad (4.3)$$

---

[7] As discussed in Sect. 2, the empirical Fuoss-Strauss plot does not yield information on the conformational change of polyions

Elastic free energy

$$F_{el}(L, l_t) = - KT \int_0^\infty 4\pi R^2 \{W(R, s) - W_0(R, s)\} \ln(W_0(R, s)) \, dR \qquad (4.4)$$

with $W_0(R, s)$ the distribution function of the uncharged chain, $l_t = l_p$, and $W(R, s)$ the segment distribution of the charged chain, $l_t = l_p + l_{pe}$. The minimum of the sum of both electrostatic interaction energy and elastic restoring energy

$$\frac{dF(L, l_t)}{dl_{pe}} = 0 = \frac{d}{dl_{pe}}(F_{DH} + F_{el}) \qquad (4.5)$$

defines the equilibrium conformation of the charged cylinder in terms of the electrostatic persistence length l_{pe}. It should be noted that these calculations yield an apparent electrostatic persistence length, as the electrostatic excluded volume is additionally calculated rather than the pure internal rigidity of the chain. It is inherently assumed that the polyelectrolyte conformation with the electrostatic excluded volume can be approximately modelled by an unperturbed chain with an increased persistence length, the difference to the intrinsic persistence length being defined as the electrostatic contribution to the chain stiffness.

4.1.3 The Local Stiffness Approximation

In a series of papers, Odijk and coworkers [246–248] and Fixmann and Skolnick [249, 250] independently derived a worm-like chain model for polyelectrolytes (OSF theory). The theory minimizes the free energy of the polyion comprising of the elastic bending energy of a worm-like chain and the electrostatic repulsion energy, approximated by a Debye–Hückel potential. The resulting radii of curvature easily translate into an electrostatic persistence length dependent on the Debye screening length κ^{-1} but, of course, being independent of the contour length L. The total persistence length l_t is given by the sum of the intrinsic persistence length, l_p, the "unperturbed" chain would have had without charges and the electrostatic part, l_{pe}, originating from the interaction of charges along the macroion:

$$l_t = l_p + l_{pe}. \qquad (4.6)$$

Finally the electrostatic persistence length reads as

$$l_{pe} = \tfrac{1}{12} l_B N^2 h(\kappa L) \qquad (4.7)$$

with N the number of charges, l_B the Bjerrum length and $h(y) \equiv h(\kappa L)$ a function accounting for end effects:

$$h(y) = e^{-y}\{y^{-1} + 5y^{-2} + 8y^{-3}\} + 3y^{-2} - 8y^{-3}. \qquad (4.8)$$

For infinite-long chains Eq. (4.7) reduces to

$$l_{pe} = \frac{l_B}{4\kappa^2 a^2} \tag{4.9}$$

with a the contour length per ionic charge $a \equiv L/N$.

For $a < l_B$, Eq. (4.9) has to be modified because counterion condensation [3, 5] reduces the charge density to a maximum value of $a \cdot \xi$ with $\xi = 1$ for $a > l_B$ and $\xi = l_B/a$ for $a \leqslant l_B$. Insertion into Eq. (4.9) yields

$$l_{pe} = \frac{l_B}{4\kappa^2 a^2 \xi^2}. \tag{4.10}$$

For dilute polyelectrolyte solutions with added salt $c_s^m > 10^{-3}$ m, the contribution of the noncondensed gegenions to the Debye screening length κ^{-1} is negligible. Then Eq. (4.10) predicts

$$l_{pe} \sim \frac{1}{c_s^m}. \tag{4.11}$$

Relations in Eqs. (4.7) and (4.9) were postulated to be valid for $l_t \gg \kappa^{-1}$ which is fulfilled if l_p is significantly larger than the Bjerrum length l_B, which holds true for most polyelectrolyte chains. In his first derivation [246], Odijk imposed the second restriction $l_{pe} < l_p$ which, however, was completely released in a subsequent paper [247]. Later, instead of utilizing the Debye–Hückel potential and counterion condensation, Fixman [251] and Le Bret [19] numerically solved the Poisson–Boltzmann equation for a curved cylinder (Le Bret for a toroid). Within numerical accuracy, Fixman's and Le Bret's results are identical and approach the OSF result at low ionic strength. At high salt levels, however, the calculated electrostatic persistence length deviates significantly from the inverse power of salt concentration into much weaker dependence on c_s (i.e. yielding larger l_{pe} values), which no longer follows a power law.

Odijk and Houwaart [247] modified the OSF approach for the electrostatic excluded volume by artificially dividing the electrostatic interaction into short-range leading to the electrostatic persistence length discussed above and into long-range causing the electrostatic excluded volume. The latter is calculated along the lines of the classical excluded volume theory for neutral polymers of chains with a persistence length l_t and an effective (hard-core) segment radius $R \approx \kappa^{-1}$. The results read as

$$Z_{el} = \left(\frac{3^{3/2}}{4\pi^{1/2}}\right) L^{1/2} \kappa^{-1} l_t^{3/2} \sim L^{1/2} \kappa^2 \sim L^{1/2}/c_s^m \tag{4.12}$$

with Z_{el} being the electrostatic excluded volume parameter, which may serve to calculate the expansion coefficient of the radius of gyration.

$$\langle S^2 \rangle = \alpha_{s,el}^2 \langle S^2 \rangle_0 \xrightarrow{\text{long chains}} \alpha_{s,el}^2 \frac{l_t L}{3} \tag{4.13}$$

with $\alpha_{s,el}^2 = 0.541 + 0.459(1 + 6.04 Z_{el})^{0.46}$.

As $l_t = l_p + l_{pe}$ already includes the electrostatic stiffening, a total expansion may be defined for long chains as

$$\alpha_{s,t}^2 \equiv \frac{3\langle S^2\rangle}{Ll_p} = \frac{l_t}{l_p}\alpha_{s,el}^2. \tag{4.14}$$

As already discussed, the intrinsic excluded volume has been neglected in the above derivation.

Very recently, the application of the OSF approach to intrinsically flexible polyions was scrutinized [252] as the free energy of bending is originally identified with the energy of one particular configuration of the chain only, thus neglecting free energy fluctuations around this configuration. Accounting for such fluctuations, however, leads to inconsistencies in the model. Thus it was argued that the OSF theory should only be applied to intrinsically stiff polyions, i.e., $\xi a \ll (l_p \cdot l_B)^{1/2}$ which requires $l_p \gg l_B$ if counterion condensation takes place. The basic criticism of the OSF model is also supported by recent molecular dynamics simulations [82–84] (see Sect. 4.2).

4.1.4 The "Fuzzy" Sphere Model

4.1.4.1 Fixman's Approach [253]

In contrast to most experiments which yield a power law dependence of $l_{pe} \sim (c_s^m)^{-1/2}$ corresponding to $\alpha_s^2 \sim (c_s^m)^{-1/2}$ in terms of the excluded volume picture for long chains (as will be shown in Sect. 4.3), the theoretical approaches discussed so far either yield an inverse scaling of $l_{pe} \sim (c_s^m)^{-1}$ (OSF) or continuously bent curves (numerical calculation and excluded volume modified OSF theory). With the keen assumption that the concentration of gegen- and co-ions within the volume of an intrinsically flexible polyion coil is enhanced as compared to the average concentration in solution, Fixman [253] long ago derived the viscosity expansion coefficient α_η^3 to vary as $\alpha_\eta^3 \sim (c_s^m)^{-1/2}$. However, quantitative comparison with experimental data required unreasonably large values of the parameter $l_p \varepsilon/\varepsilon_{eff}$ with ε the usual dielectric constant of the solvent and ε_{eff} an effective dielectric constant within the sphere modelling an enhanced counterion binding.

4.1.4.2 Numerical Calculations

On the basis of the numerical calculations described in Sect. 4.1.2, one of us [254] investigated the possibility as to whether or not the concentration of counter- or co-ions may be enhanced within the volume of a coiled polyion. No specific "binding" effect of the low molar mass ions into the polyion chain were considered, but only the simple osmotic energy, which is required to create a non-uniform ion distribution. In analogy to Eq. (4.5) the total free energy is minimized as a function of n with the additional osmotic contribution F_{os}:

$$F = F_{DH} + F_{el} + F_{os} \qquad (4.15)$$

with both F_{DH} and F_{el} decreasing functions with increasing l_{pe} and

$$F_{os} = KT \cdot \Delta n_s \cdot V_{COIL} \qquad (4.16)$$

with Δn_s the increased number concentration of salt ions within the coil and the coil volume $V_{COIL} = 4\pi/3 \, R_h^3$ or $V_{COIL} = 4\pi/3 \, R_g^3$. The numerical results indicate that the salt concentration within a coil may increase by $10^{-3} - 10^{-2}$ mol/l for $c_s = 10^{-5}$ mol/l depending on the chain length and on the coil volume definition. Of course, these results appear quite artificial and most likely do not explain the experimentally observed $l_{pe} \sim c_s^{-1/2}$ dependence, because at higher concentration the increase becomes negligible, i.e., $\Delta n_s = 0$. On the other hand preliminary experiments [255] on charged μ-gels of moderate cross-linking density, i.e., 1:5 to 1:20, reveal no swelling or shrinking of the dimensions with varying ionic strength of the solution, which is not yet understood.

4.1.5 Recent Developments

Qian and Kholodenko [256] presented a new approach to the polyelectrolyte theory based on the Feynman variational method as used in the theory of large polarons. Unfortunately, the results are presented in a form which cannot easily be utilized for the interpretation of experimental results. One interesting result, however, is derived for the "effective" persistence length as

$$l^{-1} = N^{-1}(1 + \text{const.} \ I^{0.66}) \quad \text{small } c_s^m \qquad (4.17)$$

and

$$l_{eff} = \frac{a_3}{\ln(a_4 I^{1/2})} \left(1 + \frac{a_5}{\ln(a_4 I^{1/2})} \cdot I^{-1/2} \right) \quad \text{large } c_s^m \qquad (4.18)$$

with a_3, a_4 and a_5 some constants and I the ionic strength of the solution.

4.2 Simulations

In recent years Monte Carlo simulations have advanced to monitor the conformation of single polyion chains consisting of up to several hundreds of segments or beads as a function of ionic strength and charge density.

Brender et al. discussed in a series of papers [257–259] the fraction of "kinks" in short polyelectrolyte chains (less than 48 beads) as a function of ionic strength and charge density. By variation of the temperature T the relative contribution of the electrostatic and the entropic part of the free energy was influenced, leading to a coil-like shape a high T and to a rod-like conformation at T = 0. However, the interesting results are not easily verified experimentally because the temperature range of experiments is quite restricted.

The most extensive work on single chain simulation has been presented by Christos, Carnie and Creamer [260–263] where the conformation of chains of up to 320 beads was simulated as a function of charge density and ionic strength. Due to the finite chain length, pronounced end effects leads to an increasing persistence length with increasing contour length which seems to level off at larger chain lengths.

Very recently, Barenbrug et al. [264] performed simulation of a self-avoiding walk on a cubed lattice with fixed and fluctuating charges and compared the Monte Carlo results to analytical theories. The effect of charge fluctuations is found to be almost negligible under the conditions investigated and for low charge densities the simulation, the OSF theory and the numerical calculations presented in Sect. 4.1.2 agree[8]. At high charge densities, the chain extension of the simulation on the cubed lattice becomes significantly larger than the more realistical model of Carnie et al., who utilized the carbon–carbon bond angle of 109.5. In Fig. 16 a comparison is made between the simulation results of Carnie et al., the OSF theory and the calculation by Förster and Schmidt. The observed agreement is partly misleading because it does not reflect the obvious differences between the approaches which would appear at larger chain length and lower ionic strength. Both the simulations and the numerical calculations become instable if the total persistence length approaches the contour length which in turn limits the ionic strength regime to quite a high ionic strength, because larger chains cannot yet be simulated.

Whereas most simulations monitor the conformational change of one isolated polyion chain and account for the gegen- and co-ions by electrostatic screening, Stevens and Kremer [82–84] performed molecular dynamics simulations on a multi-chain system explicitly treating the full Coulomb interactions between the charged polyion segments and the gegenions. The polyion chain is modelled as a free-jointed spring and bead model with an excluded volume in terms of a repulsive Lennard–Jones potential, which is also applied to the counterion "hard core" repulsion. Persistence length, osmotic pressure and scattering functions are evaluated for chains consisting of 16, 32 and 64 beads. The simulations impressively confirm the two scaling regimes of the osmotic pressure, $\pi \sim c_p^{9/8}$ at low, and $\pi \sim c_p^{9/4}$ at high concentrations as derived by Odijk and as experimentally observed. One important aspect emerging from the results is already briefly mentioned in Sect. 2.3.3, because in agreement with experiments and in contrast to Odijk's prediction, no chain length dependence of the cross-over concentration between the two osmotic pressure scaling

---

[8] Unfortunately the numerical calculations of the persistence length (called Schmidt theory in [264]) are not correct at both added salt concentrations of $c_s^m = 0.145$ mol/l and $c_s^m = 5.9$ mol/l which are corrected in Fig. 16.

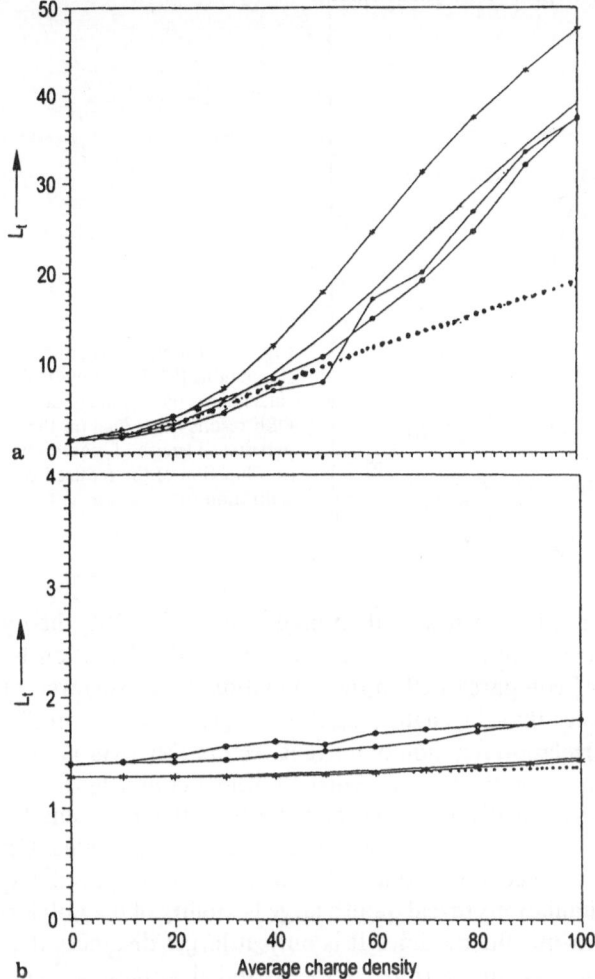

Fig. 16a, b. Comparison of different theoretical and Monte Carlo results for the persistence length l_t as a function of charge density for MC simulations with fixed (○) and mobile (●) counterions. OSF theory (—) and numerical calculations (····) (Data taken from [264]): **a** $c_s^m = 0.0145$ m; **b** $c_s^m = 5.9$ m

regimes is found. Rather, the cross-over concentration is given by a critical concentration

$$c_\pi^* \equiv (4\pi l_B \cdot a^2)^{-1} \tag{4.19}$$

with l_B the Bjerrum length and a the length of a monomer unit. Thus the cross-over concentration c_π^* does not correspond to an overlap concentration of individual polyion chains but rather reflects the concentration of polyion segments beyond which all electrostatic interactions are effectively screened. Finally, the total persistence length as a function of polyion concentration (see

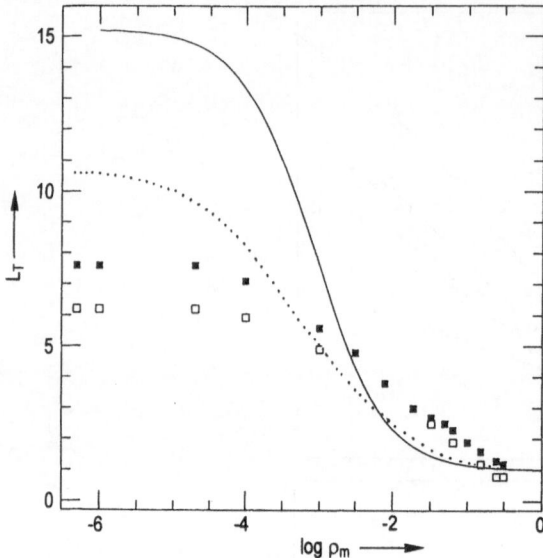

Fig. 17. Molecular dynamic simulations ($\square$, $\blacksquare$), numerical calculations (*dotted line*) and OSF results (*solid line*) on the persistence length of macroions as a function of the reduced concentration ρ_m

Fig. 17) yields much lower values at low c_p than predicted by the OSF theory and somewhat larger values at high c_p, whereas the numerical calculation by Förster and Schmidt [244] compares well to the simulation data. However, at low c_p there is some evidence that the chains might no longer follow worm-like chain statistics, i.e., the correlation function of tangent vectors $\langle \vec{R}(s)\vec{R}(s') \rangle$ does not decay exponentially, as already claimed earlier by Seidel et al. [265].

Another interesting feature is displayed by the molecular dynamics simulation concerning the local flexibility of intrinsically flexible polyion chains. On a local scale where the OSF theory introduces the "local stiffness approximation" (see Sect. 4.1.3) the simulations reveal quite a large flexibility of the polyion beads originating from entropic fluctuations. It is only at larger distances that these locally flexible blobs become correlated to expanded structures. This observation more or less confirms the basic criticism [252] of the OSF model already discussed in Sect. 4.1.3.

4.3 Experiments

The effect of ionic charges on the conformation of flexible macromolecules has been investigated by many experimental techniques, such as viscometry, static and dynamic light scattering, neutron and X-ray scattering. It is difficult to determine precisely the chain stiffness for flexible polyions, due to the following complications.

1) It is difficult to prepare well-defined and molecularly well characterized samples in terms of molar mass, molar mass distribution and chemical charge

density without any low molar mass ionic impurities (salt). This also includes the question of how many ionic impurities are present in "pure" water, the most frequently used solvent.

2) At low ionic strength conditions, the concentrations usually used for the above mentioned experimental techniques fall into the "ordinary–extraordinary" transition regime where the single chain properties are completely obscured by both strong coupling effects with gegen- or coions and by the presence of "slow modes" as discussed in Sect. 3. In order to avoid these regimes safely, the polyion concentration is restricted to

$$c_p^m < 0.1 c_s^m. \tag{4.20}$$

The condition at Eq. (4.20) is almost always violated by experimental investigations at ionic strength $c_s^m \leqslant 10^{-3}$ m requiring $c_p^m \leqslant 10^{-4}$ mol/l, i.e. $c_p \leqslant 0.02 - 0.04$ g/l for usual polyelectrolytes with monomer molar masses of 200–400 g/mol and still most violated at $c_s^m = 10^{-2}$ m as measurements are usually performed at concentrations above 1 g/l.

3) Present "state of the art" light scattering and viscosity experiments do indeed investigate polyelectrolyte concentrations as low as 1 mg/l, which would allow the experimental determination of the polyion conformation at ionic strength in the order of $c_s^m = 10^{-5}$ m without violating the conditions at Eq. (4.20). As discussed in Sect. 2, at this low ionic strength the dominating contribution to the observed experimental quantities originates from intermolecular interference of the polyions rather than from single chain properties, as is evident from "Bragg-Peaks" in the angular dependent scattering intensity. Such "Bragg-Peaks" decrease with increasing ionic strength and eventually disappear at $c_s^m \approx 10^{-4}$ m for such low polyion concentrations, i.e., $\lambda^* < 1$.

   A vanishing Bragg-Peak, however, does not necessarily indicate the observed quantities to be completely unaffected by intermolecular interaction terms. Thus, the concentration regime free of intermolecular interaction in terms of c_p and c_s^m cannot be rigorously defined on the basis of present knowledge, experiments, however, indicating that for $c_s^m \geqslant 10^{-3}$ m and $c_p^m \leqslant 0.1 \, c_s^m$, single chain properties indeed being observed.

4) Eventually at intermediate and high ionic strength the excluded volume of a polyelectrolyte chain becomes important, originating from two sources:

   i) the "intrinsic" excluded volume the macroion would have had, even in the absence of ionic charges;

   ii) the "electrostatic" excluded volume causing chain expansion by a long range electrostatic interaction which leads to the molar mass dependence of the apparent electrostatic persistence length, l_{pe}, as discussed in Sect. 4.1.3.

Whereas ii) is part of the problem, i.e., originates from the ionic charges on the macromolecule, i) is of non-ionic origin and ideally should be accounted for separately. In practice, a separation of the intrinsic and electrostatic excluded volume is not possible, because the intrinsic excluded volume is neither independent nor a monotonic function of the ionic strength.

For most polyelectrolyte solutions the total excluded volume is most pronounced at intermediate ionic strength quantitatively depending on the overall flexibility of the chain and on the thermodynamic quality of the solvent, both of which depend on the ionic strength. Qualitatively it is known that most flexible polyelectrolytes and nonionic, water soluble polymers approach unperturbed dimensions at high salt concentrations. This effect is known as the "salting out" effect, i.e., the macromolecules precipitate. Reducing the salt content improves the solvent quality, which leads to chain expansion ("intrinsic" excluded volume) and simultaneously increases the electrostatic expansion of a charged polymer. At some intermediate ionic strength, the macroion is so strongly expanded by electrostatic forces, that the intrinsic excluded volume approaches zero as the ring closure probability of the chain segments becomes very small. It is obvious that this delicate balance between electrostatic and thermodynamic forces is far beyond the scope of any analytical theory and constitutes the major obstacle in the interpretation of experimental results. Also it is worth mentioning that no simple scaling relations of molecular properties vs ionic strength are to be expected.

As will be discussed below, the complications arising from points 1) and 3) are generally appreciated in the literature in that samples are carefully prepared and measurements of the conformational properties are performed at ionic strength $c_s^m \geqslant 10^{-3}$ m. Point 4) is frequently discussed, whereas point 2) is still mostly ignored.

4.3.1 Viscometry

Viscometry is the most widely applied experimental method to determine the conformation of polyelectrolytes. The experiments are easily performed except for the measurements of the shear rate dependence, which is usually neglected (see, however, Sect. 2.3.3 for the importance at low ionic strength conditions).

In 1984, Tricot [266] summarized viscosity results of various vinylic polyacids reported in literature, analyzed the data by the Kratky-Porod worm-like chain model and compared the results with the OSF theory and the theory of Fixman [251] and Le Bret [19]. At low ionic strength, the electrostatic persistence length approximately followed a scaling relation $l_{pe} \sim (c_s^m)^{-1/2}$, in complete disagreement with the OSF theory, which postulates $l_{pe} \sim (c_s^m)^{-1}$. At high ionic strength Fixman's and Le Bret's results showed qualitatively a better agreement with experiments. Quantitatively, however, experimental values for l_p were up to 50% larger. From the measurements at $c_s^m > 2.5 \cdot 10^{-3}$ m, Tricot keenly extrapolated l_{pe}^{-1} vs $(c_s^m)^{-1/2}$ to zero salt concentration and obtained common values for l_{pe} ($c_s^m = 0$) $\approx$ 500 Å for all polyions. Because there is no theoretical justification of such extrapolations its merit is just as questionable as the Fuoss-Strauss plot.

It should be noted that almost all viscosity investigations at $c_s^m < 0.05$ m do violate the condition $c_p^m \leqslant 0.1\, c_s^m$ as discussed above and accordingly do not represent single polyion properties.

Some viscosity investigations on salt-free polyelectrolyte solutions are worth mentioning despite the dominating intermolecular electrostatic interaction discussed above. Cohen and Priel [267, 268] tried to circumvent the interaction by diluting the solution to about 1 mg/ml, well beyond the peak-concentration, and extrapolated the data to zero concentration. Anomalously high Huggins coefficients and intrinsic viscosities were obtained which led to the conclusion that, even at such high dilution, the interaction dominates in agreement with light scattering studies in a similar concentration regime.

As already mentioned in Sect. 3.3.3, Nishida and Kaji [75, 269] calculated the influence of intermolecular interaction on the intrinsic viscosity of spheres. Assuming the interaction effect on the viscosity of spheres and flexible coils to be comparable they corrected the experimental viscosity data which also exhibit a peak as a function of concentration by the calculated contribution from the interaction. The remaining part of the intrinsic viscosity shows a continuous increase as a function of ionic strength which levels off to a constant value at low ionic strength. Interpretations of these intrinsic viscosities are in qualitative agreement with dynamic Kerr effect studies [270, 271] and Tricot's data analysis.

Hodgson and Amis [157] reported dynamic viscoelasticity measurements on "salt-free" dilute solutions of protonated PVP samples in an HCl/glycol solvent. Again, a peak in the concentration dependence of the dynamic storage and loss moduli was observed independent of molar mass and measurement frequency. In contrast of Cohen et al. who postulated a free-draining behaviour of the polyelectrolyte chains at low ionic strength, the relaxation spectra could be interpreted by the Zimm normal modes for non-draining coils. The maximum was interpreted as an expansion of the coils followed by a contraction, because in these experiments the ionic strength exhibits a minimum where the maximum in the moduli occurs which makes a more detailed discussion difficult.

At intermediate and high salt concentration the complication arises from excluded volume effects. This question was investigated long ago by Nagasawa et al. [4, 272, 273] solely in terms of classical excluded volume theory, i.e., $l_t = l_p \neq f(c_s^m)$, as the concept of the electrostatic persistence length was not known at the time. The conclusion was that experimentally determined electrostatic expansion coefficients do vary as

$$\alpha_\eta^3 \sim \left(\frac{M}{c_s^m}\right)^{1/2} \tag{4.21}$$

with $\alpha_\eta^3 \equiv [\eta]/[\eta]_0$ and $[\eta]_0$ the unperturbed intrinsic viscosity in contradiction to all excluded volume theories which predict $[\eta] \sim M^{1/2}(M^{1/2}/c_s^m)^{3/5}$.

It is very interesting to note that there is one exception, being that Fixman [253] derived $[\eta] \sim M^{1/2}(M/c_s^m)^{1/2}$ under the assumption that, due to the presence of the dissociated gegenions, the ionic strength within the coiled macroion is much higher than the external added salt concentration (see Sect. 4.1.4). Particularly at high ionic strength this assumption is believed to be meaningless and the coincidence considered as accidental.

4.3.2 Static and Dynamic Light Scattering

Theory and experiments on polyelectrolytes by static light scattering, together with the viscosity results, were reviewed by Nagasawa and Takahashi in 1972 [4]. Their conclusions are still valid if the following simplifications are accepted.

1) The expansion of a polyion with decreasing ionic strength is caused by the electrostatic excluded volume only. A local stiffening (concept of the electrostatic persistence length) was not in discussion at the time.
2) No contribution of the intrinsic excluded volume is considered.
3) Electrostatic interaction is fully described by a Debye–Hückel potential.

With the invention of laser light scattering techniques, including dynamic light scattering for measurements of diffusion coefficients, new possibilities for the conformational studies of polyelectrolytes became available. However, most of these recent light scattering studies investigated salt-free polyelectrolyte solutions rather than the conformation of macroions at intermediate and high ionic strength. In principle, the precise measurements of the radius of gyration and of the translational diffusion coefficient should provide more reliable data on the conformation of polyions than simple viscosity experiments. There are only few static and dynamic light scattering investigations on the conformation of flexible polyelectrolytes reported in the literature, with qualitatively similar but quantitatively different results. It is quite safe to conclude that most of the light scattering and the viscosity data scatter for essentially the same reasons:

– at low ionic strength, the intermolecular structure interferes with the single chain properties (reported as miraculous scattering envelopes) and the condition $c_p^m \leqslant 0.1 c_s^m$ is almost always violated for measurements at $c_s^m < 10^{-2}$ m;
– the experimental results at intermediate and high ionic strength depend on subtle intrinsic excluded volume effects.

In appreciation of the first statement, light scattering investigations of the polyion conformation (and probably any other experimental technique for the investigation of conformational properties) are strictly limited to salt concentrations $c_s^m > 10^{-4}$–10^{-3} m, as even state-of-the-art instruments do not allow measurements significantly below a few mg/l.

Thus, we focus on the second statement, i.e. on the properties of macroions at intermediate and high ionic strength where we encounter the yet unsolved problem of the intrinsic excluded volume for flexible polyelectrolytes. At present, there are two approaches to tackle the problem.

1) Find a polyelectrolyte with a poor solubility, i.e., θ-conditions should be reached in the order of $c_s^m \approx 0.1$ m. As the solubility slowly increases with

decreasing salt and the polyion stiffening simultaneously becomes larger, the net intrinsic excluded volume should remain small at any ionic strength.

2) Investigate one of the common "easily prepared" polyelectrolytes such as PSS, PAA, or PMA, which have excellent solubility in aqueous salt solutions, and account for the pronounced intrinsic excluded volume in an approximate "ad hoc" manner.

Before we discuss the more recent data in terms of approaches 1 and 2, we will briefly mention some earlier investigations.

Some static light scattering investigations [274, 275] on polyacrylic acid have analyzed the shape of the particle scattering factor at ionic strength $c_s^m > 0.01$ m. It was qualitatively concluded that the experimental particle scattering factor was better matched by a flexible chain with excluded volume rather than by a worm-like chain model. Fischer, Sochor and Tan [276] were the first to analyze measured radii of gyration and viscosity data on poly-2-acrylamido-2-methylpropanesulphonate at $c_s \geq 0.01$ m in terms of a worm-like chain model. They found a significant molar mass dependence of the persistence length at all ionic strengths, quantitatively depending on the nature of the gegenions.

With a number of questionable approximations, Schmitz and Yu [278] reported an upper boundary for the electrostatic persistance length $l_{pe} \approx 250$ Å at low ionic strength for an ultra high molar mass KPAA, based solely on dynamic light scattering measurements.

We now come back to the two different approaches, described above, which deal with the excluded volume problem. Recent investigations followed the second approach and explicitly applied the Odijk–Houwaart theory, described in Sect. 4.1.3 to flexible polystyrene sulphonate, [210, 279] polyacrylamide/polyacrylic acid copolymers [280] and to some natural polysaccharides [210, 213, 281]. The investigation of Wang and Yu [279] on PSS is based on measured diffusion coefficients only, and with an approximate estimate of R_g a qualitative agreement with theory was obtained. In the work of Reed et al. [210, 213, 280], the theoretical curves qualitatively agreed with experimental results, particularly at intermediate ionic strength (0.001 m $\leq c_s^m \leq 0.1$ m) where the calculated apparent persistence length approximately scaled as $(c_s^m)^{-1/2}$. At higher ionic strength, however, the calculated curves bent downwards, i.e., yielding a much smaller apparent electrostatic persistence length as experimentally observed, even if, in the case of the polyacrylic copolymers, the intrinsic excluded volume is also taken into account. However, the reasonable agreement between theory and experiments is obtained only by optimizing both l_{pe}, and l_p, which almost always leads to unreasonably high values of the intrinsic persistence length l_p, i.e., $l_p = 31.3$ Å for polystyrenesulphonate [210] and $l_p = 37$ Å for the polyacrylic acid copolymer [280]. It should also be noted that most of these experimental results were obtained with a number of unnecessary experimental simplifications and assumptions for data evaluation, such as

1) the form factor of the polyelectrolyte chain is given by the Debye function for unperturbed Gaussian chains;

2) the slope of the reduced static scattering intensity vs q^2 is independent of concentration for $c < c_p^*$, c_p^* being the overlap concentration, i.e., no extrapolation of the slope to zero concentration was peformed. Thus, the values reported for the radius of gyration are quite uncertain;

3) the effect of small but significant polydispersities was ignored;

4) sometimes the static light scattering measurements were restricted to scattering angles $\theta \geqslant 45°$ only.

Also, the diffusion coefficient determined by dynamic light scattering was not utilized for the determination of the persistence length, because for all investigated polysaccharides the diffusion coefficient was found to be independent of the salt concentration in the investigated regime of $c_s^m > 10^{-3}$ m. Only for the polystyrenesulphonate was D_z constant for 0.002 m $\leqslant c_s^m \leqslant 0.03$ m but increased with increasing salt concentration when $c_s^m > 0.03$ m. This increase, however, was reported to be not even qualitatively compatible with the corresponding decrease observed for the radius of gyration except at very high salt concentrations $c_s^m > 0.5$ m. It was speculated that the transition from nondraining to free draining hydrodynamics might just cancel the increasing chain expansion of the macroions with decreasing ionic strength. Such an expansion, however, is in contradiction to the molar mass dependence of $D_z \sim M_w^{-1/2}$ found for all c_s^m, whereas free-draining macromolecules should exhibit a $D \sim M^{-1}$ power law. As will be shown below, these experiments are in qualitative disagreement with other static and dynamic light scattering investigations on polyvinylpyridinium salts [244], where the chain expansion was observed in both D_z and R_g, although the persistence lengths derived from both quantities also do not perfectly coincide.

Nordmeier and Dauwe [282] reported static light scattering experiments on polystyrene sulphonate at 0.005 m $\leqslant c_s^m \leqslant 2$ m and analyzed the data by a worm-like chain model. The resulting total apparent persistence length is compared to the data of [210] in Fig. 18. The agreement is quite poor, and the data of [282] do not follow at all the scaling-law (solid line in Fig. 18).

Förster et al. [244] and Beer [283, 284] investigated the chain stiffness of poly-2-vinyl-pyridiniumbenzylbromide at ionic strength 10^{-3} m $\leqslant c_s^m \leqslant 0.1$ m for different molar masses by combined static and dynamic light scattering.

As QPVP precipitates at $c_s^m \approx 0.15$ m salt, the effect of the intrinsic excluded volume is expected never to influence significantly the experimental results at any ionic strength (see approach 1).

As opposed to the investigations of the very soluble polyelectrolytes such as PSS and polyacrylic acid, no scaling behaviour of l_t or l_{pe} vs c_s^m was observed, although a lower limit for the intrinsic persistence length of $l_p = 15$ Å was adopted which equals the most likely value for the uncharged polyvinylpyridine. Rather, a continuously bent curve was found in a double logarithmic plot of l_{pe} vs c_s^m, the downwards curvature of which becomes more pronounced towards high ionic strength. At low ionic strength, the l_{pe} values determined by Förster et al. seemed to level off and become only slightly dependent on ionic strength for $c_s^m < 0.01$ m in qualitative agreement with the numerical calculations of

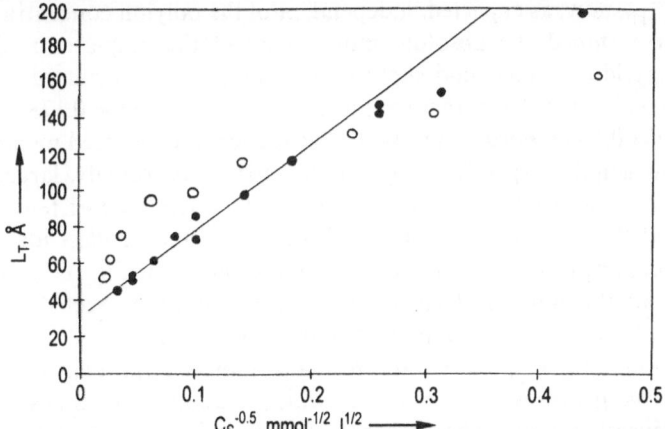

Fig. 18. Comparison of experimentally determined persistence length on NaPSS as a function of added salt, $(c_s^m)^{-1/2}$: ●: [133], $M_w = 7.8 \times 10^5$ g/mol, ○: [282], $M_w = 3.5 \times 10^5$ g/mol

Förster and Schmidt and in clear disagreement with the OSF theory[9]. Also, the R_g and R_h values could not be consistently fitted with one common persistence length below $c_s^m < 10^{-2}$ m in a qualitatively similar direction as observed by Gosh et al. for PSS. It was speculated at the time that this discrepancy between R_g and R_h might originate from an initial influence of intermolecular interference from long range electrostatic interactions. Such an explanation, however, was not satisfactory because it could not account for the observed strong increase of the apparent radius of gyration determined at very small but finite concentrations of $2 \cdot 10^{-6}$ g/l $\leqslant c_p \leqslant 2 \cdot 10^{-5}$ g/l, because, at such low polyion concentrations, the gegenions do not noticably affect the total ionic strength of the solution at $c_s^m \leqslant 10^{-3}$ m. In this work, however, the apparent radius of gyration was defined as

$$R_{g,app}^2 \equiv 3 \left[\frac{dKc/R_\theta}{dq^2} \middle/ \frac{Kc}{R_{\theta=0}} \right]_{c=c_p}. \tag{4.22}$$

Very recently, Beer [283] reinvestigated identical samples and similarly prepared ones even more carefully and could impressively demonstrate the increase of $R_{g,app}$ at low c_p to follow from the definition of the apparent radius of gyration by Eq. (4.22) if the second virial coefficient becomes anomalously large, i.e., K_c/R_θ at $c = c_p$ is much larger than $K_c/R_\theta = 0$ at $c_p = 0$. Defining an apparent radius of gyration as

$$R_{g,app}^2 \equiv 3 \left[\frac{dKc/R_\theta}{dq^2} \right]_{cp} \middle/ \left[\frac{Kc}{R_{\theta=0}} \right]_{c_p=0} \tag{4.23}$$

---

[9]Unfortunately, in the original Fig. 12 of [244] the theoretical curve for the OSF theory is not plotted correctly. The correct curve is shifted by exactly one decade to a larger persistence length which, however, does not significantly alter the conclusion

yields values of $R_{g, app}$, now, as expected, independent of the polyion concentrations. Beer also reexamined the absolute molar mass of the unquaternized "parent" polyvinylpyridines and found slightly lower molar masses caused by a somewhat larger value for the refractive index increment (dn/dc = 0.173 as compared to dn/dc = 0.165 reported in [244]). This reduces the reported molar masses and contour length, respectively, by roughly 10%. Accordingly, larger persistence lengths are evaluated, which, however, still do not consistently emerge from R_g and R_h. Also, the l_{pe} values still increase at the smallest ionic strength investigated, i.e., $c_s^m = 10^{-3}$ m. The results are shown in Fig. 19 with the OSF theory and the numerical calculations as discussed in Sect. 4.1.2. Whereas the OSF theory is seen not to match the experimental results, the numerical calculations almost perfectly fit the data if the charge density is reduced to one charge per 20 Å contour length which is about three times as small as the condensation limit. Although not understood, one explanation would be that the electrostatic interaction is overestimated by the Debye–Hückel potential.

The effect of a large intrinsic excluded volume was again investigated by Beer [283, 284] for the polyvinylpyridiniumhydrochloride (HPVP) prepared from the same "parent" polymer as QPVP samples. This was done simply by dissolving PVP in aqueous HCl at pH = 3 and adding sodium chloride in order to vary the total ionic strength. The hydrogenated polyvinylpyridines did not precipitate even in a saturated sodium chloride solution of about 6 m. The results are shown

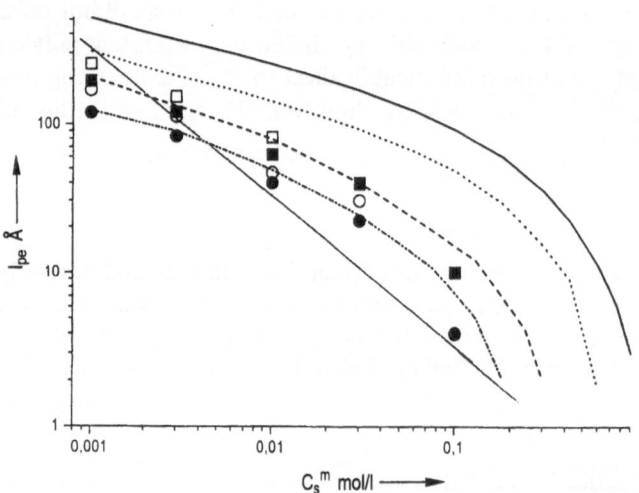

Fig. 19. Experimental and numerically calculated electrostatic persistence length of QPVP at different molar masses as a function of ionic strength. *Circles*: $M_w = 7.6 \times 10^5$ g/mol, *squares*: $M_w = 2.3 \times 10^6$ g/mol. *Filled symbols*: l_{pe} derived from R_h, *open symbols*: l_{pe} derived from R_g. *Solid and dotted lines*: numerical calculations for a charge density of 0.35 (Manning limit); *dash-dotted and dashed lines*: numerical calculations with a charge density of 0.125. *Lower solid line*: OSF theory (Taken from [283]

in Fig. 20 where l_{pe} is plotted vs c_s^m on a double logarithmic scale. The intrinsic persistence length was again taken as $l_p = 15$ Å. Now the curves do not bend downwards at high ionic strength, and lead to higher apparent electrostatic persistence lengths than observed for the QPVP samples.

At low ionic strength, it cannot be unambiguously decided whether this difference of the persistence length originates from the still quite different excluded volumes or reflects a coil-shrinking of the QPVP samples due to the hydrophobic nature of benzyl group. At high ionic strength, however, it seems safe to conclude that the frequently observed scaling $l_{pe} \sim (c_s^m)^{-1/2}$ emerges accidently from the subtle interplay of electrostatic stiffening and chain expansion due to the electrostatic and intrinsic excluded volume. Figs. 19 and 20 also show the persistence length as derived by a worm-like chain model to be molar mass dependent in agreement with the numerical calculations irrespective of ionic strength.

Mattoussi et al. [285] reported light scattering results on the precursor of poly-phenylenevinylene, the polycation polyxylylenetetrahydrothiopheniumchloride, dissolved in methanol at various ionic strengths in the range of $0.0044\text{m} \leqslant c_s^m \leqslant 0.041$ m. The solvent quality at all ionic strengths investigated (i.e., the intrinsic excluded volume) was regarded as poor. For the first time an inverse power law of $R_g^2 \sim l_{pe} \sim (c_s^m)^{-1}$ was found experimentally for a reasonably flexible polyion ($l_p = 40$ Å) in qualitative agreement with the OSF theory. However, the absolute magnitude of the observed electrostatic persistence length was found to be almost a factor 5 (!) larger than calculated by the OSF theory. In view of the broad molar mass distribution of the sample ($M_w/M_n \approx 2$), which was completely ignored in the analyses of the chain stiffness, and of the

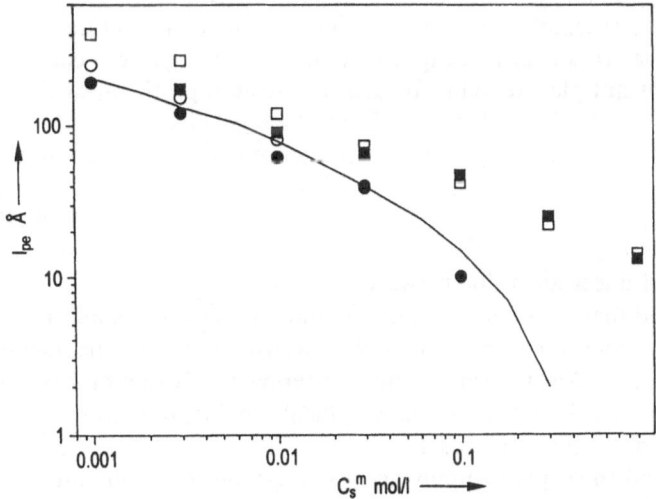

Fig. 20. Comparison of the experimentally determined persistence length from QPVP and HPVP of identical chain length L = 6900 Å; *squares*: HPVP, *circles*: QPVP; *open symbols*: l_{pe} derived from R_g, *filled symbols*: l_{pe} derived from R_h. *Solid line* as in Fig. 19.

lack of knowledge of the dissociation constants, the scientific character of this paper towards a better understanding of polyelectrolyte conformations appears questionable.

4.3.3 Neutron and X-ray Scattering

In general, the concentration domain for X-ray and neutron scattering experiments is well above 1% by weight which, for reasonably large molar masses, lies above the overlap concentration of the macromolecules, i.e., in the semidilute concentration regime. A simple scattering experiment at these high concentrations will yield either an intermolecular interference pattern or a kind of static correlation length of the transient network supposedly being formed (see discussion of polyion chain overlap in Sect. 2.3.3). Also it must be kept in mind that, except for very high salt concentrations, the presence of large polyion clusters is unavoidable, which in most experiments is not always recognized at the usually high q-values investigated. However, careful studies do report an anomalous increase of the scattering intensity at low q, which has not yet been unambiguously related to the slow mode observed by light scattering. The two phenomena, however, most likely do have the same origin. Keeping this in mind, we now focus on experiments in the high q-regime where only the local conformation of a macromolecular chain is monitored, i.e., the scattering envelopes are independent of the total chain length.

In two subsequent papers Muroga et al. [286, 287] report on X-ray scattering measurements on the sodium salt of polyacrylic and stereoregular polymethacrylic acid at various degrees of dissociation (0 to 0.85), two ionic strengths of $c_s^m = 0.1$ m and $c_s^m = 0.01$ m and polyion concentrations between 1 and 3%. The persistence length was determined graphically by the Kratky-plot I (q). q^2 vs q from the cross-over point q^* where the scattering pattern changes from the ideally constant plateau value for coils to the asymptotic linear slope characteristic for the scattering behaviour of rod-like segments.

The cross-over value q^* is related to the persistence length of the chain by

$$l_t = \frac{A}{q^*} \tag{4.24}$$

with A a constant of uncertain value between 1.9 and 2.7.

It should be noted that the interpolation procedure becomes quite uncertain for essentially two reasons: for a small number of persistence lengths per chain $(L/l_t < 50)$ the plateau value typical for the scattering behaviour of flexible segments is never reached and accordingly difficult to identify; conversely, the linear slope theoretically expected for rod-like segments at very large q sometimes lies beyond the experimentally accessible q-range or is obscured by the scattering contribution of the chain cross-section.

Within these uncertainties Muroga et al. concluded that the total persistence length for PMA and PAA of $l_t = 9$ or 13 Å, depending on the values for A in Eq. (4.24), is essentially independent of the degree of ionization, molar mass, polyion and added salt concentration. These results are in obvious contradiction to all light scattering investigations and to the neutron scattering investigations discussed below. The authors argued that the local stiffening of a polyion does not occur and explains the increase of polyion dimensions by excluded volume effects only.

First neutron scattering investigations were performed by Moan and Wolff on carboxymethylcellulose [288] and on PMA [289] with molar mass M_w = 13 000 g/mol at different polyion concentrations $0.5 \, g/l \leqslant c_p \leqslant 32 \, g/l$ and various degrees of ionization and on one fully ionized PMA sample of M_w = 370 000 g/mol at $c_p = 10$ g/l. No salt was added. For the small PMA sample, the radius of gyration of the single chain was believed to be measured, which increased with decreasing polymer concentration as $R_g \sim c_p^{-1/2}$ until for the smallest concentration a value close to the one expected for a rod-like structure was obtained. It was also demonstrated at constant polyion concentration of 10 g/l that the reduced viscosity and R_g show a similar increase with increasing degree of ionization.

There is no doubt that all these data fall into the extraordinary regime where no single chain properties are observable. Unfortunately, the small PMA sample, due to the small number of persistence lengths of the chain, did not exhibit a cross-over point q^*, in order to evaluate the persistence length in the high q-regime where the influence of the "extraordinary scattering" should largely be reduced. There is one measurement reported on the high molar mass fully ionized PMA sample at $c_p = 10$ g/l with no added salt, which yields a cross-over value $q^* = 0.019 \, \text{Å}^{-1}$ leading to roughly a ten times larger persistence length as reported by Muroga et al. Moreover, the measurements of q^* on carboxymethylcellulose show the value of q^* to decrease with decreasing c_p as $q^* \sim l_t \sim c_p^{1/2}$. Later, Plestil et al. [130] utilized the ingenious labelling technique introduced by Williams et al. [136], which allows the extraction of the single particle scattering factor even if the polyelectrolyte chains are embedded in a highly ordered lattice of interacting macroions. This procedure should concomitantly eliminate the scattering contribution of the cluster in the "extraordinary phase". The persistence length of PMA at one single polyion concentration c_p = 36 g/l was determined by the position of q^* (see Eq. 4.22) for the different degrees of ionization and for Li^+ and Na^+ as gegenions. With the constant A = 2.3, $l_t = 7.9$ Å was obtained for the nonionized samples being in good agreement with the data of Muroga et al. on isostatic PMA ($l_t = 5.0$ Å with A = 1.9, which yields $l_t = 6$ Å for A = 2.3) and on PAA ($l_t = 8.9$ Å with A = 1.9 leading to $l_t = 10.8$ Å for A = 2.3). With an increasing degree of ionization, however, the persistence length increased to $l_t = 12.5$ Å for the fully ionized samples irrespective of the gegenion, i.e., Li^+ or Na^+.

At first the labelling technique described above was applied to extract the single particle scattering factor of PSS solutions (M = 30.000 g/mol) with no added salt at different polyion concentrations $16 \, g/l \leqslant c_p \leqslant 120 \, g/l$ [134]. These measurements were repeated later [131] because in the first investigation the hydrogenated PSS was not completely matched. The extraction of the single particle scattering factor required measurements at different ratios of deuterated and hydrogenated PSS samples at fixed total polyion concentration, followed by extrapolation to zero deuterated PSS content. The resulting "single chain" scattering curves do show a downwards bending at low q in a Berry-plot, which was ignored for the determination of R_g and remained unexplained. The R_g values strongly increase with decreasing polyion concentration in essentially quantitative agreement with the theory of Le Bret [193]. At the same time another neutron scattering experiment was made on a higher molar mass PSS ($M_w = 80.000$ g/mol) [119] again utilizing the labelling technique in order to extract both the pure intermolecular structure factor and the single chain form factor. The persistence length l_t was obtained by fitting the theoretical form factor for a worm-like chain to the experimental data at intermediate and high q-values in order to avoid the ambiguous graphical interpolation in the Kratky-plot. The radius of gyration could not be obtained as the data were not taken at the required very low q-values. The results were quite similar to those of [134] as l_t increases with $c_p^{-1/2}$. Quantitative agreement with the calculation of Le Bret is observed if a reasonable value for $l_p = 12$ Å is adopted. A comparison of the persistence length in [119] and [134], however, reveals that the l_t values of the higher molar mass PSS are consistently larger if the contour length per monomeric unit is taken as 2.5 Å. Excellent agreement is obtained if the contour length/monomer of the higher molar mass polyion is reduced to 2.2 Å only, or if the molar mass were 15% smaller. Small angle neutron scattering experiments were also performed on the tetramethylammonium salt of PSS with added salt. In contrast to the results of Muroga et al., Ragnetti and Oberthür [290] came to the conclusion that the scattering envelope is perfectly fitted by a worm-like chain model with increasing persistence length towards smaller ionic strength. The authors concluded that excluded volume effects do not have any significant influence. Again, relatively high polyion concentrations of roughly 1% were measured at ionic strength 0.02 m $\leqslant c_s^m \leqslant$ 1 m.

Interestingly, a transition in scattering intensity and chain dimension was observed [290] at $c_s^m = 0.1$ to 0.5 m, which was interpreted as a conformational transition of miraculous origin. Nonetheless, the data were made to fit the OSF theory by simply adopting different intrinsic persistence lengths on either side of the transition, i.e., $l_p = 70$ Å at low c_p and $l_p = 120$ Å at high c_p. These values are much higher than those obtained for the sodium salt of PSS in salt-free solution and can hardly be explained by the size effect of the different counterions. A much simpler and more probable explanation for these results is that at $c_s^m \approx 0.1$–0.5 m the transition from the ordinary to the extraordinary regime occurs with the consequence of enhanced scattering intensities and large radii of gyration.

4.3.4 Dynamic Kerr Effect

Dynamic Kerr effect measurements are restricted to low ionic strength as already discussed in Sect. 2.3.3, which in turn raises the problem of how electrostatic interactions influence the results. First dynamic Kerr effect measurements by Oppermann [175] on PSS in salt-free solutions were interpreted in terms of the rotational diffusion of rigid rods without, however, specifying the persistence length quantitatively. In these measurements Oppermann already monitored the bimodal decay of the birefringence curves and utilized the slow relaxation time for the data evaluation. Oppermann also investigated the absolute, static values of the Kerr constant which yields the persistence length of a polyion if the absolute value of the induced dipole moment is known. Reasonable estimations were compatible with quite large persistence lengths already derived from the dynamic data.

Later, Piazza and Degiorgio [270, 271] performed dynamic Kerr effect measurements on lower molar mass PSS where the fast relaxation is not observed, because it most probably has moved out of the time window of the instrument. The resulting relaxation times were interpreted in terms of the rotational diffusion of worm-like chains according to Monte Carlo simulations of Hagerman and Zimm [291] and to Yamakawa [292]. The resulting persistence lengths are at the upper limit of experimental values derived from other techniques discussed above.

However, as long as the full decay of the field-free birefringence curves at higher molar masses and the influence of intermolecular interactions are not fully understood, the results have to be treated with caution.

5 Conclusion

This review should demonstrate that polyelectrolyte solutions are currently advancing to a fascinating topic in polymer science. The increasing effort in analytical theory, simulations and experiments is about to result in a deeper, fundamental understanding of the interacting macroions and of the conformational properties of flexible polyions. As these "simple" problems have still not been completely solved, new challenges have already emerged, such as the effect of polyion chain architecture (polyelectrolyte stars, combs, rings, µ-gels), the effect of bivalent or multivalent metal ions bound to the polyelectrolyte chain (intramolecular cross-linking, polyion coil collapse), and the complex field of surfactant–polyion and polyanion–polycation complexes. In comparison with the growing technical importance of such complex ionic systems, the fundamental academic research in this area has hardly started yet.

Note added in proof. It has very recently come to our attention that a theoretical calculation of the radius of gyration as a function of ionic strength is hidden in a paper "Adsorption of a polyelectrolyte chain to a charged surface" by Muthukumar [293] which is not acknowledged in current polyelectrolyte literature. As will be shown in a future work [294] the coincidence between the theoretical results and the experimental data of polyelectrolytes in a good solvent (see Fig. 20) is remarkable. It is interesting to note that, in the experimentally investigated regime of $\kappa R_g > 1$, Muthukumar's approach can be reduced to the old Flory excluded volume approach properly modified to polyelectrolytes.

6 References

1. Morawetz M (1961). In: Rice SA, Nagasawa M (eds) Polyelectrolyte solutions. Academic Press, London
2. Katchalsky A, Alexandrowicsz Z, Kedem O (1966). In: Conway BE, Barrada RG (eds) Chemical physics of ionic solutions. John Wiley, New York
3. Oosawa F (1970). In: Polyelectrolyes. M. Dekker, New York
4. Nagasaw M, Takahashi A (1972). In: Huglin MB (ed) Light scattering from polymer solutions. Academic Press, New York
5. Selegny E (ed) (1974) Polyelectrolytes. Reidel D, Dordrecht, NL
6. Mandel M (1988). In: Encyclopedia of polymer science and engineering, Sec. Ed., John Wiley, New York
7. Mandel M (1993) Polyelectrolytes. Reidel D, Dordrecht, NL
8. Schmitz KS (1993) Macroions in solution and colloid suspension. VCH
9. Förster S, Schmidt M, Antonietti M (1990) Polymer 31: 781.
10. Bernal JD, Fankuchen I (1941) J Gen Physiol 25: 111
11. Riley DP, Oster G (1951) Discuss. Faraday Soc 11: 107.
12. Balta-Calleja FJ, Vonk CG (1989). In Polymer science library. Vol. 8, Jenkins AD (ed). Elsevier, Amsterdam
13. Matsuoka H, Ise N, Okubo T, Kunugi S, Tomiyama, Yoshikawa Y (1985) J Chem Phys 83: 378
14. Hosemann R, Bagchi SN (1962) Direct analysis of diffraction by matter. North Holland, Amsterdam
15. Lifson S, Katchalsky A (1954) J Polymer Sci. 13: 43
16. De Gennes P-G, Pincus P, Velasco RM, Brochard F (1976) J Phys France 37: 1461
17. Odijk T (1979) Macromolecules 12: 688
18. Kaji K, Urakawa H, Kanaya T, Kitamaru R (1988) J Phys France 49: 993
19. Le Bret M (1982) J Chem Phys 76: 6248
20. Hayter JB, Penfold J (1981) Mol Phys 42: 109
21. Hansen JP, Hayter JB (1982) Mol Phys 46: 561
22. Belloni L (1985) Chem Phys 99: 43
23. Waisman E, Lebowitz JL (1971) J Chem Phys 53: 3086
24. Blum L (1975) Mol Phys 30: 1529
25. Blum L (1980) Theoretical chemistry: advances and perspectives. Vol. 5, Eyring H, Henderson D (eds) Academic Press
26. Nägele G, Klein R, Medina-Noyola N (1985) J Chem Phys 83: 260
27. Hansen JP, McDonald IR (1986) The theory of simple liquids. Academic Press, London
28. Jannink G (1986) Makromol Chem Makromol Symp 1: 67
29. Stillinger F, Lovett R (1968). J Chem Phys 49: 199
30. Hess W, Klein R (1983) Adv Phys 32: 173
31. For a review see: Sood AK (1991). In: Solid state physics. Ehrenreich H, Turnball D (eds) Vol. 45, Academic Press
32. Medina-Noyola M, McQuarrie DA (1980) J Chem Phys 73: 6279
33. Belloni L, Drifford M, Turq P (1984) Chem Phys 83: 147
34. Beresford-Smith B, Chan DYC, Mitchell DJ (1985) J Colloid Inteface Sci 105: 216
35. Schneider J, Hess W, Klein R (1985) J Phys A 18: 1221
36. Onsager L (1949) Ann NY Acad Sci 51: 627
37. Flory PJ (1956) Proc R Soc London 234: 73
38. Schneider J, Kasser D, Dhont JKG, Klein R (1987) J Chem Phys 87: 3008

39. Maeda T (1991) Macromolecules 24: 2740
40. Katchalsky A (1981) J Pure and Appl Chem 26: 327
41. Mandel M (1992). J Phys Chem 96: 3934
42. Hayter J, Jannink G, Brochard-Wyart F, de Gennes P-G (1980) J Phys Lett France 74: L451
43. De Gennes P-G (1979) Scaling concepts in polymer physics. Cornell University Press, Ithaca
44. Benmouna M, Weill G, Benoit H, Akcasu AZ (1982). J Phys France 43: 1679
45. Koyama R (1983) Physica B 120: 418
46. Koyama R (1984) Macromolecules 17: 194
47. Koyama R (1986) Macromolecules 19: 178
48. Grimson MJ, Benmouna M (1986) Chem Phys Lett 132: 55
49. Benmouna M, Grimson MJ (1987) Macromolecules 20: 1161
50. Zimm BH (1948) J Chem Phys 16: 1093
51. Vilgis TA, Benmouna M. Benoit H (1991) Macromolecules 24: 448
52. Vilgis TA, Borsali R (1991) Phys Rev A 43: 6857
53. Grimson MJ, Benmouna M, Benoit H (1988) J Chem Soc Faraday Trans 84: 163
54. Genz U, Klein R (1989) J Phys France 50: 439
55. Genz U. Klein R, Benmouna M (1989) J Phys France 50: 449
56. Khokhlov AR (1980) J Phys A 13: 979
57. Pfeuty P (1978) J Phys France Coll 12: 149
58. Khokhlov AR, Khachaturian KA (1982) Polymer 23: 1742
59. Borue V, Erukhimovich I (1988) Macromolecules 21: 3240
60. Bates FS, Fredrickson GH (1990) Ann Rev Phys Chem 41: 25
61. Joanny JF, Leibler L (1990) J Phys France 51: 545
62. Benmouna M, Vilgis TA (1991) Macromolecules 24: 3866
63. Van der Maarel JRC, Groot LCA, Hollander JG, Jesse W, Kuil ME, Leyte JC, Leyte-Zuiderweg LH, Mandel M (1993) Macromolecules 26: 7295
64. Tanford C (1961) Physical chemistry of macromolecules. John Wiley, New York
65. Wang L, Bloomfield VA (1990) Macromolecules 23: 194
66. Wang L, Bloomfield VA (1990) Macromolecules 23: 804
67. Fuoss RM (1948) J Polym Sci 3: 603
68. Fuoss RM (1949) J Polym Sci 4: 96
69. Fuoss RM, Strauss UP (1948) J Polym Sci 3: 246
70. Fuoss RM (1952) Discuss Faraday Soc 11: 125
71. Witten TA, Pincus P (1987) Europ Lett 3: 315
72. Cohen J, Priel Z, Rabin Y (1988) J Chem Phys 88: 7111
73. Borsali R, Vilgis TA, Benmouna M (1992) Macromolecules 25: 5313
74. Borsali R, Rinaudo M (1993) Makromol Chem Theory Simul 2: 179
75. Nishida K, Kanaya T, Kaji K (1990) Polymer preprints Japan 39: 3953
76. Rice, S, Kirkwood J (1957) J Chem Phys 31: 901
77. Snook I, van Mengen W (1959) J Chem Phys 31: 901
78. Gaylor K, van Mengen W, Snook I (1979) J Chem Soc Farad Trans II 75: 451
79. Berne BJ (ed) Theoretical chemistry. Vol. 6, 7
80. Schneider J, Hess W, Klein R (1986) Macromolecules 19: 1729
81. Weyrich B, D'Aguanno B, Canessa E, Klein R (1990) Faraday Discuss Chem Soc 90: 245
82. Stevens MJ, Kremer K (1993). In: Macro-ion characterization. Schmitz KS (ed) ACS Symposium Series 548
83. Stevens MJ, Kremer K (1993) Phys Rev Letters 71: 2228
84. Dünweg B, Stevens MJ, Kremer K (1994). In: Monte Carlo and molecular dynamics simulations in polymer science. Binder K (ed) Oxford University Press
85. van Megen W, Snook I (1984) Adv in Colloid and interface Sci 21: 119
86. Pusey PN, Tough RJA (1985) in Dymanic light scattering applications of photon correlation spectroscopy. Pecora R (ed) Plenum Press, New York
87. Brown JC, Pusey PN, Goodwin JW, Ottewil RH (1975) J Phys A 8: 664
88. Grüner F, Lehmann W (1979) J Phys A 12: L303
89. Grüner F, Lehmann W (1982) J Phys A 15: 2847
90. Härtl W, Versmold H, Wittig U (1984) Ber Bunsenges Phys Chem 88: 1063
91. Taylor TW, Ackerson BJ (1985) J Chem Phys 83: 2441
92. Tata BVR, Kesavamoorthy R, Arora KS (1986) Mol Phys 57: 363
93. Härtl W, Versmold H (1988) J Chem Phys 88: 7157
94. Okubo T (1989) J Chem Phys 90: 2408
95. Krause R, Nägele G, Karrer D, Schneider J, Klein R, Weber R (1988) Physica A 153: 400

96. Safran SA, Clark NA (eds) (1987) Physics of complex and supermolecular fluids. Wiley, New York
97. Kose A, Ozaki K, Takano K, Kobayashi Y, Hachisu S (1973) J Colloid interface Sci 44: 330
98. Crandall RS, Williams R (1977) Science 198: 293
99. Clark NA, Hurd AJ, Ackerson BJ (1979) Nature 281: 57
100. Hachisu S, Takano K (1982) Adv Coll Interface Sci 16: 233
101. Pieranski P (1983) Contemp Phys 24: 25
102. Ise N (1986) Angew Chem Int Ed Engl 25: 323
103. Ise N, Matsuoka H, Ito K (1989) Macromolecules 22: 1
104. Dosho S, Ise N, Ito K, Iwai S, Kitano H, Matsuoka H, Nakamura H, Okamura H, Ono T, Sogami I, Ueno Y, Yoshida H, Yoshiyama T (1993) Langmuir 9: 394
105. Sogami I (1983) Phys Lett A 96: 199
106. Sogami I, Ise N (1984) J Chem Phys 81: 6320
107. Theodoor J, Overbeek G (1987) J Chem Phys 87: 4406
108. Yamanaka J, Matsuoka H, Kitano H, Ise N (1990) J Coll and Interface Sci 134: 92
109. Yamanaka J, Matsuoka H, Kitano H, Ise N, Yamaguchi T, Saeki S, Tsubokawa M (1991) Langmuir 7: 1928
110. Maier EE, Schulz SF, Weber R (1988) Macromolecules 21: 1544
111. Schulz SF, Maier EE, Weber R (1989) J Chem Phys 90: 7
112. Hagenbüchle M, Weyrich B, Deggelmann M, Graf C, Krause R, Maier EE, Schulz SF, Klein W, Weber R (1990) Phys A 169: 29
113. Maier EE, Krause R, Deggelmann M, Hagenbüchle M, Weber R (1992) Macromolecules 25: 1125
114. Fraden S, Maret G, Caspar DCN, Meyer RP (1989) Phys Rev Lett 63: 2068
115. Drifford M, Dalbiez J-P (1984) J Phys Chem 88: 5368
116. Krause R, Maier EE, Deggelmann M, Hagenbüchle M, Schulz SF, Weber R (1989) Physica A 160: 135
117. Kaji K, Urakawa H, Kanaya T, Kitamaru R (1988) J Phys France 49: 993
118. Nierlich M, Williams CE, Boue F, Cotton J-P, Daoud M, Farnoux B, Jannink G, Picot C, Moan M, Wolff C, Rinaudo M, de Gennes PG (1979) J Phys France 40: 701
119. Nierlich M, Boue F, Lapp A, Oberthür R (1985) Colloid Polym Sci 263: 955
120. Schmidt M (1989) Makromol Chemie Rapid Common 10: 89
121. Förster S (1992) Ph.D. thesis, University of Mainz
122. Ise N, Okubo T, Kunigi S, Matsuoka H, Yamamoto KI, Ishii Y (1984) J Chem Phys 81: 3294
123. Kanaya T, Kaji K, Kitamaru R, Higgins JS, Farago B (1989) Macromolecules 22: 1356
124. Ise N, Okubo T, Yamamoto K, Kawai H, Hashimoto T, Fujimura M, Hiragi Y (1980) J Am Chem Soc 102: 7901
125. Ise N, Okubo T, Yamamoto K, Matsuoka H, Kawai H, Hashimoto T, Fujimura M (1983) J Chem Phys 78: 1473
126. Weill G (1984) Polym Commun 25: 147
127. Ise N, Okubo T, Hiragi Y, Hashimoto T, Fujimura M, Nakajima A, Hagashi H (1979) J Am Chem Soc 101: 5836
128. Cotton JP, Moan M (1976) J Phys Lett France 37: L75
129. Rinaudo M, Dumond A (1977) J Polym Lett 1: 411
130. Plestil J, Ostanevich YM, Bezzabatonov VYu, Hlavata D (1986) Polymer 27: 1241
131. Nierlich M, Boue F, Lapp A, Oberthür R (1985) J Phys France 46: 649
132. Patkowski A, Gulari E, Chu B (1980) J Chem Phys 73: 4178
133. Li X, Reed WF (1991) J Chem Phys 94: 4568
134. Williams C, Nierlich M, Cotton JP, Jannink G, Boue F, Dauod M, Farnoux B, Picot C, de Gennes PG, Rinaudo M, Moan M, Wolff C (1979) J Polym Sci Polym Lett 17: 379
135. Plestil J, Ostanevich YM, Bezzabotnow VY, Hlavata D, Labsky J (1986) Polymer 27: 839
136. Bendedouch D, Chen SH (1983) J Phys Chem 87: 1473
137. Nallet F, Jannink G, Hayter JB, Oberthür R, Picot C (1983) J Phys France 44: 87
138. Nallet F, Cotton JP, Nierlich M, Jannink G (1985) Lect Notes Phys 172: 175
139. Kaji K, Urakawa H, Kanaya T, Kitamaru R (1984) Macromolecules 17: 1835
140. Koene RS, Nicolai T, Mandel M (1983) 16: 231
141. Trommsdorf E (1931) Ph.D. thesis, Freiburg
142. Staudinger H (1932) Die hochmolekularen organischen Verbindungen Kautschuk und Cellulose. Verlag Julius Springer, Berlin
143. Kern W (1938) Z Physik Chem A181: 283

144. Conway BE, Butler JAV (1953) Nature 172: 153
145. Fujita H, Homma T (1954) J Colloid Sci 9: 591
146. Eisenberg G, Pouyet J (1954) J Polym Sci 13: 85
147. Butler JAV, Robins AB, Shooter KV (1957) Proc R Soc London A241: 299
148. Vink H (1970) Makromol Chem 131: 133
149. Vink H (1987) J Chem Faraday Trans 83: 801
150. Vink H (1992) Polymer 33: 3711
151. Rabin Y, Cohen J, Priel Z (1988) J Polym Sci Polymer Lett 26: 397
152. Cohen J, Priel Z, Rabin Y (1988) Polym Commun, 29: 235
153. Cohen J. Priel (1989) Macromolecules 22: 2356
154. Yamanaka J, Matsuoka H, Hasegawa M, Ise N (1990) J Am Chem Soc 112: 587
155. Yamanaka J, Araie H, Matsuoka H, Kitano H, Ise N, Yamaguchi T, Saeki S, Tsuokawa M (1991) Macromolecules 24: 3206
156. Yamanaka J, Araie H, Matsuoka H, Kitano H, Ise N, Yamaguchi T, Saeki S, Tsuokawa M (1991) Macromolecules 24: 6156
157. Hodgson DF, Amis EJ (1991) J Chem Phys 94: 4581
158. Kim MW, Pfeiffer DG (1988) Europhys Lett 5: 321
159. Okamoto H, Nakajima H, Wada Y (1974) J Polym Ed 12: 1035
160. Okamoto H. Nakajima H, Wada Y (1974) J Polym Sci Polym Sci Polym Ed 12: 2413
161. Rosser RW, Nemoto N, Schrag JL, Ferry JD (1978) J Polym Sci Polym Phys Ed 16: 1031
162. Hodgson DF, Amis EJ (1989) J Chem Phys 91: 2635
163. Pals DTF, Hermans JJ (1948) J Polym Sci 3: 897
164. Pals DTF, Hermans JJ (1952) Rec Trav Chim Pays-Bas 71: 433
165. Wolf C (1978) J Phys France C2-39: 169
166. Rochas C, Domard A, Rinaudo M (1979) Polymer 20: 76
167. Davis RM, Russel WB (1987) Macromolecules 20: 518
168. O'Konski CT (ed) (1976) Molecular electro-optics. Marcel Dekker, New York
169. Schnee C (1994) Ph.D. thesis, University of Bayreuth
170. Lewis J, Pecora R, Eden D (1986) Macromolecules 19: 134
171. Diekmann D, Hillen W, Morgeneyer B, Wells RD, Pörschke D (1982) Biophysical Chem 15: 263
172. Allison SA, Nambi P (1992) Macromolecules 25: 759
173. Elvingson C (1992) Biophys Chem 43: 9
174. Allison SA (1993) Macromolecules 26: 4715
175. Oppermann W (1988) Makromol Chemie 189: 927
176. Oppermann W (1988) Makromol Chemie 189: 2125
177. Krämer U, Hoffmann H (1991) Macromolecules 24: 256
178. Schurr JM, Schmitz KS (1986) Ann Rev Phys Chem 37: 271
179. Schmitz KS (1990) An introduction to dynamic light scattering by macromolecules. Academic Press, Boston
180. Ackerson BJ (1976) J Chem Phys 64: 242
181. Büldt G (1976) Macromolecules 9: 606
182. Burchard W, Schmidt M, Stockmayer WH (1980) Macromolecules 13: 580
183. Schmidt M, Stockmayer WH (1984) Macromolecules 17: 509
184. Oono Y, Baldwin PR, Ohta T (1984) Phys Rev Lett 3: 2149
185. Oono Y, Baldwin PR, (1986) Phys Rev Lett 33: 3391
186. Wang L, Bloomfield VA (1989) Macromolecules 22: 2742
187. Stephen MJ (1971) J Chem Phys 55: 3878
188. Lee WI, Schurr JM (1975) J Polym Sci 13: 873
189. Berne BJ, Pecora R (1976) Dynamic light-scattering. Wiley, New York
190. Tivant P, Turq P, Drifford M, Magdalenat H, Menez R (1983) Biopolymers 22: 2762
191. Hermans JJ (1949) Rec Trav Chim Pays-Bas 68: 859
192. Landau LD, Lifschitz EM (1983) Lehrbuch der Theoretischen Physik. Vol. 10, Akademie Verlag, Berlin
193. Akcasu AZ, Benmouna M, Hammouda B (1984) J Chem Phys 80: 2762
194. Belloni L, Drifford M, Turq P (1985) J Phys Lett 46: L207
195. Belloni L, Drifford M, (1985) J Phys Lett 46: L1183
196. Benmouna M, Vilgis TA, Hakem F (1992) Macromolecules 25: 1144
197. Schmitz KS, Schurr JM (1973) Biopolymers 12: 1543
198. Schmidt RL (1973) Biopolymers 12: 1427

199. Schmitz KS, Pecora R (1975) Biopolymers 14: 521
200. Jolly D, Eisenberg H (1976) Biopolymers 15: 61
201. Chen FC, Yeh A, Chu B (1976) J Chem Phys 66: 1290
202. Schmidt RL, Boyle JA, Majo RA (1977) Biopolymers 16: 317
203. Jamieson AM, Presley CT (1973) Macromolecules 6: 358
204. Raj T, Flygare WH (1974) Biochemistry 13: 3336
205. Mathiez P, Weisbuch G, Mouttet C (1978) J Phys Lett France 39: L139
206. Mathiez P, Weisbuch G (1979) Biopolymers 18: 1465
207. Mathiez P, Mouttet C, Weisbuch G (1980) J Phys France 42: 519
208. Mathiez P, Mouttet C, Weisbuch G (1981) Biopolymers 20: 2381
209. Grüner G, Lehmann WP, Fahlbusch H, Weber R (1981) J Phys A 14: L307
210. Peitzsch RM, Burt MJ, Reed WF (1992) Macromolecules 25: 806
211. Sedlak M, Amis EJ (1992) J Chem Phys 96: 817
212. Sedlak M, Amis EJ (1992) J Chem Phys 96: 826
213. Gosh S, Li X, Reed CE, Reed WF (1990) Biopolymers 30: 1101
214. Koene RS, Mandel M (1983) Macromolecules 16: 220
215. Koene RS, Nicolai T, Mandel M (1983) Macromolecules 16: 227
216. Koene RS, Mandel M (1983) Macromolecules 16: 973
217. Doherty P, Benedek GB (1974) J Chem Phys 61: 5426
218. Lin SC, Lee WI, Schurr JM (1978) Biopolymers 17: 1041
219. Drifford M, Dalbiez JP (1985) Biopolymers 24: 1501
220. Neal DG, Purich D, Cannel DS (1984) J Chem Phys 80: 3469
221. Nemoto N, Matsuda H, Tsunashima Y, Kurata M (1984) Macromolecules 17: 1731
222. Ramsay DJ, Schmitz KS (1985) Macromolecules 18: 2422
223. Wilcoxon JP, Schurr JM (1983) J Chem Phys 78: 3354
224. Schosseler F, Ilmain F, Candau SJ (1991) Macromolecules 24: 225
225. Sedlak M, Konak C, Stepanek P, Jakes J (1987) Polymer 28: 873
226. Smits RG, Kuil ME, Mandel M (1983) Macromolecules 26: 6808
227. Matsuoka H, Schwalm D, Ise N (1991) Macromolecules 24: 4227
228. Chu B, Wu DQ, MacKnight WJ, Wu C, Phillips JC, LeGrand A, Lantmann CW, Lundberg RD (1988) Macromolecules 21: 253
229. Lantmann CW, MacKnight WJ, Higgins JS, Pfeiffer DG, Sinha SK, Lundberg RD (1988) Macromolecules 21: 1339
230. Lantmann CW, MacKnight WJ, Sinha SK, Pfeiffer DG, Lundberg RD, Wignall GD (1988) Macromolecules 21: 1344
231. Wu DQ, Phillips JC, Lundberg RD, MacKnight WJ, Chu B (1989) Macromolecules 22: 992
232. Gosh S, Peitzsch RM, Reed WF (1992) Biopolymers 32: 1105
233. Sedlak M (1993) Macromolecules 26: 1158
234. Zero K, Ware BR (1984) J Chem Phys 80: 1610
235. Fulton RL (1978) J Chem Phys 68: 3089
236. Fulton RL (1978) J Chem Phys 68: 3095
237. Kuhn W, Künzle O, Katchalsky A (1948) Helv Chim Acta 31: 1994
238. Katchalsky A, Künzle O, Kuhn W (1950) J Polym Sci 5: 283
239. Katchalsky A, Lifson S (1956) J. Polym. Sci. 11: 409
240. Harris FE, Rice SA (1954) J Phys Chem 58: 725
241. Rice SA, Harris FE (1954) J Phys Chem 58: 73
242. Lifson S (1958) J Chem Phys 29: 89
243. Schmidt M (1991) Macromolecules 24: 5361
244. Förster S, Schmidt M, Antonietti M (1992) J Phys Chem 96: 4008
245. Koyama R (1973) J Phys Soc Jpn 34: 1029
246. Odijk T (1977) J Polym Sci, Phys Ed 15: 477
247. Odijk T, Houwaart A (1978) J Polym Sci, Polym Phys Ed 16: 627
248. Odijk T (1978) Polymer 19: 989
249. Skolnick J, Fixman M (1977) Macromolecules 10: 944
250. Fixman M, Skolnick J (1978) Macromolecules 11: 863
251. Fixman M (1982) J Chem Phys 76: 6346
252. Barrat JL, Joanny JF (1993) Europhys Letters 3: 343
253. Fixman M (1964) J Chem Phys 41: 3772
254. Schmidt M, unpublished results
255. Grottenmüller R, Schmidt M, to be published

256. Qian C, Kholodenko AL (1988) J Chem Phys 89: 2301
257. Brender C (1990) J Chem Phys 92: 4468
258. Brender C (1990) J Chem Phys ^3: 2736
259. Brender C, Danino M (1992) J Chem Phys 97: 2119
260. Carnie SL, Christos GA, Creamer TP (1988) J Chem Phys 89: 6484
261. Christos GA, Carnie SL (1989) J Chem Phys 91: 439
262. Christos GA, Carnie SL (1990) J Chem Phys 92: 7661
263. Christos GA, Carnie SL, Creamer TP (1992) Macromolecules 25: 1121
264. Barenbrug ThMAOM, Smit JAM, Bedeaux D (1993) Macromolecules 26: 6864
265. Seidel C, Schlacken H, Müller I (1994), Makromol Chemie Theory and Simulation 3: 333
266. Tricot M (1984) Macromolecules 17: 1698
267. Cohen J. Priel Z (1989) Polym Commun 30: 223
268. Cohen J. Priel Z (1990) J Polym Phys 93: 9062
269. Nishida K, Kanaya T, Fanjat N, Kaji K (1992) Polymer Prepr Japan 41: 4403
270. Degiorgio V, Bellini T, Piazza R, Mantegazza F (1990) Phys Rev Letters 64: 1043
271. Degiorgio V, Mantegazza F, Piazza R (1991) Europhys Letters 15: 75
272. Takahashi A, Nagasawa M (1964) J Am Chem Soc 86: 543
273. Noda I, Tsuge T, Nagasawa M (1970) J Phys Chem 74: 710
274. Kitano T, Taguchi A, Noda I, Nagasawa M (1980) Macromolecules 13: 57
275. Nagasawa M, Noda I, Kitano T (1980) Biophys Chem 11: 435
276. Fischer LW, Sochor AR, Tan JS (1977) Macromolecules 10: 949
277. Kitano T, Taguchi A, Noda I, Nagasawa M (1980) Macromolecules 10: 955
278. Schmitz KS, Yu JW (1988) Macromolecules 21: 484
279. Wang L, Yu H (1988) Macromolecules 21: 3488
280. Reed WF, Ghosh S, Medjahdi G, Francois J (1991) Macromolecules 24: 6189
281. Foussiac E, Milas M, Rinaudo M, Borsali R (1992) Macromolecules 25: 5613
282. Nordmeier E, Dauwe W (1992) Polymer J 24: 229
283. Beer M (1993) Diploma Thesis University of Bayreuth
284. Beer M, Schmidt M (1994) PMSE Preprints 71: 697 ACS, Washington DC
285. Mattoussi H, O'Donochue S, Karasz F (1992) Macromolecules 25: 743
286. Muroga Y, Noda I, Nagasawa (1985) Macromolecules 18: 1576
287. Muroga Y, Noda I, Nagasawa (1985) Macromolecules 18: 1580
288. Moan M, Wolff C (1975) Polymer 16: 776
289. Moan M, Wolff C (1975) Polymer 16: 781
290. Ragnetti, Oberthür RC (1986) Colloid Polym Sci 264: 32
291. Hagerman P, Zimm BH (1981) Biopolymers 20: 1481
292. Yoshizaki T, Yamakawa H (1984) J Chem Phys 81: 982
293. Muthukumar M (1987). J Chem Phys 86: 7230
294. Muthukumar M, to be published

Editor: Prof. Wegner
Received: July 1994

Kinetics of the Free-Radical Emulsion Polymerization of Vinyl Chloride

I. Capek
Polymer Institute, Slovak Academy of Sciences, Dúbravská cesta 9,
842 36 Bratislava, Slovakia

Dedicated to Dr. M. Lazár, D.Sc. on the occasion of his 65th birthday

The aim of this review is to summarize and discuss the kinetic data of the emulsion polymerization and copolymerization of vinyl chloride. The current understanding of the kinetics of free-radical polymerization of conventional monomer is briefly described and kinetic data of radical polymerization and copolymerization of vinyl chloride in the presence of hydrophobic and hydrophilic additives are summarized. Effects of the initiator type and concentration, the reaction conditions and the type of diluent are evaluated. Variation of kinetic and molecular weight parameters in the heterogeneous polymerizations with emulsifier type and concentration are discussed.

List of Abbreviations and Symbols

a	root-mean-square end-to-end distance per square root of the number of monomer units of the polymer
a_s	the surface area of an emulsifier molecule
AIBN	2,2′-azobisisobutyronitrile
AP	ammonium peroxodisulfate
BA	butyl acrylate
BMA	butyl methacrylate
BPO	benzoyl peroxide
BQ	p-benzoquinone
CA	chloranile
C_M	constant for chain transfer to monomer
CMC	critical micellar concentration
D_{ab}, D_{ij}	mutual diffusion constant
D_p	diffusion constant of a macroradical chain
D_m	diffusion constant of monomer
$D_{i, cmd}$	center-of-mass diffusion constant
EA	ethyl acrylate
EDTA	ethylenediamine tetraacetic acid
ESR	electron spin resonance
DCP	diacetyl peroxydicarbonate
E	emulsifier
E_o	overall activation energy
E_p	activation energy for propagation
E_a	activation energy for decomposition of initiator
E_t	activation energy for termination
f	initiator efficiency
FS	formaldehyde sulphoxylate
HLB	hydrophilic/lipophilic balance
I	initiator
I˙	initiator radical
IPN	interpenetrating polymer network
k_d	decomposition rate constant of initiator
k_{dw}	decomposition rate constant of initiator in water
k_d'	decomposition rate constant of initiator in the presence of emulsifier
k_i	initiation rate constant
k_p	propagation rate constant of the growing radical
$k_{p, rd}$	reaction diffusion controlled rate constant
k_t	termination rate constant of the growing radicals
k_{tc}, k_{td}, k_{tpr}	termination rate constant for combination, disproportionation and primary radical termination
k_{tr}	chain transfer rate constant

k_{re}	re-entry rate constant
$k_{t,tr}$	termination rate constant for translation diffusion
$k_{t,seg}$	termination rate constant for segmental diffusion
k_{tw}	termination rate constant in water
k_e	radical entry constant
k_{des}	desorption (exit) rate constant
k_a	absorption rate constant
K	lumped rate constant
LCB	the long chain branching
M	monomer
$M^·$	monomeric radical
$[M]_{eq}$	equilibrium monomer concentration in particle
$M_i^·$	growing radical with the initiator fragment (j the number of monomer units in oligomer radical before it enters the particle, i = 0, 1, 2, ... j)
$M_j^·$	continuous phase radical
M_n	number-average molecular weight
M_w	weight-average molecular weight
M_v	viscosity-average molecular weight
MMA	methyl methacrylate
MWD	molecular weight distribution
N_A	Avogadro's number
N	number of particles per unit volume
N_w	number of particles in water
N_n	number of particles with n radicals
n	average-number of radicals per particle
P	dead polymer
$P^·$	polymeric radical
P_o	saturation pressure
PEO	poly(ethylene oxide)
PVC	poly(vinyl chloride)
p_s	adsorbed emulsifier
PSD	particle size distribution
$R^·$	primary radical
R_i	rate of initiation
R_p	rate of polymerization
R	rate of primary particle formation
R_d	decomposition rate of initiator
r	monomer reactivity ratio
s	solubility
S	total surface of polymer particles
SDS	sodium dodecyl sulfate
St	styrene
VAc	vinyl acetate
VD	vinylidene chloride

VC	vinyl chloride
VCM	vinyl chloride monomer
V_p	total volume of latex particles
$[VC]_{eqw}$	saturation concentration of vinyl chloride in water
X_n	number-average degree of polymerization
x_f	fractional conversion
x_c	critical (fractional) conversion
X	saturation degree
x	reaction order
ϕ	monomer volume fraction in the polymer
ρ_A	rate of absorption of radicals into polymer particles
ρ_m	density of monomer
ρ_p	density of polymer
$\omega_{m/p}$	weight ratio of monomer and polymer
ω_p	polymer weight fraction
σ	Lennard-Jones diameter of monomer
γ	surface tension

1 Introduction

Poly(vinyl chloride) (PVC) is one of the most widely produced polymeric materials in use today. It is commercially produced by four major processes: suspension, bulk, emulsion and solution. An industrially important method of production of PVC is emulsion polymerization. There are a lot of data regarding the kinetics and mechanism of emulsion polymerization of vinyl monomers. However, relatively little work has been done on the kinetics of vinyl chloride emulsion polymerization and much less on the emulsion copolymerization. Concerning the preparation of copolymer latexes of vinyl chloride monomer, there are only patents [1–3].

Vinyl chloride differs from monomers like styrene, methyl methacrylate or vinyl acetate principally by the insolubility of the polymer in its monomer. Its key feature is that poly(vinyl chloride) (PVC) is insoluble in its monomer, but it is slightly swollen by its monomer. Polymerization of vinyl chloride differs also from the conventional heterogeneous polymerization of monomers, such as acrylonitrile, in so far that vinyl chloride partly swells its polymer but acrylonitrile does not. It differs also from the conventional emulsion polymerization of unsaturated monomers in which the polymer particles are swollen by their monomer.

Emulsion polymerization of unsaturated monomers where the monomer and polymer are mutually insoluble, are of considerable industrial importance, but there is no generally accepted theoretical framework to discuss them. A major problem in the mechanistic investigation of any multiple – phase polymerization system is the number and complexity of the reactions which must be taken into account. For example, in such particles the formation of reaction loci and the growth of polymer chains occur simultaneously and each of these processes is itself complex. In quantifying the kinetics of heterogeneous polymerizations, where monomer and polymer are mutually immissible (insoluble), it is essential to take proper account of the compartmentalized nature of the system.

For the derivation of the rate expressions and polymer quality equations, one can assume that the polymerization of VC takes place in several stages. During the first stage, which extends from zero to about 0.1% conversion, the reaction mixture consists mainly of pure monomer, the concentration of polymer being less than its solubility limits. During this stage, the nucleation of monomer-swollen emulsifier micelles takes place. Stage 2 extends from the point of the appearance of the polymer-rich phase (0.1%) to a fractional conversion x_c (0.7–0.8) at which the separate monomer phase disappears. During this stage, the reaction mixture consists of the monomer-rich and polymer-rich phases and monomer droplets. The latter saturates both phases with monomer. It is suggested that polymerization proceeds in both phases at different rates and is accompanied by transfer of monomer from the monomer droplets to polymer particles and from the monomer-rich phase to polymer-rich phase, so that the composition of the latter is kept constant. The disappearance of the separate monomer phase (droplets) is associated with a pressure drop in the reactor. In the conversion range, $x_c < x < 1.0$, we have once again a homogeneous system consisting of the copolymer phase swollen with monomer. During this stage, the monomer weight fraction decreases in the polymer phase as the total monomer approaches a final limiting value.

We begin by describing the current understanding of the kinetics of polymerization of vinyl chloride with a high water-solubility. In particular, we note the importance of desorption of radicals and the formation of less reactive radicals for vinyl chloride. These concepts will then be drawn upon as we try to understand the behavior of a monomer insoluble in its polymer in polymerization.

2 Kinetics of Conventional Monomer Polymerization

2.1 General

The following reaction schemes can be proposed for solution polymerization of vinyl chloride monomer (VCM) initiated by the radical initiator (soluble in monomer or solvent) [4]

initiator decomposition

$$I \xrightarrow{k_d} 2R^{\cdot} \tag{1}$$

$$R^{\cdot} \xrightarrow{k_{d1}} R^{\prime\prime} + y \tag{2}$$

initial propagation step

$$R^{\cdot} + M \xrightarrow{k_{p1}} M_i^{\cdot} \text{ (or } P_i^{\cdot}) \tag{3}$$

$$R^{\prime\prime} + M \xrightarrow{k_{p1}} M_i^{\cdot} \tag{4}$$

subsequent propagation

$$M_i^{\cdot} + M \xrightarrow{k_p} M_{i+1} \tag{5}$$

termination

$$M_i^{\cdot} + M_j^{\cdot} \xrightarrow{k_{tc}} P_{ij} \tag{6}$$

$$M_i^{\cdot} + M_j^{\cdot} \xrightarrow{k_{td}} P_i + P_j \tag{7}$$

$$M_i^{\cdot} + R^{\cdot} \xrightarrow{k_{tp}} P_i \tag{8}$$

chain-transfer

$$M_i^{\cdot} + M \xrightarrow{k_{trm}} P_i + M^{\cdot} \tag{9}$$

$$M_i^{\cdot} + P \xrightarrow{k_{trp}} P_i + P^{\cdot} \tag{10}$$

$$M_i^{\cdot} + I \xrightarrow{k_{tri}} P_i + I^{\cdot} \tag{11}$$

$$M_i^{\cdot} + S \xrightarrow{k_{trs}} P_i + S^{\cdot} \tag{12}$$

The emulsion polymerization of vinyl chloride initiated by a water-soluble initiator principally includes the kinetic events given by Eqs. (1)–(12) and also some further steps such as the entry of radical into the particle

$$M_j + \text{particle} \xrightarrow{k_e} \text{entry} \tag{13}$$

and the radical desorption (re-entry)

$$\text{an active particle} \underset{k_{re}(+R^{\cdot})}{\overset{k_d'(-R^{\cdot} \text{ or } M^{\cdot})}{\rightleftharpoons}} \text{an inactive particle} \tag{14}$$

Here, I is the initiator, $R^{\cdot}$ the primary radical, $R^{\cdot\cdot}$ the secondary radical, M the monomer, $M^{\cdot}$ the monomeric radical, P dead polymer, $P^{\cdot}$ the polymer radical, $I^{\cdot}$ the initiator radical, $M_i^{\cdot}$ the growing radical with the initiator fragment (soluble or insoluble in the continuous phase, j the number of monomer units in oligomer radical before it enters the particle, $i = 0, 1, 2, \ldots, j$), $M_j^{\cdot}$ the continuous phase free radical, k_d the first-order initiator decomposition rate constant. k_{d1} the decomposition rate constant of initiator (or the primary radical), k_{p1} or k_{p2} the propagation rate constant of a primary radical, k_p the propagation rate constant of the growing radical, k_t the average termination rate constant of the growing radicals, t_{tc}, k_{td}, k_{tpr} the termination rate constant for combination, disproportionation and primary radical termination, k_{trm}, k_{trp}, k_{tri} and k_{trs} the chain-transfer constants to monomer, polymer, initiator and solvent, respectively, k_e the radical entry rate constant, k_d' the desorption rate constant and k_{re} the re-entry rate constant.

The first microscopic process, the formation of primary radicals by the decomposition of initiator governs the rate of initiation and starts the growth of the macromolecule. The rate of initiation is known to be a function of the initiator efficiency (f), the initiator decomposition constant (k_d) and the initiator concentration ([I]) and can be expressed by the following equation

$$R_i = 2f k_d[I] \tag{15}$$

The rate of polymerization, R_p of the conventional solution or bulk polymerization can be expressed by the following equation [5]

$$R_p = k_p[M][R^\cdot] = k_p(2f\,k_d/k_t)^{0.5}[M]^{0.5}[I]^{0.5}$$
$$= (k_p/k_t^{0.5})[M]R_i^{0.5} \tag{16}$$

where [M] is the monomer and [R$^\cdot$] radical concentration.

The rate of the emulsion polymerization is known to be a function of the radical (primary particle) formation, R. The rate of primary particle formation can be expressed by the following equation

$$R = R_i N_A \tag{17}$$

where N_A is the Avogadro's number.

The rate of polymerization of the classical emulsion (the initial reaction system is a multi-phase system) can be expressed by the following equation [6]

$$R_p = k_p[M]_{eq}\,Nn/N_A \tag{18}$$

where $[M]_{eq}$ is the equilibrium monomer concentration in the particle, N the number of particles per unit volume and n the average-number of radicals per particle.

It is well-known [6, 7] in emulsion polymerization that it is this "compartmentalization" which gives rise to greatly different characteristics of polymer generation in a latex system from those in bulk or solution. The micellar model [6, 7] involves 1) the creation or entry of polymerizing free radicals which become active at the locus of polymerization, 2) aqueous-phase events and 3) first- and second-reaction order loss of these free radicals. In an emulsion polymerization system where polymer and monomer are mutually soluble, the polymerizing locus is the whole particle. If monomer and polymer are mutually insoluble, most evidence [8–12] seems to indicate that the polymerization locus is the particle/water interfacial region.

An important characteristics of emulsion polymerization is that, unlike other polymerization processes, the polymerization rate and molecular weight can be increased at the same time. This is due to the compartmentalization of the system that reduces the probability of mutual termination of propagating radicals. The behavior of this compartmentalized system depends on the rate of exchange of species between the elements of the system. The main ingredients of an emulsion polymerization system include monomer, dispersant, emulsifier, and initiator. Water is commonly used as the dispergant. A water-insoluble monomer can be dispersed in water by means of an oil-in-water emulsifier and polymerized with a water-soluble initiator.

A conversion curve for the emulsion polymerization system is characterized by its sigmoidal shape. A few separate regions are distinguished in the conversion – time curves (see Fig. 1 [13]) [14]. Initially, the emulsion polymerization is very slow (interval 1). As polymerization proceeds, the rate of polymerization gradually increases, reaching an approximately constant value of the rate of

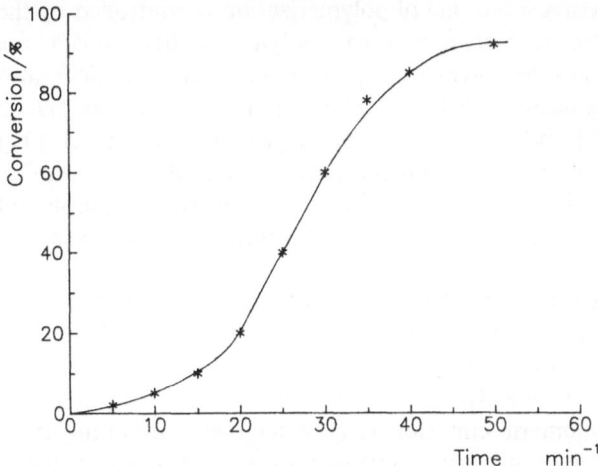

Fig. 1. Variation of the monomer conversion in the emulsion copolymerization of methyl methacrylate (MMA) and ethyl acrylate (EA) with reaction time [13]. Recipe: 150 g water, 20 g MMA, 80 g EA, 2.5 g Tween 40, 2.5 g Spolapon AOS, 60 °C. $[(NH_4)_2S_2O_8] = 2.92 \times 10^{-3} \, mol \, dm^{-3}$

polymer formation within a long conversion region (interval 2). This behavior is the direct result of the generation of new particles. The increase in the number of particles is accompanied by the increase of the rate of polymerization. As soon as the number of particles is fixed through the depletion of the free emulsifier micelles the polymerization becomes constant. Finally, the polymerization rate falls as the monomer droplets disappear, and consequently the monomer within the polymer particles becomes exhausted. In this system, the first interval is located within the conversion region of 0–30%, interval 30–70% and the last interval above 70% conversion. The rate of polymerization determined in the interval 2 is the maximum rate.

In interval 1, polymer particles are formed by the entry of radicals from the continuous phase into micelles accompanied by chain propagation. During this stage, the number of particles gradually increases with conversion. This stage is highly dependenting on initiator and monomer type and concentration, degree of agitation, presence of oxygen and on other reaction conditions. Beyond this interval, the free emulsifier micelles disappear. In contrast, the interval 2 is highly reproducible with the fixed number of polymer particles. At this stage a polymeric radical may grow undisturbed for a long time until another radical enters the particles or a chain transfer event takes place.

In intervals 1 and 2, the reaction mixture consists of three phases: the aqueous phase, droplets of monomer and monomer-swollen polymer particles. Beyond interval 2, when conversion is high enough, the monomer droplets disappear, and the reaction system becomes a two-phase system. At this stage, the unconverted monomer is present only in polymer particles and in water (if it is soluble).

For most emulsion systems, the rate of polymerization is controlled by the rate of entry and exit of free radicals to and from polymer particles, not by the rate of monomer diffusion to the polymerization sites. The entry of radicals into the polymer particles has been treated as a collisional process [14] as well as a diffusional process [15] and a colloidal process [16]. Nomura et al. [17] pointed out that radical desorption from the polymer particles and micelles plays an important role in particle formation and numerous examples of deviations from the Smith-Ewart kinetic model have been attributed to radical desorption.

A number of mechanisms and models have been proposed for latex particle formation in emulsion polymerization systems [18–20]. These include particle formation by entry of a free radical into a micelle [6, 7] or by homogeneous nucleation of oligomeric free radicals in the aqueous phase [18, 21–23] or within microdroplets of the monomeric emulsion [24]. After their formation, these primary particles may simply grow by conversion of monomer into polymer within these particles, or undergo coagulation [22, 25].

It has recently been demonstrated [26] that a most sensitive test of the theories describing the nucleation mechanism in an emulsion polymerization system is provided by the determination of the time of evolution of the full particle size distribution (PSD) of the latex sometime after the cessation of nucleation. This implies that the rate of production of new latex particles must be an increasing function of time for much of the nucleation period, although the rate ultimately drops off quite rapidly near the completion of nucleation. Any one-step micellar entry or homogeneous nucleation mechanism cannot be operative in this system because both mechanisms (micellar and homogeneous nucleation) predict that the rate of production of new latex particles decreases monotonically with time or at least remains constant.

The rate of production of latex particles was explained in terms of a two step coagulative nucleation model [26, 27]. The first-formed primary particles undergo coagulation to form latex particles. The small primary particles differ from mature latex particles in at least two important respects; first, they are colloidally unstable, undergoing coagulation with other primary particles; second, they polymerize very slowly. The colloidal instability of such small particles arises from their small size and concomitant curvature of their electrical double layers. The low rate may arise from the reduced swelling of particles by monomer, due to their hydrophilic character or/and their small particle size [28] and rapid desorption of any free radicals from primary particles arising from their small size [29]. This model is expected to be operative in polymerization of monomers with restricted mutual solubility of monomer and polymer. In such a case, the growth of particles is expected to be affected by the coagulation of small and large particles.

After emulsifier exhaustion, the remaining primary particles coagulate until only latex particles remain; since coagulation is rapid, this will lead to rapid decrease in the nucleation rate. This time can be estimated from the expression derived by Smith and Ewart [6] for the time when the total surface area of the particles equals to that which can be covered by the emulsifier.

The coagulative nucleation mechanism [26, 27] was derived to explain the decrease of the exponent of [E] with increasing emulsifier concentration. At high [E], the nucleation time is of long duration since the new primary particles are readily stabilized. As a result, more latex particles are formed and eventually outnumber the very small primary particles after long periods of time. Thus primary/latex particle collisions become more and more frequent and fewer latex particles are formed. The nucleation rate approaches zero after these long periods and N_c approaches a constant at high [E]. It should be noted that the exponent predicted by Smith and Ewart, 0.6, is constant over all values of [E] and fails to predict the larger exponent of [E]. The decrease in the exponent of [E] with increasing emulsifier concentration, however, should be also discussed in terms of 1) the electrostatic repulsion between negatively charged surfaces of polymer particles and negatively charged radicals, 2) the electrolyte effect and 3) the contribution of polymerization in emulsifier monomer droplets (the transfer of emulsion systems to microemulsions in which the rate of polymerization is much lower).

2.2 Initiation

The initiation process, similar to other free-radical vinyl polymerizations, involves the chemical decomposition of unstable peroxides – azocompounds, or persulfates – into free radicals which can react rapidly with monomer to begin the propagation of polymer chains [4]. In the case of a water-soluble initiator, the radical concentration in polymer particles is related to the initiator concentration in water and the radical capture efficiency of latex particles. The radical capture efficiency of monomer droplets is very small and, therefore, their contribution to overall polymerization process is negligible. Thus, the small surface area of monomer droplets and/or high concentration of radicals in monomer droplets disfavor the growth events. Using an oil-soluble initiator, the radical concentration in particles and monomer droplets is related to the initiator concentrations in both phases. The initiator concentration between these phases is usually expressed in terms of an initiator partition coefficient.

It is known that values of f at low conversion are slightly less than 1.0 [30, 31], i.e., f mostly varies in the range 0.5–0.6. Russell et al. [32] have reported that f decreases slowly with conversion down to ω_p (the polymer weight fraction) ≤ 0.8 and then falls steeply with $\omega_p \geq 0.8$).

Noyes [33, 34] has also suggested that, at higher conversions of monomers, chemical barriers to recombination of the two free radicals are very small whereas the activation energy for propagation is much higher. Hence, initiator fragments are virtually trapped next to each other and can barely, if at all, diffuse out of the cage of its environment. Thus, it is highly likely that, at high conversions, initiator moieties undergo "primary termination", which means the geminate recombination of initiator fragments that have not escaped from the cage in which they were formed. Furthermore, it is clear that as ω_p increases and

D_i (the diffusion constant of the initiator) decreases, the fraction of initiator radicals escaping from primary recombination decreases and, hence, the overall initiator efficiency, f, decreases. The "secondary termination" process [33, 34] (a process whereby free radicals that have separated from an encounter subsequently in the course of diffusion, re-enter each other and undergo geminate recombination) is not operative at high conversions while the microdomains with the radicals are separated and their diffusion to reach each other has a low probability.

The "secondary termination" may be operative, however, in the heterogeneous polymerization where monomer-rich microdomains characterized by low viscosity contain a large amount of initiator. Here, the microdomains are separated from each other by the polymer shell (wrapping effect) and, therefore, radical deactivation takes place by participation of mobile radicals.

Stickler et al. [35] predicted that at high conversions (in bulk systems) a large population of polymer chains of relatively small degrees of polymerization is formed. Later, Russell et al. [32] came to the same conclusion which they explained by the following consideration: At high conversion, the propagation will continue until the oligomer free radical is of sufficient length to be entangled in the surrounding polymer matrix, at which point the propagating chain is effectively immobilized on the time scale of propagation. This entanglement length is supposed to be of the order of 100 monomer units. The probability that a separating pair of initiating free radicals will undergo several propagation steps and then recombine will, therefore, be greatly enhanced compared with the situation at lower conversions. Since propagation is thought to be diffusion controlled in this high conversion regime, the initiator efficiency should depend on conversion. A marked decrease in f at high ω_p was observed. At critical conversion, the two free radicals are thus very likely to be entrapped in close proximity to each other and so to undergo geminate recombination. Hence, once diffusion and propagation become comparable, the initiator efficiency is expected to decrease dramatically.

Complete conversion of acrylates (e.g., butyl) and methacrylates (e.g., methyl) to polymer at 50 °C or 60 °C over a few hours may be obtained in emulsion systems [32, 36–38]. In the bulk polymerization of MMA, complete conversion can be achieved after more than 10 days (at 80 °C) [39]. The halting of bulk polymerization at high conversions cannot be ascribed to the drastic decrease in the value of k_p, while the polymerization in particles in a high conversion regime (monomer droplets are depleted) proceeds under the same viscosity conditions. The radicals and/or reaction loci. Russell et al. [32] propose that this experimental phenomenon is due to a large decrease in the initiator efficiency, f. This conclusion results from the fact that in emulsion systems initiation occurs in the aqueous phase (the viscosity of the continuous phase is low and, therefore, does not change with conversion), whereas in bulk systems initiation takes place in the monomer/polymer phase, which becomes very viscous at high conversions.

The complete conversion in emulsion polymerizations of unsaturated monomers which do not dissolve their polymers is expected to be similar to that for bulk systems. This results from the fact that polymerization in such systems proceeds in the aqueous phase and on the surface of polymer particles.

After depletion of monomer droplets and a decrease in the amount of monomer in the water, the rate of initiation decreases due to the increase of the water-phase termination and the decrease of radical capture efficiency of latex particles.

It has been shown that the radical capture efficiency for a monomer with high water-solubility such as acrylonitrile and, therefore, also for vinyl chloride is very high (close to 100%) [40].

Thus, at high conversions in bulk systems the cage effect is operative, whereas in the emulsion or dispersion systems the cage effect does not appear (if a continuous phase-soluble initiator is used). Besides, in emulsion systems, radicals that enter immediately propagate and are transformed into large and entangled ones. Therefore, the concentration of small radicals (derived from initiator) is very low. In bulk systems, there are two populations of radicals (mobile and entangled) and in the dispersion systems only one (entangled) population. The fraction of mobile radicals (monomeric) formed by chain transfer to monomer is the same in both systems but its contribution is different.

Shen et al. [41] have suggested that at higher conversion the radicals are wrapped in the more "rigid" micro-surroundings. Under very high conversion conditions, the wrapping layer becomes tight and dense by which the diffusion of radicals of initiator molecules is restricted. Primary radicals cannot escape from the cage surrounded by the dense matrix, so that the initiation efficiency tends to zero and the system reaches the limiting conversion. The wrapping layer of a radical is formed gradually and its density increases as the chain is growing. The probability of a radical becoming wrapped is mainly dependent upon its chain length besides the concentration of polymer in the system. The diffusion barrier in the micro-region is determined by the thickness and density of the wrapped layer. This concept of a wrapping layer indicates nonequality of radical reactivities and correlates the reactivity with its chain length.

Mechanism of radical formation and initiation at high conversion was discussed by Capek et al. [42] who investigated the dispersion copolymerization of PEO macromonomers initiated by oil (monomer)- and ethanol/water-soluble initiators. The formation of radicals in the continuous phase (from an ethanol/water-soluble initiator, 2,2'-azobisisobutyramide, VA) is mainly responsible for the growth and higher radical activity of polymer particles. In these systems (initiator is located only in the continuous phase), high rates and large molecular weights of polymer are observed. If the initiator (benzoyl peroxide, BPO) can diffuse from the continuous phase to the monomer-swollen polymer particles (after nucleation of particles), low rates and molecular weights of polymer are observed. BPO molecules or their radical pairs are supposed to react with entangled propagating radicals in a volume in which the cage effect is operative. The BPO-growing radical reactions terminate the growth events and simultaneously release the single radicals. The transfered mobile radicals formed by a cage effect mechanism take part in propagation as well as in termination.

Thus, the growth of macroradicals in particles is transferred by the reaction BPO-growing radical to propagation of mobile radicals termination. This mechanism controls the concentration and the ratio of long and short chain radicals and keeps the high conversion polymerization system in an active form. This approach was used in Ref. [42] to explain how the mobile radicals at high conversion are formed where the cage and wrapping models fail.

2.3 Propagation

A polymerization reaction occuring in a bulk or monomer/solvent system can be divided into two separate steps [4]: 1) the diffusive step, in which the reactants diffuse towards each other and 2) the chemical reaction step, in which the reacting species overcome an activation barrier. Each of the reaction events considered here is likely to be controlled by either a chemical or physical process, and sometimes by both. Denoting by $k_{p,dif}$ and $k_{p,chem}$ the rate constants of these steps, the overall rate constant k_p for that reaction is as follows

$$1/k_p = 1/k_{p,dif} + 1/k_{p,chem} \tag{19}$$

The rate of polymerization is usually controlled by the rate of only one of the above two processes. In the low and intermediate conversion regime, $k_p \sim k_{p,chem}$, while in the very high conversion regime, $k_p \sim k_{p,dif}$. The propagation rate constant, k_p, is expected to be independent of both molecular of both molecular weight fraction of polymer (ω_p), at least until the glass transition is approached. At very high conversion, k_p is expected to be dependent on ω_p. The only method which has been successfully applied to determine k_p is the ESR method.

Zhu et al. [43], who used ESR spectroscopy to study the bulk polymerization of MMA at $25\,°C$, observed two distinct free radical signals: 1) nontrapped, liquid-state, propagating free radicals and 2) trapped, solid-state, nonpropagating free radicals. The rate parameter k_p is defined with reference to propagating free radicals. If ESR spectroscopy detects a significant population of nonpropagating free radicals, then from equation (16) it is clear that this technique will give values of k_p that are too low.

Variation of the propagation rate constant with conversion has two regimes. In the range of low and medium conversion ($\omega_p < 0.8$), k_p is nearly constant or slightly decreases with conversion. Beyond this point ($\omega_p \sim 0.84$), k_p decreases strongly. In this conversion regime, the polymerization proceeds under monomer starved conditions. Here, the polymer matrix becomes glassy and the transport of monomer is diffusion controlled. It was found experimentally that k_p decreases by only 2 orders of magnitude from its value at low conversion [36]. The mechanistic interpretation of this result is that propagation becomes diffusion controlled above $\omega_p > 0.84$.

Values for k_p (ω_p) for MMA emulsion polymerization at 50 °C have been reported by Ballard et al. [34]. They were many orders of magnitude different from those values suggested by bulk studies [43].

At conversions well above $\omega_p \geq 0.8$, the value of k_p corresponds to that found in the absence of chemical kinetics control. The reaction diffusion constant [44] is denoted by $k_{p,rd}$ and may be estimated from the Smoluchowski expression [45]

$$k_{p,rd} = 4\pi D_{ab}\sigma N_A \tag{20}$$

where D_{ab} is the mutual diffusion coefficient of the reacting species, N_A is Avogadro's constant, and σ is the radius of interaction of the reactants. For the diffusion controlled propagation, an appropriate value for σ is the Lennard-Jones diameter of the monomer such as MMA. Thus, $D_{ab} = D_m + D_p$, where D_p is the diffusion constant of a macroradical chain end. Because the major contribution to the motion of the chain end at high ω_p comes from propagation, it follows that

$$D_p = k_p(M)a^2/6 \tag{21}$$

Here, M is the monomer concentration and a is the root-mean-square end to end distance per square root of the number of monomer units of the polymer.

It has long been recognized that some bulk polymerizations virtually stop at high conversions of monomer to polymer well before complete conversion [46]. Since the growing chains are likely to be entangled with the matrix polymer at such high polymer concentrations and are, therefore, capable of undergoing only extremely slow center-of-mass diffusion (transition), it is assumed that the origin of this vanishing rate cannot be due to increasing termination rate constant k_t. It has, therefore, been proposed that k_p must diminish markedly in this high-conversion range.

The increase of polymer concentration sometimes decreases the solvent power of the reaction medium which causes shrinkage of the active or dead polymer chains. Under such conditions, a denser polymer matrix is formed in which the average space available for movement of a chain radical contracts. At the beginning, the primary radicals propagate very quickly, and then these radicals propagate with monomer diffusing into the radical micro-region, so that the polymer accumulates and wraps up the radical in the micro-region. As the polymerization proceeds, the micro-surroundings becomes more and more rigid. The thickness of the wrapping layer is dependent upon the chain length of the radical itself. At the same time, the monomer, the initiator, and the short radicals are still in the liquid micro-surroundings. It is suggested that the radicals wrapped in the more "rigid" micro-surroundings can terminate very slowly because they can diffuse through the wrapping layer only with difficulty [41]. In the final stage, the wrapping layer becomes very dense and thick, so that the diffusion of monomer is restricted by it, and k_p becomes very small or tends to zero. This approach seems to be acceptable for heterogeneous polymerization.

2.4 Termination

In the general termination reactions Eqs. (6 and 7), the growing radicals may terminate either by combination to form a larger macromolecule, or disproportionation to form two separate dead polymer molecules [4].

In the general termination equation, Eq. (6 and 7), it is accepted that the termination rate constant, k_t, is diffusion controlled and, therefore, it is a strong function of chain length for short chains and a rather weak function for long chains.

The termination is controlled by diffusion because of the very low activation energy of the chemical reaction. North and Reed [47] divided the termination in homogeneous polymerization into two separate steps, i.e., a diffusion step and the actual chemical reaction. The former is a consecutive process of translation and segmental diffusion. Thus, the termination rate constant, $k_{t,dif}$, can be expressed as follows

$$1/k_{t,dif} = 1/k_{t,tr} + 1/k_{t,seg} \tag{22}$$

where $k_{t,dif}$ is the overall termination constant for translation diffusion, $k_{t,tr}$ the rate constant for translation diffusion and $k_{t,seg}$ the rate constant for segmental diffusion.

Variation of the termination rate constant k_t with conversion contains several different regimes. In the low conversion range, the value of k_t slightly increases with ω_p due to the fact that the segmental diffusion is the rate determining process. At higher conversions, the termination rate constant is diffusion controlled and decreases with ω_p up to the critical conversion at which the polymer chains are entangled [47–49]. Beyond this region, the termination is controlled by the reaction diffusion [44]. As soon as the propagation reaction becomes diffusion controlled, the termination rate decreases and approaches zero.

Here, the Smoluchovski equation (20) can also be used, where D_{ij} (or D_{ab} in Eq. (20)) is the mutual diffusion constant for species of degree of polymerization i and j, respectively, and separation at which termination occurs is r_{ij}. The separation at which termination proceeds between two monomeric free radicals is of the size of a monomeric unit [46], a value which is close to the van der Waals radius for species of the size considered here.

Experimental diffusion studies suggest that for polymerizing systems beyond low conversions, $D_{i,cmd} \propto M^{-1}$ for a small molecular weight of the polymer and $D_{i,cmd} \propto M^{-2}$ for large molecular weight chains [50]. The short radicals are mobile by virtue of their relatively unhindered (center-of-mass) translation diffusion. Long chain radicals are supposed to be less mobile, their ends can be thought of as moving entirely by reaction diffusion and the termination between any two entangled radicals should proceed by the reaction diffusion.

It is suggested that primary radical termination (characterized by Eq. (8)) is an important cause of non-ideal kinetics of vinyl chloride polymerization [51]. Bamford et al. [52] suggested that the influence of primary radical termination on vinyl polymerization is limited to initiator systems which produce primary radicals $(R^{\cdot})$ that have relatively long half-lives and so do not readily decompose to give secondary radicals $(R^{\cdot\prime})$. The intercept of the plot of $1/R_p$ against $1/[I]^{0.5}$ on the vertical axis a term $\{k_t/[k_p^2 k_i^2/(k_{tpr} \sigma)]\}$, where σ is the geometrical mean constant for the rate constants for the recombination of primary radicals and mutual termination of growing polymer chains} is a measure of the extent of primary radical termination.

Deb and Meyerhoff [53–55] proposed a method for examining the influence of primary radical termination on vinyl polymerization systems where the decomposition of the primary radicals is in competition with the initiation reaction. They proposed that 1) mutual termination of $R^{\cdot}$ and $R^{\cdot\prime}$, 2) primary radical transfer, and 3) termination of growing polymer chains by $R^{\cdot\prime}$ are negligibly small and so the extent of primary radical termination, $k_{tpr}/k_i k_p$, can be obtained from the slope of the plot of log $R_p/[I][M]$ against $R_p/[M]^2$. Although this approach has proved useful in quantifying the extent of primary radical termination, some effects should also be discussed. At low monomer concentration, however, abstraction of hydrogen atoms by both the primary and secondary radicals can become significant. Kinetic orders of the concentrations [initiator] and [monomer] less than 0.5 and larger than 1.0 indicate the existence of the primary radical termination.

The physical state of the system not only greatly changes with proceeding polymerization but also becomes microscopically heterogeneous after the onset of auto-acceleration. A lot of experimental data indicated that the inequality of the radical reactivity results from differing physical states of the micro-surroundings of radicals. Although the chemical process is the same, the diffusivity is different in the different micro-surroundings.

Bamford, Jenkins and co-workers [56] successfully interpreted the reactivity of trapped radicals by the occlusion theory for the heterogeneous polymerization of acrylonitrile. According to this theory, the probability of a radical becoming occluded by coalescence and the diffusion barrier within the precipitated polymer coils depends on their number and size, respectively.

Stable radicals trapped in the polymer matrix participate in termination but not in the chain growth. The polymer matrix saturated with stable radicals can act as a radical trapper in which deactivation of primary radicals takes place. Stable radicals may be deactivated, however, through the reaction with primary radicals.

Deactivation of polymer particles is discussed in terms of the first-order radical loss process [57, 58] through the chain transfer to monomer (a chain-transfer agent) a desorption of transferred radicals, and by the formation of long-lived or trapped radicals.

When monomer and polymer are mutually insoluble, the loss of radical activity may proceed in several ways; 1) Burnett and Melville [59] attributed the abnormal behavior of heterogeneous polymerization of vinyl monomer to the enhanced termination of macroradicals by coagulation of colloidal particles containing growing radicals. Thomas and Pellon [60] have found that the rate of the heterogeneous polymerization of acrylonitrile increased with the 0.8th power of [AIBN], the overall activation energy was of 140–148 kJ mol^{-1} and the molecular weight varied inversely with the 0.2 power of [AIBN]. To account for these observations, a termination mechanism is proposed that involves the possibility of unimolecular chain termination by a process of "burial". This process is conceived of as a mechanism by which a growing chain may become shielded from further growth by coiling or embedding itself in the solid phase.

Thus, the growing free radicals become buried in the glassy matrix of the polymer particles and is unable to propagate. Thereafter, the buried free radical may undergo termination with another free radical. This second step is not rate determining as far as polymerization is concerned, since propagation has already ceased. Bamford and Jenkins[8–10] have shown that the characteristic kinetic features of heterogeneous polymerizations are attributed to occlusion of free radicals by precipitated polymer, and they discussed the factors influencing the degree of occlusion and its effect on the individual rate constants. The mean degree of occlusion is a function of the extent of reaction and the state of the polymer phase, and the same is true of the rate constants. As the reaction proceeds, both propagation and termination constant were reported to decrease; the change in termination constants greatly predominates and accounts for the initial increase in rate with time and several other phenomena peculiar to heterogeneous polymerizations. Termination by "burial" is supposed to occur only when about 1% of the radicals have formed. According to Bamford and Jenkins, this small fraction could not significantly affect the kinetics of polymerization in the manner described by Thomas and Pellon [60].

1) Bemford's and Jenkins's approach to the first-order radical loss process is based on the hypothesis that the propagating end of the macroradical becomes inaccessible for the monomer. The free radical may grow, so that it becomes surrounded by a glassy polymer matrix through which no further monomer can diffuse. Thus, the further addition of monomer is suppressed on steric grounds due to the absence of backbone rotations in the glassy state. Termination with another mobile radical can take place, but this would be on a much slower time scale, and thus the loss of free radical activity would appear kinetically first order.

2) Another mechanism that has been suggested [12] for the first-order loss is the geminate free radical recombination; here, it is postulated that free radicals from the aqueous phase may undergo attachment to the particles in pairs, their recombination being slow because they are immobilized in a glassy precursor particle formed by aqueous-phase propagation. These radical pairs could undergo termination once they reach the surface of the particle, which is likely to be less glassy than the precursor.

3) A transfer mechanism [61] suggests that termination of radicals proceeds at or near the particle surface. One could suppose that transfer of free radical activity to a monomer unit is a rate-determining step, which would imply first-order kinetics; this monomeric free radical could then undergo rapid

(non-rate-determining) termination on the surface of the particle or/and desorption.

4) The phase separation of growing chains decreases the radical activity of the polymerization system known as unimolecular termination (the first order radical-loss process). This event strongly suppresses both termination and propagation events. The result is a decrease in the polymerization rate and the molecular weight of the polymer. Polymerization systems are characterized by limiting conversion caused by the decreased penetration of monomer into polymer coils and/or surroundings of reaction loci.

5) Diluents, that are very poor solvents (or at very high ω_p) and bring about immediate precipitation of the polymer as it forms, represent a heterogeneous reaction system in which at least some of the active polymeric radicals co-precipitate with inactive polymer [62]. Besides, the branched structure of growing radicals and/or the presence of unsaturated polymer chains may decrease the radical activity of the system (the shielding effect and the long-lived radicals).

Variations in the radical reactivities of termination and propagation in physical macro-states were ascribed to the wrapping effect by Shen et al [41]. The authors reported that the value of k_t dropped only by one order of magnitude over the 0–0.5 ω_p range. This reduction of k_t is supposed to be only due to the variation of the physical macro-state of the system rather than the physical state of the microsurroundings. In other words, the reactions at this state are mainly macro-diffusion controlled, including the translation and segment diffusion processes. At $\omega_p \geq 0.5$, the k_t decreases more strongly. This is ascribed to the micro-diffusion which starts to play a role. At $\omega_p \geq 0.7$, the matrix becomes more rigid thus converting the propagating radicals into the wrapped ones. Hence, micro-diffusion becomes the controlling step of termination instead of macro-diffusion. Thus, the macro-diffusion depends upon the matrix, whereas micro-diffusion depends upon the micro-surroundings. The authors [41] suggest that the radicals in different micro-surroundings have unequal reactivity and mobility owing to the wrapping effect.

The termination rate constant k_t is thus a function of the following variables: (1) the polymer weight fraction ω_p, (2) the molecular weight of the dead polymer (M_m), (3) the number-average degree of polymerization of the two mutually terminating chains, (4) the nature of the continuous phase, (5) the size and chain density of the polymer coils, (6) the chain density around the growing end and unsaturated unit and (7) the mutual solubility of polymer and monomer (solvent) and the degree of growing radicals.

2.5 Chain-Transfer

Transfer to monomer and polymer reactions are very important in VC polymerization because of their frequency relative to the chain transfer propagation reaction controls the molecular weight. Transfer to the monomer causes a shift of the molecular weights to lower values. On the other hand, transfer to the polymer can lead to the formation of branched (larger) polymers. It is clear that if both transfer events are active then the molecular weight distribution broadens.

Transfer to monomer and intramolecular (backbiting) reactions are considered to be responsible for the formation of the short side chains observed in PVC. Note that transfer reactions do not change the total number of live polymer chains and, therefore, the rate of polymerization and the number-average molecular weight will not be affected.

The double bond formation can arise from both transfer to monomer and termination by disproportionation. These bonds are called "terminal double bonds" and are located at or close to the ends of polymer chains. The various anomalous and unsaturated structure found in the polymer product may be attributed to the chain transfer processes.

Chain transfer reactions mostly proceed by abstraction of a monovalent atom such as hydrogen or a halogen. The scission of a bond carbon – oligovalent (e.g., H) atom is of interest for the introduction of endgroups into a polymer produced in a free radical reaction. Radically induced vinyl monomer polymerization with the possibility of chain transfer to a polymer of different chemical structure present in the reaction mixture leads to graft copolymers if bond scission occurs outside the main chain, no matter whether a single atom or a grouping is abstracted. Quite a different result is obtained if a radical attack involves a bond in the main chain of the polymer, if this bond scission occurs at a monovalent atom, which must be at the chain end, there is block copolymer formation. If bond scission occurs inside the polymer backbone, either block or random copolymers are produced [63].

If the remaining polymer radical is active enough to initiate polymerization of the vinyl monomer, then there is the possibility of block or random copolymer formation. In the case of formation of less reactive radicals, the decrease in the polymerization rate and the molecular weight of the polymer is observed.

The radical polymerization of unsaturated monomers with the possibility of chain transfer to a monomer gives rise to monomeric radicals with the same or alternatively less reactivity than propagating radicals. In the former, the rate is constant and the molecular weight of the polymer varies. In the latter (the degradative chain transfer), both the rate and the molecular weight of the polymer decrease [64] and the reaction order with respect to the [initiator] increases. This process involves a intramolecular transfer which produces a radical of reduced propagation activity. Park and Smith [65] have reported monomer and initiator orders for degradative chain transfer processes occuring with monomer, initiator, and solvent. Degradative chain transfer to initiator and solvent increased the reaction order with respect to the monomer concentration above 1.0 and degradative transfer to monomer decreased its reaction order to below 1.0. Degradative chain transfer to monomer increased the reaction order with respect to the initiator concentration above 0.5 and degradative chain transfer to solvent decreased the reaction order with respect to the initiator concentration below 0.5 Polymerization of branched monomers and/or formation of branched polymers favor the degradative chain transfer processes due to a hindered propagation.

A chain-transfer constant (the ratio of the transfer rate constant to that for propagation) is generally determined using the Mayo model [66]. This method is accurate if k_t is sufficiently low that not many dead polymer chains are formed by bimolecular termination.

3 Vinyl Chloride Polymerization

3.1 Effects of Initiator and Emulsifier

3.1.1 General

According to the micellar theory [6, 7] the rate of polymerization (in the stationary interval 2) is proportional to the 0.4 power of the concentration of initiator and the 0.6 power of the concentration of emulsifier [4]. It was later shown by Roe [23], Fitch [22, 25, 67] and Hansen and Ugelstad [18] that this behavior is also consistent with homogeneous nucleation. Indeed, identical exponents can be predicted by virtually any mechanism which assumes that (1) coagulation does not occur, (2) nucleation ceases when the surface area of the latex particles is equal to the total surface area capable of being occupied by the emulsifier molecules, and (3) the rate of production of free radicals is uniform. Here, determination of exponents for [I] and [E] fails to discriminate between competing micellar and homogeneous nucleation theories.

Smith and Ewart [6] assume that the particles are formed by entry of primary radicals into the monomer-swollen emulsifier micelles or polymer particles where they initiate or terminate the growth events. Roe [23] assumes that the particles are formed outside the micelles. Particle formation stops when the surface of the particles has grown to such a size that the emulsifier concentration in the aqueous phase is below a critical point, somewhat lower than CMC. Fitch et al. [22, 25, 67] have proposed a mechanism which implies that primary particles are formed in the aqueous phase by precipitation of oligomer radicals above a critical chain length. Such primary particles are colloidally unstable, undergoing coagulation with other primary or with large polymer particles and polymerize very slowly.

The basic principle of the Fitch theory is that the formation of primary particles will take place up to a point where the rate of formation of radicals in the aqueous phase is equal to the rate of disappearance of radicals by capture of radicals by particles already formed. According to the Fitch theory of homogeneous particle nucleation, the addition of emulsifier does not lead to any increase in the number of primary particles formed. It only leads to an increased stability of the primary particles, preventing them from coagulating with a resulting decrease in the final number of particles formed. At low emulsifier concentrations, particles are nucleated mostly in the aqueous phase. At higher emulsifier concentrations, particles are nucleated both in water and in the micelles. We accept that the initiation of polymerization and the growth of active polymer chains occurs first in the aqueous phase. The growth of macroradicals in water occurs as long as these macroradicals remain soluble in water. The precipitation of such macroradicals leads to the formation of primary particles. Thus, the primary initiation must occur in water, even though the monomer is sparingly soluble in water. The oligomer radicals which have entered start the secondary initiation in the monomer-swollen emulsifier micelles. Primary charged radicals do not initiate the polymerization in the monomer-swollen emulsifier micelles or the monomer-swollen polymer particles due to the strong electrostatic repulsion experienced between the charged monomer-swollen micelles or polymer particles and charged primary radicals (such as $SO_4^{\cdot-}$). Generally, the oligomer radicals start the polymerization in micelles or polymer particles while the electrostatic repulsion between the growing radicals and the particles is weak. The oligomer radicals orienting their active ends towards micelles and charged ends towards the aqueous phase can be effectively captured by micelles or polymer particles. After its entry, the charged end of macroradical will be at the particle-water interface, while the active end grows inside the polymer particle or terminates by collision with another macroradical inside the micelle or polymer particle. Charged groups on the polymer particle surface increase the stability of polymer particles.

It is suggested that the initiation of emulsion polymerization of unsaturated monomers is a two-step process. It starts in the aqueous phase by the primary free radicals derived from the water-soluble initiator or secondary derived from the emulsifier molecules. The second step occurs in the monomer-swollen micelles or particles and is caused by the water-soluble or water-insoluble oligomeric radicals which have entered [68]. In the homogeneous one-phase reaction system, both the formation of primary radicals and the growth of the polymeric chain take place in one phase.

According to the micellar and the homogeneous nucleation theories [6, 7, 22, 23, 25, 27], the formation of new particles stops as soon as the micellar emulsifier disappears, and the number of particles (N), which grow during the polymerization process, corresponds to the equation

$$N \propto [E]^{0.6} \times [I]^{0.4} \tag{23}$$

where [E] and [I] are the concentrations of emulsifier and of initiator in the aqueous phase, respectively.

Several examples have appeared supporting the idea of homogeneous nucleation of particles or a two-step initiation process of monomers which have a very low solubility in water (e.g. styrene, hexyl acrylate,...).

The situation, characterized by Eq. (23), represents an idealized state. In many cases, the concept of the classical kinetic model of emulsion polymerization has been modified because it was found to be inapplicable to monomers with higher water solubility [25, 67, 69]. The polymerization of partly water-soluble monomers was assumed to proceed not only in the micelles or particles but also in the continuous aqueous phase.

Emulsion copolymerization is an important process in the manufacture of widely used products such as paints, foams, coatings, fibers and others. Despite this, many mechanistic aspects of emulsion copolymerization remain unclear, although they have industrial importance.

Copolymerization is a very important tool for investigating emulsion polymerization mechanism. In other words, not only the rate of polymerization and the particle size, but also copolymer and particle composition can be used to derive the mechanism. The pair of monomers with different solubilities is considered as a model system for such investigation. The fraction of monomers in the aqueous phase and in polymer particles may be controlled by the nature of monomers and the monomer feed composition. The composition of the copolymer is a function of the monomer feed composition in particles and is only slightly affected by the monomer composition in the aqueous phase. This follows from the much higher rate of polymerization in articles compared to that in the homogeneous (aqueous) phase.

3.1.2 Initiator

The emulsion polymerization of vinyl chloride is usually carried out by the thermochemical initiation with water-soluble initiators such as peroxodisulfate, hydrogen peroxide and/or the redox systems.

Gerrens et al. [70, 71] investigated the kinetics and mechanism of the emulsion polymerization of vinyl chloride (VC) initiated by potassium peroxodisulfate in the presence of anionic emulsifier (sodium-α-oxyoctadecane sulfonate). The rate is proportional to the 0.8th power of the initiator concentration. The high reaction order on the initiator concentration is ascribed to the degradative chain transfer [53–55] to the VC monomer and unsaturated poly(vinyl chloride) (PVC). The transferred monomeric VC and PVC radicals are less reactive due to resonance stabilization. This decreases the bimolecular termination and favors the first radical loss process. As the polymerization advances, the reactivity of growing radicals decreases. For example, increasing conversion from 1 to 3% increases the lifetime of growing radicals from 1.7 to 20 s [72]. Thus, the growing radicals spend most of their existence in a coiled state due to the non-solvent action of the monomer. Occlusion of radicals decreases the radical activity of the reaction system. The authors [70, 71] also reported that the average number of radicals per particle is very low. This was ascribed to the chain transfer and the formation of less reactive radicals. In addition, they found that the number of particles is independent of the initiator concentration.

The dependence of the particle number versus conversion was described by a curve with a maximum (9.2×10^{18} particles/dm^3) at 50% conversion [70, 71]. At low conversion ($\sim 10\%$) the number was 6×10^{18} particles/dm^3 and at high conversion 7×10^{18} particles/dm^3. These data show that aggregation of particles takes place.

Meehan et al. [73] reported that the exponent on the initiator concentration was 0.4. The addition of a small amount of a radical scavenger (a retarder) led to an increase of the reaction order from 0.4 up to 0.9. Gerrens et al. [70, 71] discussed this behavior in terms of the degradative chain transfer. As a result of the degradative chain transfer (exponent above 0.5), the expression for the termination of macroradicals (R$^\bullet$) by monomers (allyl acetate [74] and by vinyl chloride [75]) changes from

$$R_t = k_t [R^\bullet]^2 \text{ to } R_t' = k_t' [R^\bullet] [M] \tag{24}$$

where M denotes the monomer or/and unsaturated polymer (group) concentration. Thus, the total rate of polymerization is expected to be proportional to [Initiator]$^{1.0}$ and independent of the monomer and the particle concentration:

$$R_p = (k_e [I] k_p)/(k_t' N_A) \tag{25}$$

Under such conditions, the rate of polymerization is independent of the polymer particles. The number of particles is independent of or varies very slightly with the initiator concentration. The molecular weights of poly(vinyl chloride) were very small and nearly independent of initiator concentration. Thus, the degradative chain-transfer to VC or unsaturated PVC depresses the effects of initiator and particle concentrations on polymerization behavior.

Peggion et al. [76] have reported that the rate of emulsion polymerization of vinyl chloride is proportional to the 0.5th power of the initiator concentration (with anionic emulsifier (SDS) above its CMC). The reaction order thus differs slightly from that proposed for the conventional emulsion polymerization of vinyl monomers. These data were discussed in terms of the bimolecular termination of growing radicals. The number of particles was found to be independent of the initiator concentration. The data were supposed to be affected by the restricted penetration of radicals and monomers into the polymer phase and separation of the growing radicals from the monomer phase and chain transfer events. The trapping of radicals increases with conversion and so the reaction order with respect to [initiator] increases [77]. The authors [76] suggested that the emulsion polymerization of VC is similar to "a heterogeneous polymerization" [56] in which the formation of occluded radicals increases the order with respect to [initiator].

The same group [76] carried out polymerizations of vinyl chloride in the presence of anionic emulsifiers (Aerosol AY and Aerosol MA at concentrations below CMC) varying peroxodisulfate concentrations. The reaction orders with respect to [initiator] 0.66 and 0.7, respectively, have been observed for the emulsifiers. The high reaction order indicates that the bimolecular termination mechanism is suppressed while the first order radical loss is operative. The formation of less stable primary particles favors the association of particles and so the fraction of burried radicals. The formation of less reactive radicals through the chain-transfer events and by the precipitation of growing radicals is believed to increase the reaction order with respect to [initiator].

Variations of the number of particles at various stages of polymerization were followed with the type of emulsifier [76]. SDS and Aerosol AY emulsifiers were used. In the first system (0.013 g emulsifier/100 g water, [peroxodisulfate] $= 1.23 \times 10^{-3}$ mol dm^{-3}), the number of particles varied only slightly between 6 and 80 or 90% conversion, i.e., from 1.2 to 1.3×10^{14} particles/dm^3. A somewhat stronger variations in the particle number was observed with Aerosol AY. Here, the number of particles increased from 4.6×10^{14} to 7.0×10^{14} particles/dm^3 in the conversion rage 15–85%. Unexpectedly, in both systems, the number of particles slightly decreased with the peroxodisulfate concentration. For example, with increasing [initiator] from 1.23 to 6.2×10^{-3} mol dm^{-3}, N decreased from 4.0×10^{14} to 2.2×10^{14} particles/dm^3.

Ugelstad et al. [78] also reported that the reaction order with respect to the initiator (peroxodisulfate) concentration is 0.5. The value 0.5, however, was not ascribed to the bimolecular termination but to the desorption of radicals from particles. A rapid desorption and reabsorption of radicals in the particles are suggested to be responsible for the observed behavior, including the observation that, at a given moment, only a small fraction of the particles contain a radical. Desorption is known to decrease the number of radicals per particle much below 0.5, even if desorbed radicals are quantitatively reabsorbed by the polymer particles. A radical desorbed from a particle may reenter a particle already containing a radical increasing termination. In the case of rapid desorption and reabsorption of radicals, the total rate of absorption of radicals into the particles is expressed as $\rho_A = k_a [R^{\cdot}]_w$, where k_a is the absorption rate constant and $[R^{\cdot}]_w$ the concentration of radicals in water. The rate of desorption from particle is expressed as $\rho_d = k_{des} n N_n$, where N_n is the number of particles with n radicals. The following expression for the rate of polymerization under a rapid desorption and reabsorption of radicals in the particles was suggested.

$$R_p = \frac{k_p [M]_{eq} R_i^{0.5}}{N_A} \left[\frac{(V_p k_{des} + N_w k_{tp}) k_a^2}{2 k_{tp} k_{des} k_a^2 + 2 k_{tw} k_d^2 (V_p k_{des} + N_w k_{tp})} \right]^{0.5} \quad (26)$$

where V_p is the total volume of latex particles, N_w the number of particles in water, k_{tp} the rate constant for termination in the latex particles and k_{tw} the rate constant for termination in water. If termination in water is negligible compared to that in the particles, the rate expression is

$$R_p = \frac{k_p [M]_{eq}}{N_A} (R_i)^{0.5} \left(\frac{V_p}{2 k_{tp}} + \frac{N_w}{2 k_{des}} \right)^{0.5} \quad (27)$$

or

$$R_p = \frac{k_p [M]_{eq}}{N_A} (R_i)^{0.5} \left(\frac{V_p N_A}{2k_{tp}} + \frac{N_w^{1/3} V_p^{2/3}}{2k_{des}} \right)^{0.5} \tag{28}$$

If the radicals are terminated mainly in water, then

$$R_p = \frac{k_p [M]_{eq}}{N_A} \frac{R_i^{0.5}}{2k_{tw}} \frac{k_a}{k_{des}} \tag{29}$$

According to Ugelstadt et al. [78] as well as Friis and Hamielec [79], Eqs. (27–29) describe the rate of vinyl chloride emulsion polymerization well up to the pressure drop. Nilsson et al. [80], however, observed (using a highly sensitive calorimeter reactor) that the approximation of the suppressed water-phase termination is not entirely justified especially at higher conversions.

Ugelstadt et al. [78] exclude the effect of the water-phase polymerization on the polymerization behavior. The number of latex particles is reported to be independent of conversion and the initiator concentration. The reaction order with respect to the number of particles ($R_p \propto N$) increases from 0.05 to 0.15 in the range investigated (1.0×10^{16} to 1.0×10^{19} particles per dm^3 water).

Numerical solutions of the dN/dt equations gave the following expression [78]

$$N \propto [E]^{0.94} [I]^{0.04} \tag{30}$$

Above the CMC, the theory fits well with the experimental results. For example, with increasing [peroxodisulfate] from 3.0×10^{-3} to 24.0×10^{-3} $mol\, dm^{-3}$ decreased N from 4.7 to 4.3×10^{16} particles/dm^3 (2 g of Aerosol DBM/dm^3, 50 °C). On the contrary, below CMC the model deviates from the observed data. This was ascribed to increased effect of homogeneous nucleation of particles.

In most polymerizations, the rate of polymerization was found to increase with conversion. The marked increase in the rate was observed at ca. 70–80% conversion. The number of particles passed through the maximum, 14×10^{16} particles/dm^3 (at 40% conversion), with conversion. At low conversion, N was found to be 11×10^{16} particles/dm^3 and at high conversion 9×10^{16} particles/dm^3 (8 g Aerosol DBM, [peroxodisulfate] = 6.0×10^{-3} $mol\, dm^{-3}$, 50 °C).

Sörvik et al. [81, 82] carried out a batch emulsion polymerization of VC at subsaturated pressure. Here, the reaction orders with respect to the water – soluble initiator is 0.5 and the number of particles increased only slightly with [initiator]. In the seeded emulsion polymerization of VC, the same results were obtained. This behavior was discussed in terms of the equilibrium between association and nucleation of particles.

Giskenhaug [83] has reported that the emulsion polymerization of VC is influenced by the water-phase polymerization. The half order with respect to initiator, as well as the slight effect of the particle concentration on the rate of polymerization was discussed in terms of the bimolecular termination of growing radicals. The following expression for the rate of the water-polymerization was derived

$$R_{pw} = k_p [VC]_{eqw} (R_i/2k_{tw})^{0.5}/N_A \tag{31}$$

where $[VC]_{eqw}$ is the saturation concentration of vinyl chloride in water. The good agreement between experimental data and those predicted by Eq. (31) was reported. This was taken as an evidence of the importance of the water-phase polymerization. By inserting the rate of VC polymerization, the number of particles N, the equilibrium monomer concentration $[VC]_{eq} = 6 \, mol \, dm^{-3}$ [70] and $k_p = 3.6 \times 10^7 \, dm^3 \, mol^{-1} \, h^{-1}$ [84] into Eq. (18) the average number of radicals per particle was estimated to be in the range from 10^{-1} to 5×10^{-4}. The low number of radicals in the polymer particles was ascribed to the chain-transfer to VC monomer and unsaturated polymer chains which led to the formation of fewer reactive (transferred) radicals.

Mork and Ugelstadt [85] investigated the emulsion polymerization of vinyl chloride initiated by peroxodisulfate/bisulphite/Cu^{2+}/citrate redox system. The rate of polymerization was found to be of first order with respect to the concentration of bisulphite and independent of the peroxodisulfate concentration.

The effect of the concentrations of citrate and copper ions was found to be strongly dependent on the ratio [citrate]/[Cu]. In borax and phosphate buffer, the rate of polymerization was found to be proportional to $[Cu]^{0.25}$ $[citrate]^{-0.5}$ at ratios larger than about ten. When the ratio approaches unity, an abrupt increase in the rate was observed. The authors [85] reported that 1) no radicals were produced by the reaction of peroxodisulfate with bisulphite in the complete absence of trace impurities of metal ions. The activity of such impurities may be completely suppressed by complex formation with trisodium citrate and 2) under a given set of experimental conditions, the rate of redox initiated polymerization may be adjusted within wide limits to the desired value by addition of the proper amount of citrate.

The kinetics of emulsion polymerization of vinyl chloride initiated by the redox initiator system (peroxodisulfate ammonium as initiator/Na_2SO_3 as the activator/$CuSO_4$ as the promoter) was studied by Karakas and Orbey [86]. Figure 2 shows the conversion-time data for the free-radical emulsion polymerization of VC initiated by peroxodisulfate at constant concentration of the activator and/or promoter. The conversion curves consist of three intervals (see Fig. 2). Initially, the rate of polymerization is slow. As polymerization proceeds, the rate gradually increases, reaching a constant value within a long conversion region (from 20 up to 70%). The low rates of polymerization below 20% conversion result from the nucleation and association of particles and desorption of radicals. The deviation of VCM polymerization (the linear conversion curves) from the expected one (the S-shaped conversion curves) may be discussed in terms of the different location of reaction loci. The reaction order with respect to the initiator concentration was found to be 0.65 supporting the monomolecular termination mechanism. Here, the concentration of peroxodisulfate potassium was low and changed from $5.8 \times 10^{-4} \, mol \, dm^{-3}$ to

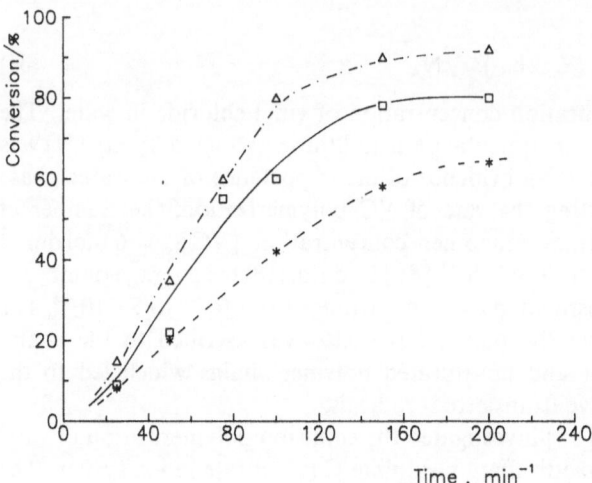

Fig. 2. Variation of the total monomer conversion in the emulsion polymerization of vinyl chloride with reaction time and with initiator $((NH_4)_2S_2O_8$, AP) concentration [86]. Recipe: [VC] = 8.5 mol dm^{-3}, [NaSO$_3$] = 0.21 × 10^{-3} mol dm^{-3}, [CuSO$_4$.5H$_2$O] = 0.5 − 1.0 × 10^{-5} mol dm^{-3}, 50 °C. [SDS] = 1.8 × 10^{-2} mol dm^{-3}. [AP] × 10^3/(mol dm^{-3}) = (∗) 0.58, (□) 0.99, (△) 1.4

1.4 × 10^{-3} mol dm^{-3} while the amounts of other reaction ingredients were kept constant. Under such conditions, the primary radical termination is negligible but the first radical loss process becomes important. As soon as one or more of the species of redox system were used in low concentrations, the low rates and conversions were observed. On increasing the total concentration of all substances (≥ 10^{-3} mol dm^{-3}), the rates of polymerization and final conversions increase. Thus, one may conclude that the initial concentrations should be kept above a minimum value (ca 5 × 10^{-4} mol dm^{-3}). The rates of polymerization increased significantly when relative ratios of [activator]/[initiator] and [activator]/[promotor] were kept the same. Small amounts of all initiation components should be used because these materials are included in the structure of the final product and high concentrations of them may lead to undesirable product properties.

Figure 3 illustrates variations of the particle size with the peroxodisulfate concentration and conversion. The size of the particles is independent of the initiator concentration. The shape of the curve of the dependence of particle size on conversion shows that the particles grow in stages 2 and 3. The increase in the particle size is more pronounced in the range of high conversions (stage 3). Thus, the polymerization of VC in the aqueous phase generates new small particles by homonucleation even at high conversion. Due to the depletion of free emulsifier, these new particles formed are unstable. A possible explanation for the increase of particle size beyond the depletion of monomer droplets is that a coalescence of small and/or large particles takes place. Absorption of radicals by polymer particles was reported to be diffusion controlled and, therefore, the

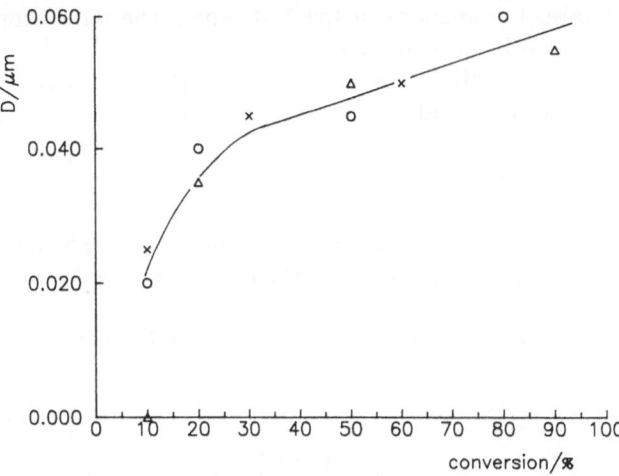

Fig. 3. Variation of the average particle diameter with the initiator (AP) concentration and conversion [86]. $[AP] \times 10^3/(mol\, dm^{-3}) = (\bigcirc)$ 0.62, $(\times)$ 1.0, $(\triangle)$ 1.4. Other conditions are given in the legend to Fig. 2

rate of initiation and the average number of radicals per particle were strongly influenced by absorption and desorption events. In the present system, the particle growth was suggested to proceed with absorption of radicals and subsequently with the propagation reactions inside the particles. Similar results were reported in Refs. [87–90] in which the size of the PVC particles was found to be independent of the initiator concentration.

Boieshan [91] has compared the emulsion polymerization of vinyl chloride initiated by 1) potassium peroxodisulfate with that initiated by 2) the redox system (potassium peroxodisulfate/$FeSO_4$/ethylene diamine tetraacetic acid (EDTA)/Na-formaldehyde sulphoxylate (FS)). In both systems, the rate of polymerization was found to increase with increasing initiator concentration. In the former case (with potassium peroxodisulfate), the rate of polymerization is proportional to $[I]^{0.61}$. The reaction order 0.61 was reported to be the result of two contributions: a) biradical termination of growing radicals for which $R_p \propto [I]^{0.5}$ and b) the first order radical loss process for which $R_p \propto [I]^{1.0}$. The overall reaction order is the sum of both contributions and its value is located in the range from 0.5 to 1.0. At very low conversion, the polymerization proceeds in the aqueous phase where the growing radicals are deactivated in mutual bimolecular reactions. As the polymerization proceeds, the concentration of less reactive radicals increases and so does the reaction order with respect to [initiator]. Chain transfer together with "burying" events are supposed to govern the termination mechanism.

Thus, for the first stage of the polymerization reaction, when there is no autoacceleration, x (the reaction order) = 0.5, and the radicals disappear by recombination. For the autoacceleration stage (x = 0.61) there are not only biradicalic reactions but also terminations of the first order. The latter stop reactions by "burying" radicals in the solid phase.

With a redox system, the reaction order 1.0 was obtained. This was attributed to deactivation of macroradicals, $P^{\cdot}$, by the iron ions;

$$P^{\cdot} + Fe^{3+} \xrightarrow{\quad k_t' \quad} P + Fe^{2+} \tag{32}$$

The VC monomer and unsaturated PVC may act in a similar way. The high reaction order (1.0) may also result from the catalytic affect of Fe^{2+} ions on the decomposition of initiator.

By the graphic method, the order with respect [Fe], [FS], and [EDTA] was determined

$$R_p \propto [Fe]^{0.34} [EDTA]^0 [FS]^{0.66} \tag{33}$$

The reaction order with respect to [Fe] and [FS] results from the oxidation-reduction process, i.e. the influence of the mobile electrons. The exponent value 0 for EDTA concentration confirms the idea that EDTA does not take part in the initiation.

It has been reported that if the quantity peroxodisulfate is sufficient to maintain the reaction with Fe and FS, its concentration in the aqueous phase has practically no influence. However, peroxodisulfate is necessary in the redox initiation system as a raw material from which the initiating free radicals are formed, but the polymerization rate and the final conversion depend mainly on the concentration of FS and Fe. That is why the determination of the reaction order with respect to [peroxodisulfate] is not easy. Maintaining, however, the optimum constant ratio of the redox system components, the dependence of the rate on the initiator concentration was studied. It was found that:

$$R_p \propto [\text{peroxodisulfate}]^{1.0} \tag{34}$$

Neelson et al. [92] followed the effect of emulsifier (sodium dodecyl sulfate) concentration on the decomposition rate of initiator and the rate of polymerization. The decomposition rate of potassium peroxodisulfate increased with increasing the emulsifier concentration. For example, with increasing $[SDS]/(\text{mol dm}^{-3})$ 0, 1.74, 3.47, 6.95, 13.89, and 24.3 increased $k_d \times 10^6/s$ 2.9, 3.6, 12.3, 16.9, and 21.9 (50 °C, [VC] = 11.5 mol dm^{-3}). For the catalitic decomposition of peroxodisulfate caused by emulsifier the authors derived the following semiempirical expression

$$k_d = k_{dw} + k_d' [E]^m \tag{35}$$

where k_{dw} is the rate constant for decomposition of initiator in water and term $k_d' [E]^m$ denotes the decomposition of initiator caused by its interaction with emulsifier. Thus, the extrapolation of decomposition rate constants obtained for different emulsifier concentrations to zero emulsifier concentration gives the value of k_d which is close to that (for pure water) reported by Kolthoff and Miller [93]. Neelson et al. suggested that the increased decomposition of initiator results from the presence of the initiator-saturated interface water/micelles and/or water/polymer particle which favors the decomposition of the initiator. The increased decomposition is caused by the interaction of initiator with emulsifier which favors the scission of the 0–0 bond of initiator. Also, new emulsifier transferred radicals are formed which also start the growth of polymer chains.

Decomposition of peroxodisulfates in the aqueous sodium dodecyl-sulfate solutions (below and above CMC) with and without poly(vinyl chloride) particles and/or vinyl chloride monomer was investigated by Georgescu et al. [94, 95]. They also observed the increase of peroxodisulfate decomposition in the presence of emulsifier (Fig. 4). In contrast, the decomposition rate decreased with increasing particle concentration. The dependence of the initial decomposition rate of potassium peroxodisulfate vs the emulsifier concentration is described by a curve with a maximum at CMC. The catalyzed decomposition of initiator was ascribed to the interaction of initiator with free emulsifier molecules or with emulsifier micelles. The effect of particles was ascribed to the decrease of the water-soluble fraction of emulsifier caused by adsorption of emulsifier on the polymer particle surface. The relationship of the decomposition rate constant vs the emulsifier (sodium dodecyl sulfate) concentration with and without poly(vinyl chloride) particles is described by a curve of the same shape but with different absolute values of k_d; k_d is lower in the presence of PVC particles (Fig. 4). The most intensive decomposition of initiator occurs at concentrations close to the CMC. Decomposition of peroxodisulfate recorded with vinyl chloride and poly(vinyl chloride) is faster than in pure water and slower than in emulsifier solutions. Variations in the decomposition rate results from the

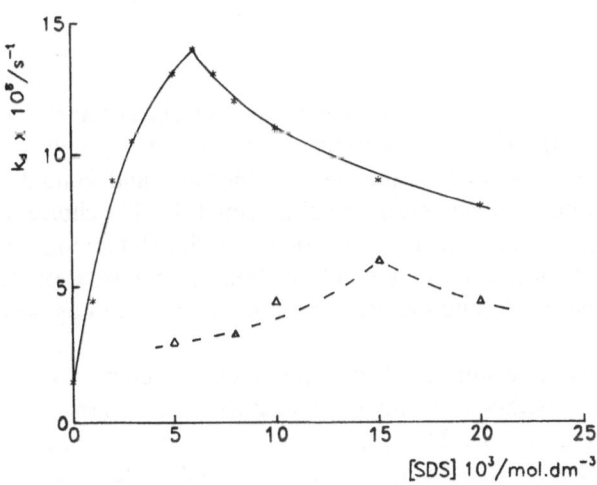

Fig. 4. Variation of k_d with the initial concentration of SDS [94]. $[K_2S_2O_8] = 1.8 \times 10^{-3}$ $mol\,dm^{-3}$, pH $= 7 - 9$, 50 °C. (∗) Without PVC particles, (△) With PVC particles (5 wt% solids in the aqueous phase)

interaction between hydrophobic chains of poly(vinyl chloride) and sodium dodecyl sulfate as shown in Ref. [96]. The following equation is derived for the decomposition rate of peroxodisulfate:

$$R_d = 1.5 \times 10^{-6} A + 4.7 \times 10^{-6} A B^{0.54} - 5.3 \times 10^{-5} A^{0.54} C^{0.45} \qquad (36)$$

where A represents the initial concentration of peroxodisulfate, B represents the concentration of dissolved SDS and C represents the concentration of micelar SDS. Equation (36) showed that the maximum value of the R_d corresponds to the CMC value. In the presence of micelles (competetive interaction), the decomposition of initiator also becomes slower. The dissolved emulsifier and the free-emulsifier surface of the polymer particles increase the decomposition rate of the peroxodisulfate. The radical intermediate SDS˙ was suggested to be responsible for the induced peroxodisulfate decomposition. The micelle and the adsorbed emulsifier on the particle surface decrease the decomposition rate. The decreased decomposition rate was explained by the trapping of the SDS in micelles or on the polymer surface by hydrophobic interactions.

The relative significance of the primary radical termination in the VC polymerization was tested by kinetic analysis of the polymerization rate data and analysis of the samples recovered at 10% conversion for combined initiator fragments [97]. The rate of polymerization was measured for 0.56, 2.7, and 3.5 mol dm^{-3} VC at 60 °C using a wide range of initiator concentration. The reaction order on [dibenzoyl peroxide] was found to be 0.34, 0.42 and 0.51, respectively. The mathematical treatment of the primary radical termination based on kinetic data (at low monomer) gave the values of $k_{tpr}/k_i k_d' \sim 5 \times 10^5$ indicating the importance of the primary radical termination.

3.1.3 Emulsifier

To prepare stable PVC latexes by emulsion polymerization of VC, considerable attention must be paid to the choice of a proper emulsifier. The emulsifier greatly affects the reaction kinetics and the physico-chemical and colloidal properties of the final polymer product and/or the dispersion [98]. The choice of the proper emulsifier is a complex problem because of its manifold functions, e.g. the surface tension of the aqueous solution, emulsification or solubilization of oil-soluble monomers or additives, and the protection of latex particles against flocculation [99].

Greth and Wilson [100] successfully applied, in the emulsion polymerization of unsaturated monomers, a method of classifying emulsifiers based on the HLB (hydrophilic/lipophilic balance) value. They plotted the most important properties of the emulsion polymerization system, i.e., the latex stability, particle size, emulsion viscosity, and rate of polymerization against the HLB value of emulsifiers used. The dependence is described by a curve with maximum or minimum at a certain value of HLB, as it is expected from the micellar model [6, 7].

Testa and Vianello [101] applied the HLB method to the emulsion polymerization of VC. They plotted the rate of polymerization and some properties of the latex against the HLB of anionic emulsifiers or their blends with non-ionic emulsifiers (Empicol (HLB = 40), sodium laurate (20.8), Atlas G 3300 (11.7) and Span 20 (8.6)). No bell-shaped curves were obtained, as one would have expected. The properties of polymer latexes and/or the rate of polymerization seemed to be rather dependent upon the amount of the emulsifier.

Figure 5 shows the dependences of the rate of polymerization and particle size. The rate of polymerization increases with increasing HLB values but in a different way for each blend. Also, the latex stability increased. The average particle size decreases linearly with increasing HLB values. The particle size distribution becomes narrower with increasing HLB values. The results are clearly different from those for the conventional emulsion polymerization of unsaturated monomers [100]. The difference results from the reduced solubility of PVC in VC monomer which influences the properties of the latex particles especially their solubilization properties on which the HLB approach is based. The monomer saturated PVC shell containing the reaction loci seem to be responsible for such behavior.

The authors [101] came to the conclusion that PVC insolubility in its monomer changes the latex properties to such an extent that it does not follow the emulsification laws on which the HLB method is based. This does not mean that the HLB approach is not useful in the vinyl chloride polymerization but its role is somewhat reduced. The polymerization loci are emulsifier monomer droplets or monomer-swollen micelles – their number is a function of the HLB nature.

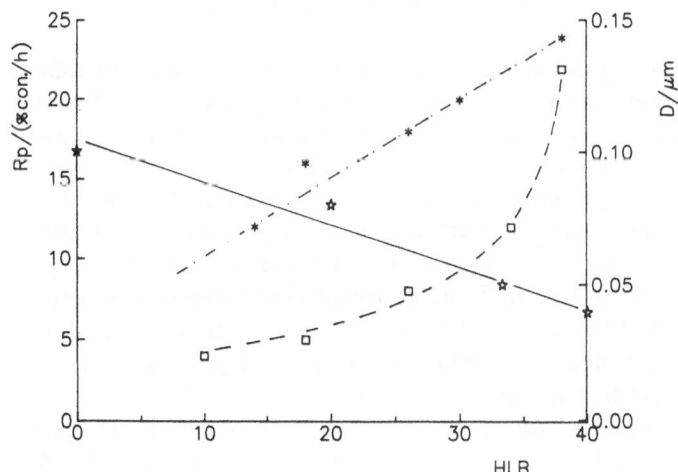

Fig. 5. Variation of the rate of polymerization and the average particle diameter with emulsifier blends and HLB. Blends of Empicol/Span 20 (E/S) and Empicol/Atlas G 3300 (E/A) were used [100]. $[K_2S_2O_8] = 1.8 \times 10^{-2}$ mol dm^{-3}, the monomer/water weight ratio = 0.7, the total amount of emulsifier 0.5 wt%. (∗) E/S, (□) E/A, (☆) E/S (D)

Peggion et al. [76] investigated the effects of type and concentration of emulsifier on the kinetics of the emulsion polymerization of VC. The polymerization rate was found to be either independent of or only slightly dependent on the emulsifier concentration. The sensitivity of the polymerization to the emulsifier concentration decreases with increasing conversion. At low conversion, the rate of polymerization slightly increases with increasing emulsifier (Aerosol MA and Aerosol AY) concentration while at medium and high conversion the rate is constant. Here, the emulsifier concentration above the CMC was used (2 or 3 times the CMC). At low conversions, the small particles are saturated with monomer and act as the reaction loci. An increase in the number of particles increases the rate of polymerization. At high conversions, the large particles contain a large fraction of the unsaturated polymer-rich phase which may act as a radical scavenger. Here, the increase of emulsifier concentration or the number of particles is counteracted by the deactivation role of the polymer-rich phase. The gel effect starts at high conversion, i.e. about 80%. At this critical conversion, the monomer droplets disappear and pressure decreases.

The particle number varies considerably (100 times with Aerosol AY and Aerosol MA, 10 times with Aerosol DBM and DSS) as the amounts of emulsifiers are varied (on the same scale) whereas the corresponding polymerization rates changed 2 or 3 times at the most with Aerosol AY and Aerosol MA and were constant or slightly increased with DSS and Aerosol DBM. These results indicate that the polymerization in particles is not dominant. The number of particles was found to increase quite linearly with the concentration of emulsifier (SDS or Aerosol DBM). In contrast, it increases markedly when the concentration of emulsifier approaches the CMC for Aerosol AY and Aerosol MA.

Ugelstadt et al. [78], however, reported that the rate of polymerization increased relatively strongly with increasing the emulsifier (anionic) concentration.

Gerrens et al. [70, 71] also investigated the effect of concentration of anionic emulsifier (sodium α-oxyoctadecane sulfonate) on the rate of polymerization, degree of polymerization, and number of particles. In the range of low emulsifier concentration (below 5 g/dm^3 of water), the number of particles increased proportionally to the 1.2 power of the emulsifier concentration, whereas the rate was independent of the emulsifier concentration. At high emulsifier concentration (above 10 g/dm^3), the rate as well as the number of particles was proportional to the 0.6 power of the emulsifier concentration. Molecular weights of PVC increase with increasing emulsifier concentration at both low and high concentrations. At high emulsifier concentrations, the small particles are formed in which the growth events are more pronounced.

Small particles are supposed to consist of the one monomer-rich phase whereas the large particles consist of two monomer-rich and polymer-rich phases. The monomer-rich phase favors the growth events whereas the polymer-rich more termination.

Effects of type and concentration of emulsifier on the emulsion polymerization of vinyl chloride were also studied by Boieshan [91]. The author reported that: 1) the size of the PVC particles and 2) the weight ratio between the aqueous and monomer phase decrease with increasing emulsifier concentration and 3) the size of latex particles increases on increasing the saturation extent of the PVC phase with its monomer. The type of emulsifier influences the latex particle formation in 2 ways: 1) by the area a_s occupied by a molecule emulsifier on the polymer surface and 2) by its adsorption characteristics. For example with potassium palmitate, which has a smaller a_s ($0.25\ nm^2$) than sodium alkyl sulphonate (E 30, $a_s = 0.34\ nm^2$), much larger latex particles are obtained even when the same molar concentration of emulsifiers is used. The total surface area of polymer particles increased with increasing conversion. The most intensive increase in the surface area was observed in the range of low conversions at most up to 30 or 40% in which the rate of polymerization is the most sensitive to the emulsifier concentration.

The influence of emulsifier on the kinetics of polymerization is not effected by the type of initiator. The same values for the reaction orders with respect to [E-30] were reported with peroxodisulfate (0.16) and the redox system (0.14). The reaction order with respect to [palmitate] and [E-30] with the redox system was reported to be 0.24 and 0.16, respectively.

For the autoaccelerated kinetics, the reaction order decreased and at a certain stage a negative value was found. This means that the rate of polymerization is independent of [E]. The autoacceleration increases as the concentration of emulsifier decreases, and was more pronounced in large particles. When the latex particles are small (d < 40 nm) at $[E] > 7.0 \times 10^{-2}\ mol\ dm^{-3}$ the autoacceleration does not take place.

The author [91] has also followed the physical significance of the $[E]^x$ value and the correlation between this value and the adsorbed emulsifier on the latex particles was observed.

For the E-30 emulsifier the following equation was derived

$$S_p = (a_s N_A)[E]^{0.26}/k \tag{37}$$

Similarly, for potassium palmitate

$$S_p = (a_s N_A)[E]^{0.26}/k \tag{38}$$

where S_p is the adsorbed emulsifier on latex particles, a_s the surface area of the emulsifier molecule, and k constant. One can see that there is a good agreement between the model (Eqs. 37 and 38) and experimental data.

Karakas and Orbey reported [86] that the rate of polymerization of vinyl chloride was nearly independent of the emulsifier (ammonium stearate) concentration. On the other hand, the number of particles ($N \propto [E]$) and the limiting conversion increased with increasing ammonium stearate concentration.

The low sensitivity of the rate polymerization to variation of emulsifier type and concentration was reported in Ref. [87].

Neelsen et al. [102] have investigated the structure of PVC obtained by the emulsion polymerization of vinyl chloride initiated by potassium peroxodisulfate in the presence of sodium dodecyl sulfate. Polymerizations were carried out with the labelled initiator (potassium peroxodisulfate) and emulsifier. The analysis of polymer led to conclusion that the ionic end groups of polymers were formed from covalently bound fragments of the emulsifier and only partly of initiator. Incorporation of ionic groups into the polymer was suggested to be a result of: 1) chain transfer to emulsifier and formation of emulsifier radicals (a reaction of primary, oligomer, chloro or propagating radicals with emulsifier molecules) and 2) initiaton of polymerization by the emulsifier radicals. The reactions of radicals with emulsifier in the monomer saturated particle surface leads to the formation of the emulsifier radicals. The decrease of the molecular weight of PVC with increasing emulsifier concentration was explained by the participation of emulsifier in the chain transfer events of growing radicals to emulsifier (Table 1). The trend of the molecular weights vs conversion dependence indicates that the transform of reaction loci from the aqueous phase to the particle phase (shell) proceeds from the start of polymerization. The decreased termination in particles favors the growth events and so the increase in molecular weight.

The slight variation of ionic groups in polymers with different initiator types and concentrations (potassium peroxodisulfate or other initiators carrying no ionic groups) was taken as an evidence of an active role of the emulsifier in the chain transfer or termination events. This behavior was a function of emulsifier type. For example, the fraction of emulsifier fragments in the polymer molecule increased in the following order of emulsifiers used:

$$\text{potassium dodecanate} \; < \text{SDS} < \text{alkyl sulfonate E-30} \tag{39}$$

Boieshan [91, 95] considers that the superficial layers of the emulsifier in the particle surface are important in the emulsion polymerization of vinyl chloride. The author suggests that the polymerization of vinyl chloride is "a homogeneous process" in which the number of the latex particles is only a secondary factor. The adsorbed layer zone of emulsifier saturated with monomer and oligomer radicals governs the overall rate of polymerization. The monomer saturated zone favors the growth events as well as the decomposition of

Table 1. Variation of the molecular weight of PVC with emulsifier concentration and conversion [102][a]

[SDS] × 10³ (mol dm⁻³)	3.2	11	17	20	35	46	53	66	85
					$M_n \times 10^{-4}$				
14.0	4.5		5.9			5.9		6.0	6.0
3.5		6.3							
1.7					6.3		6.3		
0[b]				6.4					
14.0[c]	5.7								

[a] $[K_2S_2O_8] = 5.9 \times 10^{-3}$ mol dm⁻³, 50 °C, [b] $[K_2S_2O_8] = 5.5 \times 10^{-3}$ mol dm⁻³,
[c] potassium dodecanate

initiator. The result is the high rates and the formation of high molecular weight polymers.

Tauer et al. [103] studied the emulsion polymerization of vinyl chloride initiated by a water-soluble initiator with and without emulsifier. The conversion–time curves of an emulsifier-free emulsion polymerization of VC resemble those of the precipitation polymerization. Here, the rate of polymerization of VC without emulsifier is lower than that with emulsifier. Variation of the rate of polymerization and the particle size with initiator type and concentration are summarized in Table 2. These results indicate that the rate of polymerization increases with increasing initiator concentration. As expected, the particle size decreased with increasing initiator concentration. The polymerization system contained large and small particles. As expected the size of polymer particles decreases with increasing emulsifier concentration. The particles size distribution increased with increasing emulsifier concentration and was the narrowest at zero emulsifier concentration. Coalescence of particles increases the particle size and the coagulum yield and decreases the particle concentration. The addition of a small amount of electrolyte accelerates coalescence as well as the formation of large particles and coagulum.

Deviations in the emulsion polymerization of VC from the expected behavior were discussed by Chedron [104] in terms of the aqueous polymerization and its competition with the micellar polymerization. The monomer-saturated aqueous phase and precipitation of growing radicals should support the importance of the aqueous-phase polymerization. The molecular weights of poly(vinyl chloride) slightly decreased with emulsifier concentration and increase with conversion. The trend of the molecular weights vs conversion dependence was similar to that in Ref. [102].

Peggion et al. [76], however, concluded that the strong deviations from the micellar theory cannot be ascribed to the solution polymerization which competes with the micellar polymerization and suggest the following reaction mechanism: The radicals produced by decomposition of the initiator react with the monomer dissolved in water. The polymer radicals separate from the

Table 2. Variation of the rate of polymerization and the particle diameter with the initiator type and concentration [103]

Nr	Initiator		R_p (% conv. h^{-1})	$D_{n,s}$ (nm)	$D_{n,1}$
	Type	(g dm^{-3})			
1	PS	1.49	15	161	473
2	PS	1.49	16.8		
3	PS	1.49	17.5	379	683
4	PS	0.4	8.5		
5	H_2O_2/Fe^{2+}	2.21/0.06	14.0	110	336
6	PS	1.6	22.0		

[SDS] $= 1.4 \times 10^{-2}$ mol dm^{-3}

solution at a very low polymerization degree. The formed nuclei are protected by the emulsifier and can grow to latex particles on consuming the monomer adsorbed on their surface. When the number of particles has become large enough to adsorb the emulsifier completely, the polymer radicals produced continuously in the aqueous phase precipitate on the preformed polymer particles, as they have no longer sufficient emulsifier to be stabilized. From here on, the number of particles does not change during the reaction (6–100% conversion). The polymerization in the latex particles goes on consuming the monomer adsorbed on their surface. The monomer-saturated particle shell is continuously supplied by the monomer from the aqueous solution and the monomer droplets. Thus, the particle surface zone is expected to be a reaction locus.

The influence of the emulsifier type (anionic and nonionic) and concentration on the emulsion polymerization was investigated in Ref. [105]. It was reported that the dependence of the rate of polymerization on the alkyl chain length (from C_{10} to C_{18}) of sodium alkylsulfates was described by the curve with a maximum at C_{16} ($R_p = 6.3 \times 10^{-4}$ mol dm^{-3} s^{-1}). The rates of polymerization with C_{10} and C_{18} were reported to be 2.1×10^{-3} mol dm^{-3} and 1.2×10^{-3} mol dm^{-3}, respectively. A similar trend was observed with the molecular weights, i.e., $M_v \times 10^{-4}$ was 7.6, 11.7 and 4.7 for C_{10}, C_{16}, and C_{18}, respectively.

In the case of alkylbenzene sulfonates, -sulfates and salts of fatty acids, there was no great difference in R_p and M_v. Also here, the rate of polymerization and the molecular weight passed through a maximum. However, these differences were very small.

The rate of polymerization was reported to decrease strongly going from Tween 20 ($R_p = 28.1 \times 10^{-4}$ mol dm^{-3} s^{-1}) to Tween 80 ($R_p = 2.2 \times 10^{-4}$ mol dm^{-3} s^{-1}). The molecular weights, however, decreased only slightly, i.e., from 6.1 to 3.7×10^4. Similar trends were observed with Spans (20–80), R_p decreased from 11.5×10^{-4} mol dm^{-3} s^{-1} to 0.56×10^{-4} mol dm^{-3} s^{-1}, respectively. In this system, however, the molecular weights decreased more strongly, i.e., from 10.5 to 3.6×10^4.

In all these emulsifier systems, the polymerization proceeded quantitatively.

The contact between oligomer radicals with the latex particles is supposed to take place in the emulsifier layer [106]. The overall polymerization rate of VC is supposed to be sum of two concentrations: 1) a water-phase polymerization and 2) the polymerization on the surface of particles.

Goebel et al. [107] have reported that the particle concentration reaches a maximum at very low conversion (below 1%). This agrees well with the concept of an emulsifier-free emulsion polymerization of conventional monomers in which the maximum number of particles below 5% appears. Beyond this critical conversion, the number of particles starts to decrease due to the particle association. The strong association of particles results from the low stability of small particles. The same results were found by other groups [76, 108, 109] in which a strong flocculation of latex particles led to a decrease of particle concentration with conversion.

In the emulsion polymerization of VC, the formation of polymer particles may occur by a coagulative nucleation process [26, 27]. In favor of this approach is the mutual insolubility of VCM and PVC. Here, the primary particles formed undergo coalescence whereby large "true" polymer particles are formed. In addition, the formation of primary particles during polymerization favors the growth of particles by association of small with large particles. The latex particles were also formed after the disappearance of the micelles [92]. Thus, the polymerization in the monomer-saturated aqueous phase produces the new particles. The precipitation of oligomer radicals within the monomer droplets may increase the number of particles. In favor of this hypothesis is the formation of less reactive (occluded) radicals.

Barriac et al. [110] reported that the number of particles increases with increasing conversion. Thus, the coagulative nucleation seems to be operative. The surface active oligomers formed throughout the polymerization (after the depletion of free emulsifier) should be responsible for this trend, i.e., they take part in stabilization of primary particles. The formation of emulsifier should be responsible for this trend, i.e., they take part in stabilization of primary particles. The formation of emulsifier radicals via chain transfer reactions also increases the stability of new-formed particles.

Goebel et al. [88] have found that the particle concentration decreases with conversion and the decrease is inversely proportional to emulsifier concentration. The surface tension parallels the number of particles and both abruptly increase with conversion up to 10% conversion. The higher the emulsifier concentration, the higher the number of particles which are observed. Keeping a constant concentration of emulsifier, the number of particles decreases with increasing monomer concentration. When the monomer pressure decreased by one third, the number of particles increased twice. This was ascribed to variation of the emulsifier adsorption activity with the monomer pressure. The particle concentration remains constant if the small amount of inhibitor is added. The growth of polymer particles was reported to proceed by the propagation of monomer as well as by the association of polymer particles.

The research of the influence of the emulsifier type and concentration on the emulsifier reaction order $x(N \propto [E]^x)$ is summarized in Table 3. In most cases, the reaction order x is much higher than 0.6 (expected). Thus, the strong particle association at low emulsifier concentration and/or "trapping" of emulsifier in the interface is discussed. The reaction order above 1.0 is taken as an indication of low stability of particles. By "trapping" of emulsifier, the amount of emulsifier available for the formation of particles is lower than the initially charge. Under such conditions, the association increases and the rate decreases. Table 3 shows that sodium alkyl sulfonates should be the most efficient emulsifiers for stabilization of PVC particles especially at low emulsifier concentration.

If a seed is used, a capture of radicals proceeds also by seed particles from the start of polymerization. The product of the number of particles, N, and the seed particle radius, r(nm), is considered to govern the particle formation. When the product Nr was increased from 10^7 to 5×10^8 (in the seeded emulsion polymeriz-

Table 3. Influence of the emulsifier type and concentration on the emulsifier reaction order x (N $\propto$ [E]^x) [88]

Emulsifier	[E] (g dm^{-3})	x	Ref.
SDS	0.3–30	1.0	110
Aerosol	0.3–30	1.33	110
SDS	0.66–4.0	1.0	76
Aerosol AY	8–20	3.6	76
Aerosol MA	5–13	4.5	76
Oxyoctadodecyl sulfonate	5	1.2	70
	10	0.5	70
RArSO$_3$Na	2–25	1.5	111
RSO$_3$Na	2–25	1.5	111
RSO$_4$Na	2–25	1.5	111
C$_{12}$H$_{15}$SO$_4$Li	2–25	1.5	111
Aerosol AY	10–45	2.5	111
Aerosol MA	10–45	2.5	111
C$_{11}$H$_{23}$COONa	5–30	1.0	112
C$_{12}$H$_{25}$SO$_4$Na	0.123–3.97	1.3	78
Aerosol DBM	2–32	0.8	78

ation of styrene), the fraction of new particles decreased from 1 to 0 [113]. In the seeded subsaturation polymerization of vinyl chloride, a growing radical will either precipitate or be captured by a seed particle. The eventual formation of new particles should, therefore, be determined by the number and size of the seed particles. In the emulsion polymerization of VC, the limiting combinations of particle size and its number correspond to Nr = 3.3×10^9 and 2.6×10^{11} [114].

Butucea et al. [115] investigated the effect of the initiator and emulsifier concentration on the kinetic, colloidal, and molecular weight parameters of the emulsion polymerization of VC. The rate of polymerization increased with increasing peroxodisulfate concentration independent of the emulsifier type (SDS, sodium dibutyl sulfosuccinate (SDB) or the emulsifier blend SDS/fatty alcohols (FA) with alkyl chains $C_{16} - C_{18}$. As expected the particle size decreases with increasing [peroxodisulfate].

The rate of polymerization was reported to increase with [emulsifier], e.g., in the [SDS] range from 1.7×10^{-3} to 1.0×10^{-2} mol dm^{-3} increased R_p (mol dm^{-3} s^{-1}) from 2.7 to 3.8 and the number of particles from 64×10^{16} to 78×10^{16} particles/dm^3, respectively. Increasing FA fraction in the emulsifier blend (SDS/FA) lowered both the rate of polymerization and the number of particles, i.e., varying the ratio FA/SDS from 3:1 to 6:1 the rate of polymerization decreased from 3.3×10^{-4} mol dm^{-3} s^{-1} to 2.5×10^{-4} mol dm^{-3} s^{-1} and the number of particles from 0.6×10^{16} to 0.3×10^{16} particles/dm^3, respectively ([peroxodisulfate] = 1.5×10^{-3} mol dm^{-3} s^{-1}).

In small polymer particles with an average diameter about 10 nm, the polymer with the degree of polymerization ≤ 600 was formed. In the large particles with a diameter of 500 nm, the degree of polymerization is above 700.

All polymerizations proceeded under the same reaction conditions. Thus, the molecular weight of PVC increases with increasing the particle size. In small particles, the number of radicals per particle $n < 0.005$ and $k_t > 5 \times 10^8 \, dm^3 \, mol^{-1} \, s^{-1}$ and in large polymer particles $n \cong 1.0$ and $k_t \cong 1.10^9 \, dm^3 \, mol^{-1} \, s^{-1}$ were observed. The difference in kinetic parameters may be attributed to desorption and occulusion of radicals. Generally, desorption events are very operative in small particles. The formation of occluded or buried radicals or the fraction of less reactive radicals is more operative in large particles.

Tauer and Petruschke [116] investigated the effect of PVC-seeded polymer particles on the kinetics of the emulsion polymerization of vinyl chloride. Figure 6 shows that the rate of polymerization increases with increasing concentration of seed particles. The kinetic data summarized in Tables 4–6 indicates that the polymerization is strongly influenced by the emulsifier and seed particle concentrations. From these results it follows that the rate of polymerization is a function of: 1) the particle size of seeded PVC particles, 2) the emulsifier concentration, and 3) the number of particles or the product $N_s D_s$. For all polymerizations, the following rate expression was derived

$$R_p = A N_s^{0.5} + B \qquad (40)$$

where A and B are constants (see Table 5). The author reported that the formation of new particles is governed by the number of seed polymer particles. With increasing N_s the number of new polymer particles decreases. Beyond the critical number of particles the nucleation of particles ceases. Under such conditions, all primary particles are effectively captured by the present seed

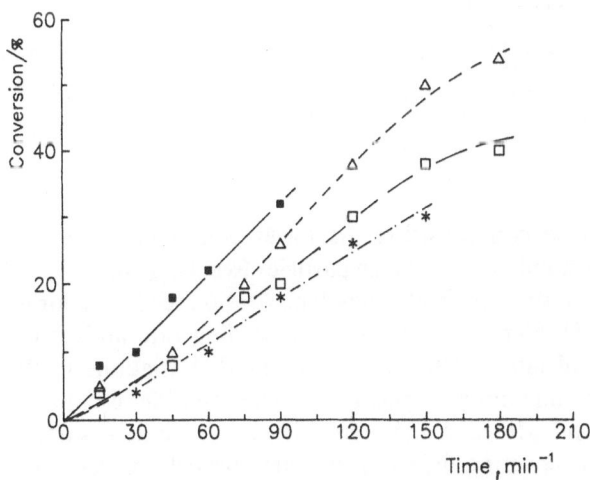

Fig. 6. Variation of the monomer conversion in the emulsion polymerization of vinyl chloride with the number of seeded PVC particles [116]. $D_s = 154 \, nm$, $N_s \times 10^{-16}/(dm^3)$: (∗) 0.18, (□) 0.43, (△) 0.85, (■) 1.5. Other conditions are given in the legend to Table 4

Table 4. Variation of the rate of polymerization in the emulsion polymerization of vinyl chloride with the number of seeded PVC particles [116][a]

$D_s = 35$ nm			$D_s = 154$ nm		
[SDS] 10^3 (mol dm^{-3})	$N_{s1} \times 10^{-16}$ /dm^3	R_p (g PVC/dm^{-3} h^{-1})	[SDS] 10^3 (mol dm^{-3})	$N_{s2} \times 10^{-16}$ /dm^3	R_p (g PVC/dm^{-3} h^{-1})
0.23	1.3	62.0	0.53	0.18	120
2.27	12.7	94	1.22	0.43	139
4.44	25.4	133	2.44	0.85	164
13.4	76.3	165	3.66	1.3	150
22.2	127.0	205			
36.6	209.4	337			

[a] $[VC] = 11$ mol dm^{-3}, 50 °C, $[K_2S_2O_8] = 5.9 \times 10^{-3}$ mol dm^{-3}

Table 5. Variation of A and B with the seeded particle size, D_s [116]

D_s (nm)	$A \times 10^7$ (g dm$^{-1.5}$ h^{-1})	B (g dm^{-3} h^{-1})
35	1.18	49.3
77	2.29	46.5
92	4.76	51.8
154	8.55	81.9

Table 6. Critical seed particle size and number [116]

dN/dt	$[E] \times 10^3$ (mol dm^{-3})	$N_s D_s \times 10^{-18}$ (nm dm^{-3})
0	$3.69 \leq [E] \leq 2.29$	≥ 3.71
0	$[E] < 3.69$	< 3.17
0	$[E] < 22.3$	< 3.71

particles. The higher particle concentration the lower emulsifier amount is needed to keep a constant number of polymer particles (see Table 6).

The mechanism of the continuous emulsion polymerization of vinyl chloride was discussed in Ref. [117]. Here, the effects of the feeding rate and/or the reaction time on the size of latex particles were followed. Variations in the particle size or number have an undulating character. The emulsifier is continuously fed into the reactor and adsorbed on the free sites on the particle surface. Under these conditions, the covering degree of the latex particles is very high (close to 100%) as it was shown in the surface tension measurements. The surface tension (σ) of the PVC latex during the continuous emulsion polymerization of VC varies between 38 to 46 mN/m [118]. From the surface tension isotherms of the emulsifier, the surface area occupied by an emulsifier (E-30, sodium alkyl monosulfonate) molecule $a_s = 22.3 \times 10^{-20}$ m^2 was estimated. Using the soap titration method [119], the average value $a_s = 34 \times 10^{-20}$ m^2 was estimated. This value is 1.5 times greater than that estimated from the surface tension isotherm. The difference is considered to derive from the different molecule pattern of the emulsifier. In the batch polymerization, the molecules of emulsifier are in a rarefied (unorganized) state occupying a greater surface compared to those in the continuous polymerization [118] (the state of maximum packing).

The research group [118] has shown that the particle size and number and the total surface area (S) of particles have an undulating character. This phenomenon was explained by a compensating tendency of the contradictory processes.

In disperse systems, the free energy tends to decrease:

$$dF = \gamma \, dS + S \, d\gamma < 0 \tag{41}$$

As the polymerization proceeds, the total surface are of the particles increases, i.e., the free surface emulsifier molecules are adsorbed and the surface tension (γ) grows. The result is an increase of both terms in Eq. (41). In such a situation, the response of the system consists of phenomena that compensate for the increase of dS and dγ; i.e., the particles agglomerate. The agglomeration process, once started, can lead to a decrease of the total surface area below the emulsifier covering limit. In this case, the continuously fed emulsifier remains available and accummulates in micelles, and the surface tension falls. Entry of radicals into monomer-swollen micelles initiates the growth events. Here, the small particles with a large. total surface area are formed. Growing continuosly, the new particles attract all emulsifier present in the solution. The formation of new particles stops. Again, the surface tension increases, as well as the total surface area of the system, until a new wave of particle agglomeration appears. These processes – the particle formation and agglomeration – occur simultaneously. At different moments, one or the other may dominate.

The oscilation amplitude around average values is not uniform polymerization process on polymerization conditions. If some of the parameters are altered (feeding rate, temperature, emulsifier or initiator concentration), the system responds immediatelly by an increase in the oscillation amplitude. In contrast, when all parameters are kept constant, the oscillations become weaker.

The dispersion degree of the latex particles during the continuous emulsion polymerization of vinyl chloride depends on the emulsifier concentration in the system. Latex particle sizes vary around some average values in connection with the surface area which can be covered with emulsifier in a state of maximum packing.

By homogenization of hexadecane/VC mixture in the aqueous solution of emulsifier, droplets are formed which are stabilized against "degradation by diffusion" [69]. Similarly, hexadecane or slightly water-soluble initiators [120] can be homogenized with emulsifier and VC, and the polymerization of these emulsified droplets gives a stable latex. Two modes of "monomer droplet" polymerization are defined: 1) formation of stable monomer/solute droplets by high shear, and 2) formation of a complex emulsifier systems or solute droplets subsequently swollen by migration of monomer.

Dewald et al. [121] have applied both approaches to prepare stable monomer/solute droplets or microemulsions. Large particles are formed by homogenization of a mixture of monomer and a water-insoluble initiator. Small droplets are formed by swelling of emulsifier (SDS)/fatty alcohol (C_{16}–C_{18} linear saturated alcohol) miniemulsion droplets by monomer. Polymerizable miniemulsion droplets are formed when alcohol in excess is used (2:1 emulsifier: alcohol).

The same group [122] have investigated the particle and droplet growth in the miniemulsion polymerization of VC as a function of conversion and seed particles. Comparison of results ((particle size vs conversion) of all systems, the monomer- and initiator-containing droplets (I) and monomer- and alcohol-containing droplets (A)) indicates that the seed growth and droplet growth profiles are relatively independent of the initiator type, differing mainly in the final particle size. The miniemulsion polymerization of VC initiated by an oil-soluble initiator gave final particles (at 90% conversion) with a diameter around 1100 nm and 500 nm for (I) and (A) droplets. Under the same conditions, the final particle diameters were found to be 800 nm and 600 nm for a partly water-soluble and a fully water-soluble initiator, respectively. In the case of a polymer seed, the most extensive increase in D with conversion was observed with a water-soluble initiator. The results with an oil-soluble and a partly water soluble initiator are nearly the same. Figure 7 shows the dependence of the

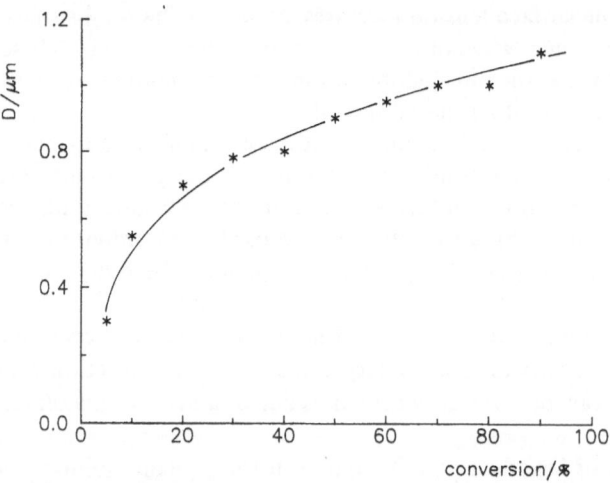

Fig. 7. Variation of the particle diameter in the miniemulsion polymerization of vinyl chloride with conversion [122] 1wt% SDS, 1wt% C_{16}–C_{18} linear saturated alcohol, initiator: DCP (dicetyl peroxydicarbonate)

particle size on the conversion is for the oil-soluble initiator and is the same for all systems studied. Summarizing, the authors have suggested that decomposition of an oil-soluble initiator begins in the droplets. At low conversion, these droplets release radicals to the aqueous phase. Droplets (A and I) then readsorb the radicals and these initiate growth events in the large polymer particles. Initiable droplets grow quickly. Beyond 30 or 40% conversion, the growth of particles proceeds similarly to that for a conventional emulsion polymerization. Polymerizations utilizing partially water-soluble and a fully water-soluble initiators showed that an intensive seed growth occurs significantly later. Authors have reported that the droplet formation and particle growth can by controlled by intensive homogenization of initiators (of significant water solubility) accompanied by equilibrating the initiator between phases, variation of amounts or ratios of emulsifier and alcohol (which affect the number of droplets capable of undergoing initiation), initial seed size, and the feeding rate of the reaction components.

3.2 The Monomer Parameter

Vinyl chloride monomer differs from monomers like styrene, methyl methacrylate or vinyl acetate principally in the insolubility of the polymer in the monomer. Moreover, vinyl chloride is highly solubile in water. The water solubility of VC and a water-phase polymerization are partly responsible for deviations from the micellar model. The reaction order with respect to [emulsifier] is known to decrease with increasing water solubility of the monomer [123]

The solubility, s, of VC in water can be expressed as [124]

$$s = 8.8 \times 10^{-3} (P/P_0) (\text{kg VC/kg water}) \tag{42}$$

where P_0 is the saturation pressure of VC at the particular temperature.

In the emulsion polymerization of vinyl chloride, the monomer is distributed between emulsifier micelles, polymer particles, monomer droplets, vapor and water. At ca. 80% conversion, the pressure begins to drop. This point marks the disappearance of monomer droplets. The reaction pressure depends directly on the extent of saturation of the reaction media with monomer and temperature. This evolution is normal if one considers that the solubility of vinyl chloride in PVC at saturation is about 0.25–0.3 kg VC/kg PVC and in water 8.8 g VC/l dm³ [124]. These data show that the polymerization of vinyl chloride may proceed in water as well as in polymer particles. The polymerization in the particles is more important due to the restricted termination. The polymerization of vinyl chloride in water forms primary particles which are either transformed to large particles or are adsorbed by large particles.

If one inserts the experimental values of the solubility of vinyl chloride in PVC (0.25–0.3 kg/kg PVC) [124] in Eq. (35) [34],

$$[VC]_{eq} = (\rho_m \phi_m / M_{vc}) \times 10^3 \ \text{mol dm}^{-3} \tag{43}$$

where M_{vc} is the molecular weight of the monomer, ϕ_m is the monomer volume fraction in the polymer particle and ρ_m is the density of monomer, the equilibrium monomer concentration in the monomer-swollen PVC particle is 4.0–4.4 mol dm^{-3}. The quantity ϕ_m can be expressed as

$$\phi_m = \omega_{m/p}\rho_p/(\rho_m + \omega_{m/p}) \tag{44}$$

where $\omega_{m/p}$ is the weight ratio of monomer and polymer in the particle. If one inserts the values of $\omega_{m/p}$ 0.25 or 0.3 in Eq. (44), ϕ_m is calculated to be 0.29 or 0.33. Solubility (or swelling) of VC in PVC was also reported to be 0.42 kg VC/kg PVC [70] and 0.25 kg VC/kg PVC [125].

The calculated values of $[VC]_{eq}$ 4.0–4.4 mol dm^{-3} are in a good agreement with those reported by Schneider et al. [126] (see Table 7). This table shows that the [VC] in particles (at monomer saturation pressures) is 4.5 mol dm^{-3} and decreases with decreasing pressure. The same trend follows the solubility of vinyl chloride in water. Gerrens et al. [70] reported the equilibrium monomer concentration to be as high as 6.0 mol dm^{-3} at 50 °C (estimated from the variation of VC pressure with conversion). Butucea et al. [115] estimated the value of the equilibrium monomer concentration 4.5 mol dm^{-3} which is close to that reported by Schneider et al.

The swelling of PVC latex particles is almost independent of temperature. The maximum uptake of VC, however, depends on the polymerization temperature. The PVC particles prepared at 11, 50 and 90 °C showed maximum uptakes of VC 0.22, 0.29 and 0.37 kg VC/kg PVC, respectively. Effects of particle and emulsifier concentrations have also been studied. It was reported that with increasing particle size the uptake of VC increased. Experimental results show that with increasing emulsifier concentration (or extra addition of emulsifier) the solubility of VC increases. It is suggested that VC is solubilized in the surface layer of adsorbed emulsifier [124].

The similar results were reported in Ref. [80]. Table 8 shows that with increasing emulsifier concentration the uptake of VC and the interfacial tension increase.

Table 7. Monomer concentration parameters in the vinyl chloride – swollen poly(vinyl chloride) particles and in the aqueous phase at saturation and subsaturation pressures at 50 °C [126]

P_{vc}[1] (MPa)	$[VC]_{eq}$[2] (mol dm^{-3})	$[VC]_{vc}$[3] (mol dm^{-3})	f_2[4]
0.79[5]	4.5	0.176	1.13
0.53	1.84	0.102	1.05
0.4	1.25	0.07	1.03

[1] The pressure of VC monomer, [2] the equilibrium VC monomer concentration in the PVC particles, [3] the VC monomer concentration in water, [4] the swelling factor of polymer particles related to variations of radius of particles, [5] pressure of saturated vapors.

Table 8. The effect of extra addition of ammonium laurate (AL) on uptake of vinyl chloride at 50 °C [80]

Increase in AL concentration $(g\,dm^{-3})$	Increase in solubilization (g VC/kg PVC)	γ^1 $(mM\,m^{-1})$
1.84–3.2	5.0	4.4
1.84–4.55	5.9	9.0
1.84–5.92	8.0	13.6

Sörvik et al. [81, 82] carried out the emulsion polymerization of VC at subsaturated pressures. They reported that the reaction order with respect to monomer concentration is ca. 1.0. The polymerization proceeds under monomer-starved conditions in which the rate of polymerization is proportional to the monomer concentration at the reaction loci. The rate of polymerization linearly decreases with decreasing pressure (without any maximum). The dependence of polymerization rate on the equilibrium monomer concentration, $R_p \propto [M]_{eq}$ can be expressed as the dependence on the monomer concentration in water because this dependence is a function of the saturation degree X and, therefore, we can write $R_p \propto [M]_{eqw}$ or $R_p \propto [8.8X]$, respectively.

The same group [114] has also reported that M_n increases with increasing $[M]_{eq}$. From experimental results, the following dependence was obtained $M_n \propto [M]_{eq}^{0.6}$. This relationship is ascribed to variation of the chain transfer to monomer (C_M). The chain transfer constant to monomer, C_M is constant at saturation pressures and decreases with decreasing pressure at subsaturation conditions.

Emulsion polymerization of vinyl chloride was also carried out at subsaturation conditions by Butucea et al. [127] who followed the effect of monomer, emulsifier and initiator on the polymerization behavior. Some results from these investigations are summarized in Table 9. This table shows that the rate of polymerization increases with increasing concentration of all these reaction components. From these results the following semiempirical equation for the dependence of the rate on the initiator, monomer (in water) and emulsifier concentrations was suggested.

$$R_p = K8.064 \times 10^{12} e^{-16\,989.4/RT} [M]_w^{2.3} [I]^{0.37} [E]^{0.43} \qquad (45)$$

where K is the lumped rate constant which includes the constant k_i, k_p, k_t and k_{tr} and $[M]_w$ (= [8.8 X]) the monomer concentration in water. The value of K was estimated to vary from 3×10^{-3} to 1.5×10^{-2}, and depends on experimental conditions.

Effects of monomer concentration on the polymerization were determined in a series of experiments using a constant seed and initiator concentration [114]. From the log-log plot of R_p against V_p (the total volume of the monomer-swollen polymer particles, see Eqs. (27–29)), the order of R_p with respect to

Table 9. Variation of the rate of polymerization with concentration of emulsifier, initiator and monomer [127]

Monomer[a] Temp. (°C)	[VC] (g dm³)	Saturation degree, X	R_p (g dm⁻³ h⁻¹)	Emulsifier[b] SDS × 10² (mol dm⁻³)	R_p (g dm⁻³ h⁻¹)	Initiator [K₂S₂O₈] × 10³ (mol dm⁻³)	R_p (g dm⁻³ h⁻¹)
48	8.48	0.963	32.5	5.5	23.4	0.55	18.3
45	7.75	0.881	23.4	4.2	20.5	0.74	20.8
40	6.81	0.773	18.25	3.0	16.5	0.93	22.0
35	5.94	0.675	13.5	1.7	13.3	1.5	27.1
30	5.16	0.587				2.2	30.2

[a] $[K_2S_2O_8] = 1.1 \times 10^{-3}$ mol dm⁻³, $[SDS] = 5.5 \times 10^{-2}$ mol dm⁻³, [b] Temp. 50 °C, X = 0.88, $[K_2S_2O_8]$ = 1.1×10^{-3} mol dm⁻³, 50 °C, $[SDS] = 5.5 \times 10^{-2}$ mol dm⁻³.

V_p was obtained. The exponent slightly decreased with increasing monomer concentration in the particles (it decreased from 0.5 to 0.4). Equations (27, 28 and/or 29) predict that the limits should be equal to 1/2 and 1/3. The experimental values indicate that the term containing termination and/or desorption rate constants, respectively, completely dominates. It is well-known from several phenomena that termination decreases with decreasing molecular mobility of the reaction ingradients (with decreasing monomer concentration in gel). Ugelstadt and Hansen [15] expect that k_{des} will decrease only slightly at concentrations near saturation. A more remarkable decrease is expected at low monomer concentrations (or at high conversion). Friis and Hamielec [79], also expect k_{des} to decrease fairly rapidly after the pressure drop.

The relation between R_p and $[M]_{eq}$ under constant reaction conditions ($\sim 10^{18}$ particles per dm³) is described by a curve with a plateau at higher monomer concentrations. This rate of polymerization increases with increasing $[M]_{eq}$ and V_p. At higher V_p, a maximum on the curve R_p vs. $[M]_{eq}$ is obtained. At lower V_p (≤ 0.1), plateau was found. N was found to be proportional to the reaction order with respect to V_p. This behavior was ascribed to a decrease in k_t. Nilsson et al. [124] and Goebel et al. [88], however, reported that the particle did not influence the maximum rate in the batch polymerizations.

The effect of the monomer concentration on the rate of VC polymerization was followed in Ref. [97]. The authors found that the reaction order with respect to [VC] was higher than 1.0. This behavior was discussed in terms of the primary radical termination.

Hjertberg et al. [114] attributed the increase of V_p (with decreasing monomer concentration) to the stronger decrease of k_t compared to that of k_{des}. This is reasonable, as termination involves macroradicals, while only the fraction of small radical species is desorbed from the particles. Termination in particles (k_t) completely dominates when the monomer concentration has decreased to about one third of the saturation value. At higher concentrations, the influence of the term involving k_{des} is larger, but this term never completely dominates polymerization. The reaction order of R_p with respect to V_p was found to decrease with increasing particle concentration. With $N = 1 \times 10^{18}$ dm^{-3}, the exponent varied from 0.41 to 0.47. Using a lower particle concentration 5.3×10^{16} dm^{-3}, the exponent slightly increased from 0.47 to 0.5. In both cases, the monomer pressure P/P_0 varied from 0.9 to 0.7. This difference may be ascribed to variation of monomer concentration or extent of monomer saturation of the particle shell.

Ugelstadt et al. [78] studied effects of seed particles and pressure on the emulsion polymerization of VC ($N = 1.0 \times 10^{17}$ dm^{-3}). A maximum polymerization rate was observed at a pressure of about 87% of the saturation value. The rate increased by a factor of at least 2 compared to the rate at saturation pressure [114]. Under subsaturation pressures, the polymerization rate passes through a maximum during the pressure drop. The increase was explained by a reduction of the transfer rate constant k_{tr}. The same happens with the desorption rate constant k_{des} [78]. Liegeois [128] drew the same conclusions and found that molecular weight decreased with increasing pressure.

The rate of polymerization decreases strongly with decreasing monomer concentration after the pressure drop ($P/P_0 \leq 0.8$). Here, the order with respect to $[M]_{eq}$ becomes well above 1. This is not expected, if other parameters such as N and V_p are kept constant and k_i is independent of the monomer concentration [129]. The termination (k_t) and desorption rate (k_{des}) constants decrease and so does the total exponent of $[M]_{eq}$. The propagation constant is believed to be constant except for very high conversions [130]. A decrease in k_p (at glass transition region) could perhaps explain the rapid decrease in R_p while changing P/P_0 from 0.8 to 0.5.

The glass transition of the polymer – monomer gel is reached when the monomer concentration has decreased to about 3.5 wt% at 55 °C [114]. According to Berens [131], the gel should contain about 6% vinyl chloride at the lowest pressure used, $P/P_0 = 0.53$. A decrease in k_p before the glass transition is in accordance with the behavior in the bulk polymerization of acrylates or methacrylates [132].

The actual concentration of monomer in the polymer particles (gels) in a diffusion controlled polymerization (at subsaturation pressure and at low agitation speed) equals the equilibrium value given by the value P/P_0 [133]. The apparent P/P_0 values, as well as the corresponding concentrations of monomer in the polymer gel are related to different agitation speeds. For example, with increasing the agitation speed from 500 to 1500 rmp, the monomer concentration increases from 6.1 to 25 g VC/100 g PVC [134].

The total mass flow of monomer, as measured by the monomer consumption, consists of two parts; first, the monomer is consumed by the polymerization reaction and, second, the polymer formed (particles) will absorb additional monomer according to the apparent P/P_0 value. The consumption of monomer is the driving force for the monomer transport which depends on the number of particles as well as the volume of the monomer swollen particles (or polymer), V_p. In a reaction controlled system (at high agitation speed), the rate of polymerization increases with V_p and the particle concentration.

The emulsion polymerization of unsaturated monomers depends strongly on monomer concentration at high conversion. When conversion is greater than the critical one at which the monomer droplets dsappear, the polymerization rate and the molecular weights are expected to decrease with conversion due to the decrease of the monomer concentration [6].

Jaeger et al. [126] investigated the molecular weights of poly(vinyl chloride) at different monomer pressures. Under monomer saturation conditions, the molecular weights of polymer were constant or slightly increased (at low conversions) as the polymerization advanced. At subsaturation pressures, the molecular weights increased with increasing conversion up to the total consumption of monomer. The trend of the molecular weights in the latter case may be attributed to the decrease of the chain transfer to monomer.

Under monomer-saturation conditions ($P_{VC} = 0.79$ MPa), the molecular weight, M_W, was reported to be ca. 1.1×10^4 (the medium and large conversion range). At subsaturation conditions, the molecular weights were lower, e.g., at $P_{VC} = 0.53$ MPa $M_w = 5.0 \times 10^3$ (the same reaction conditions, [peroxodisulfate] $= 5.9 \times 10^{-3}$ mol dm^{-3}). Variations of the moecular weights with VC pressure used the authors to calculate the chain transfer constant $C_{VC} = 3 \times 10^{-3}$. Insignificant variations of the molecular weights with [peroxodisulfate] was ascribed to the strong chain transfer to monomer which counteracts effects of other kinetic parameters.

The rate of polymerization was reported to increase with increasing weight ratio water/monomer. This shows that the water-phase polymerization becomes important. The higher the amount of VC that is dissolved in water, the higher is the rate of polymerization observed [87]. The rate of water-phase polymerization was found to be proportional to the monomer saturation degree of the aqueous emulsifier solution [87].

Under subsaturation conditions, variation of monomer concentration also influences other parameters such as the volume fraction of polymer in particles, rate constants and saturation of reaction media. Kinetic studies are more complicated and call for knowledge of the relations between different parameters or constants and the monomer concentration.

At high conversions, the rate of polymerization was found to increase with conversion. The decrease in termination counterbalances that of propagation (the decrease in $k_p[M]_{eq}$). If this difference is large enough, the overall polymerization rate increases (the gel effect).

The effect of monomer concentration on the polymerization rate and particle size was reported in Ref. [88]. With increasing monomer pressure (P_{VC}/P_{VC}°) from 0.51 to 1.0 increased the rate of polymerization from 60 to 129 (g dm^{-3} h^{-1}). Besides, the addition of VC during polymerization led to the abrupt decrease in the number of particles. It was found that the number of particles decreased with increasing pressure, e.g., from $P_{VC} = 0.53$ MPa to 0.79 MPa decreased the number of particles from 20×10^{18} particles/dm^3 to 5×10^{18} particles/dm^3 ([SDS] $= 1.4 \times 10^{-2}$ mol dm^{-3} and [peroxodisulfate] $= 2.6 \times 10^{-3}$ mol dm^{-3}).

3.3 The Temperature Parameter

The effect of temperature on the emulsion polymerization of VC rarely been studied. It is known that the polymerization temperature strongly influences not only the rate of polymerization but also the structural deffects in the PVC matrix. Reactions forming defects are most sensitive to temperature. The activation energies for these reactions are much higher than for the propagation reactions, and therefore they are more sensitive to polymerization temperature. Hjertberg and Sorvik [135] investigated effects of the particle seed and temperature at subsaturation pressure on the dehydrochlorination rate. The minimum dehydrochlorination rate was at 55 °C. The dehydrochlorination rate was also found to be a function of conversion at subsaturation pressure (Hamielec et al. [136]). Here, the dehydrochlorination rate increased linearly with temperature without any minimum.

The overall activation energy (E_0) for the solution polymerization may be expressed as follows;

$$E_0 = E_p + E_t/2 + E_d/2 \tag{46}$$

where E_p, E_t, and E_d is the activation energy for propagation, termination, and decomposition of initiator, respectively. For the emulsion polymerization of monomers with high water solubility, such as VC, the overall rate of polymerization may be expressed by the following expression

$$R_{p0} = R_{p1} + R_{p2} = k_p[M]_{eq} n N/N_A + k_p/k_t^{0.5}[M]_{eq}(2fk_d[I])^{0.5} \tag{47}$$

where R_{p1} is the rate in particles and R_{p2} in the continuous phase. The activation energy for the homogeneous polymerization (R_{p2}) of styrene, acrylates or methacrylates is the range 90–100 kJ mol^{-1} [137] which is very close to that for the homogeneous polymerization of VC [70, 71] (see later). The polymerization of styrene and acrylates in emulsions is characterized by the activation energy 30–55 kJ mol^{-1} [137]. The activation energy of polymerization of conventional (homogeneous or heterogeneous) polymerization of conventional VC for conventional initiators (AIBN, benzoyl peroxide or persulfates) was found to be 101 kJ mol^{-1} [70, 71]. In the emulsion polymerization of VC monomer, the activation energy of polymerization was estimated to be around 42 kJ mol^{-1} [137]. These data indicates that the emulsion polymerization is governed by the kinetic events proceeding in the latex particles. The water-phase polymerization should contribute mainly to the formation of primary particles.

The number of particles, however, is nearly independent of temperature despite the fact that the rate of initiation increases. This indicates that the chain transfer events and the phase separation become dominant.

The overall activation energy E_0 was reported to depend on the type of initiator [91]. In the redox system (peroxodisulfate/reducing agent), the activation energy was found to be 22.9 kJ mol^{-1}. On the other hand, in the thermochemical initiation (only peroxodisulfate was used) the activation energy was 74–80 kJ mol^{-1}. Thus, the former is lower by a factor of 3. In the gel effect stage, the activation energy was reported to be a little higher.

Meelson et al. [92] determined the activation energy E_d for the decomposition of peroxodisulfate in the presence and absence of emulsifier (SDS). In the absence of emulsifier, the activation energy was found to be 140 kJ mol^{-1} and in the presence of SDS 120 kJ mol^{-1}.

3.4 Other Parameters

Boieshan [91] found that by increasing the stirring rate the particle diameter increased, e.g., from 66 to 76 nm, when the stirring rate changed from 300 to 1000 r.p.m. Here, the particle number increases rapidly up to a critical conversion of ca. 20% until the micellar emulsifier disappears – then the number of particles slightly increases.

Hjertberg [133] investigated the effect of agitation on the emulsion polymerization of VC initiated by a water-soluble initiator (at subsaturated pressure) using PVC latex as a seed. Using a propeller, the polymerization rate increased with increasing the agitation up to ca. 1,000 rpm, and then it reached a constant value. Simultaneously, the molecular weight and thermal stability increased. At slow agitation, the rate of polymerization is diffusion controlled, while at higher speed the polymerization becomes reaction controlled. Thus, below a certain level of agitation, the insufficient transport of monomer lowers the concentration of monomer in polymer particles (gels). As a consequence, the polymerization rate decreases until it balances the rate of monomer transfer. When the agitation rate decreases from 1500 to 500 rpm, or the pressure falls from $P/P_0 = 0.97$ to 0.53, the monomer concentration decreases by a factor of 3. That the polymerization rate is indendent of agitation at higher speeds was taken as an evidence that the system becomes reaction controlled.

Hjertberg and Sörvik [114] observed that the rate of emulsion polymerization of vinyl chloride increased with stirring up to 1000 rpm. As the monomer is charge as vapor, intensive agitation is needed to avoid a diffusion-controlled regime.

Neelson et al. [138] investigated the emulsion polymerization of vinyl chloride in the presence of inhibitors. The used p-benzoquinone and a stable radical 2,2',6,6'-tetramethylpyperidine-N-oxide at concentrations $\leq 1 \times 10^{-5}$ mol dm^{-3}. After the consumption of inhibitor, the conversion vs. time curve was the same shape as that without inhibitor. In some experiments the inhibition period varies with the emulsifier and initiator concentration. The inhibitor efficiency decreased with increasing concentration of emulsifier. The solubilization of inhibitor is expected to decrease the amount of inhibitor available for reactions with radicals. For this reason, the inhibitor acts more efficiently at the low emulsifier concentration as it was reported in Ref. [139].

From these experiments, the following equation for the rate of initiation was derived.

$$R_i = k[I][E]^b \tag{48}$$

where

$$k_{i,eff} = k[E]^b \tag{49}$$

where b, is equal to 0.25 at 50 °C and 0.45 at 60 °C

or

$$R_i = k_{i,extr}[I] + k_{i,eff}[I][E]^c \tag{50}$$

where $k_{i,extr}$ is the rate constant extrapolated to $[E] = 0$.

The same group [138] also derived a semiempirical expression for the rate of initiation on the basis of results of emulsion polymerization of VC in the presence of the labelled peroxodisulfate

$$R_i = k[I][E]^c \tag{51}$$

where

$$k_{i,eff} = k[E]^c \tag{52}$$

where c = 0.2 for 50 °C. They reported that the rate constant $k_{i,eff}$ obtained by the inhibition method was approximately 2 times smaller than that obtained by the analysis of radioactive polymer. This difference may be attributed to the formation of byproducts which can again take part in the deactivation steps. The emulsification of inhibitor also decreases the available amounts of inhibitor for reactions with radicals. The rate of initiation estimated from the radioactive OSO_3^- end groups was stated as the real value.

Neelsen et al. [140] investigated the effect of NaCl on the emulsion polymerization of vinyl chloride in the presence and absence of emulsifier (SDS) or inhibitor (p-benzoquinone (BQ) or chloranile (CA)). Under reaction conditions 2 [BQ] = [CA], the same induction periods and the conversion curves were obtained. The addition of NaCl (a coagulating agent) caused the decrease of the rate of polymerization and the increase of the particle size. The addition of NaCl led to an increase in the induction period. Here, the same induction period and the same shape of the conversion curves were obtained and [BQ] = [CA]. The different behavior of BQ in the presence of NaCl was explained by the reaction of Cl˙ with BQ which leads to the chlorination of BQ. Cl˙ radicals are generated by the reaction of peroxodisulfate with NaCl (a labelled salt). The addition of NaCl increased flocculation of particles which led to the formation of larger particles and lower rates of polymerization. The dependence of the rate of polymerization on the NaCl concentration was expressed by the following equation.

$$R_p \propto [Cl^-]^{-0.15} \tag{53}$$

Prediction and control of molecular weight averages, the number of short- and long-chain branches, and terminal double bonds per polymer molecule is of considerable importance to the PVC industry since the low thermal stability of PVC has been linked to the formation of some branched and unsaturated molecular structures mainly generated by chain transfer to monomer. In vinyl chloride polymerization, the molecular weight distribution (MWD) and molecular weight averages are actually controlled by transfer to monomer and are almost independent of initiator concentration and monomer conversion up to a conversion of about $\omega_p \sim 0.85$. In this conversion range, the MWD of polymer will be given by the polydispersity index close to 2. However, as the monomer conversion increases, the relative rates of the various free-radical reactions that control the molecular structure of PVC chains change due to the appearance of strong diffusion control. In this conversion region, the number of long and short-chain branches increases as well as the production of low molecular weight polymer chains. This might result in an increase of the polydispersity index from 2 to a value as high as 3 and cause a deterioration in the thermal stability of the PVC.

Tauer et al. [141] investigated variations in the particle size and number with conversion at the begining of polymerization. The experimental results shown in Fig. 8 show that the particle size increases linearly with conversion but the number shows a maximum at certain (critical) conversion. The latter dependence is discussed in terms of strong aggregation between primary particles. The authors also suggested a model to describe variations of particle concentration with conversion and obtained good agreement between the experimental and theoretical kinetic data. In addition, the calculation showed that the average number of radicals per particle was much lower than 0.5. These data indicate that the chain transfer to monomer and desorption of monomeric

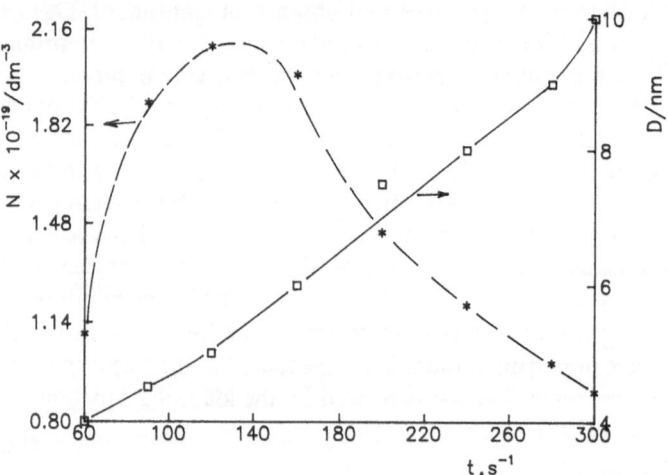

Fig. 8. Variation of the number and diameter of primary particles at the beginning of the emulsion polymerization of vinyl chloride with reaction time [141]. $[SDS] = 1.4 \times 10^{-2}\,mol\,dm^{-3}$, $[K_2S_2O_8] = 5.9 \times 10^{-3}\,mol\,dm^{-3}$, $[Na_3PO_4] = 2.63 \times 10^{-3}\,mol\,dm^{-3}$, 50 °C

radicals are operative. They found that the rate of polymerization and the number of particles reach the maximum even at very low conversions. The number of particles at the maximum was estimated to be 2×10^{19} dm^{-3}.

The kinetics of emulsion polymerization of vinyl chloride is described by kinetic parameters such as rate constants of the initiation, k_i, propagation, k_p, termination, k_t, and desorption of radicals, k_{des}. Concerning the k_p value, one may admit that a value of 1×10^4 $dm^3 mol^{-1} s^{-1}$ at $50\,°C$ is generally accepted for vinyl chloride, at least in processes of saturation with monomer [142]. Variation of the other kinetic parameters was more complicated. The value k_i depends on the nature and concentration of emulsifier and co-emulsifier utilized and also on monomer [92, 143, 144].

3.5 Branching

For the homogeneous radical polymerization, the following relation is valid for the number–average degree of polymerization X_n [145, 146]

$$1/X_n = \{(k_t f k_i [I])^{0.5}\}/k_p [M]_{eq} + C_M \tag{54}$$

where C_M is the constant for chain transfer to monomer. It was shown [147] that k_t/k_p is the same in homogeneous and heterogeneous polymerization of vinyl chloride and therefore Eq. (44) can be used for the emulsion polymerization of vinyl chloride in both phases [147].

The well-known fact that the concentration of initiator has a very small influence on the PVC molecular weight may be ascribed to the dominating event – the chain transfer to monomer. Hjertberg and Sorvik used a plot of $1/X_n$ against $[I]^{0.5}$ to estimate $C_M = 1.3 \times 10^{-3}$ [114]. This value is in a good agreement with that for conventional vinyl chloride polymerization [111].

The assumption of constant C_M is based on the classical reaction scheme for radical polymerization, e.g., for the chain transfer to monomer:

$C_M = k_p/k_{tr}$

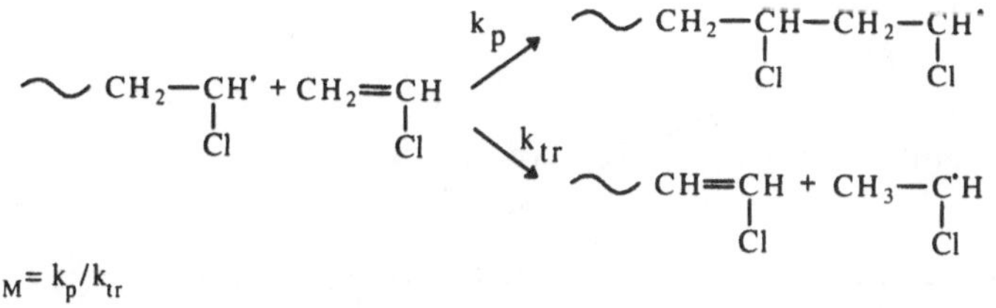

Scheme 1.

As both reactions are of first order with respect to monomer and macroradicals, C_M should be independent of the monomer concentration. One would expect the ratio k_{tr}/k_p to remain constant at decreasing monomer concentration as both reactions involve a monomer and a macroradical.

Based on free-volume ideas, Hamielec et al. [148] suggested that the ratio k_{tr}/k_p should increase appreciably with conversion after the pressure drop. This is in accordance with the experimental observation between the molecular weight and the monomer concentration.

With the recent knowledge concerning the structure of end groups and short chain branches, it is obvious that the simple mechanism for chain transfer to monomer presented in Scheme 1 is not valid. Instead, a more complicated mechanism (Scheme 2) is now favored [114, 149, 150]. The first step in chain transfer is a head-to-head addition (k_1) instead of tail-to-head (k_p). The ratio between the two reaction rates at level A should be reasonably independent of $[M]_{eq}$. The following step in transfer (k_2) is 1,2-Cl migration. The competing reaction (propagation, k_3) at this level (B) would be monomer dependent. However, according to results reported in Ref. [151], the concentration of the resulting head-to-head structure is negligible, below 0.1/1000 VC [151], and k_3 can be assumed to be zero. At level C, the macroradical can be transformed into an inactive chain by loss of a chlorine atom (k_5) which subsequently reacts

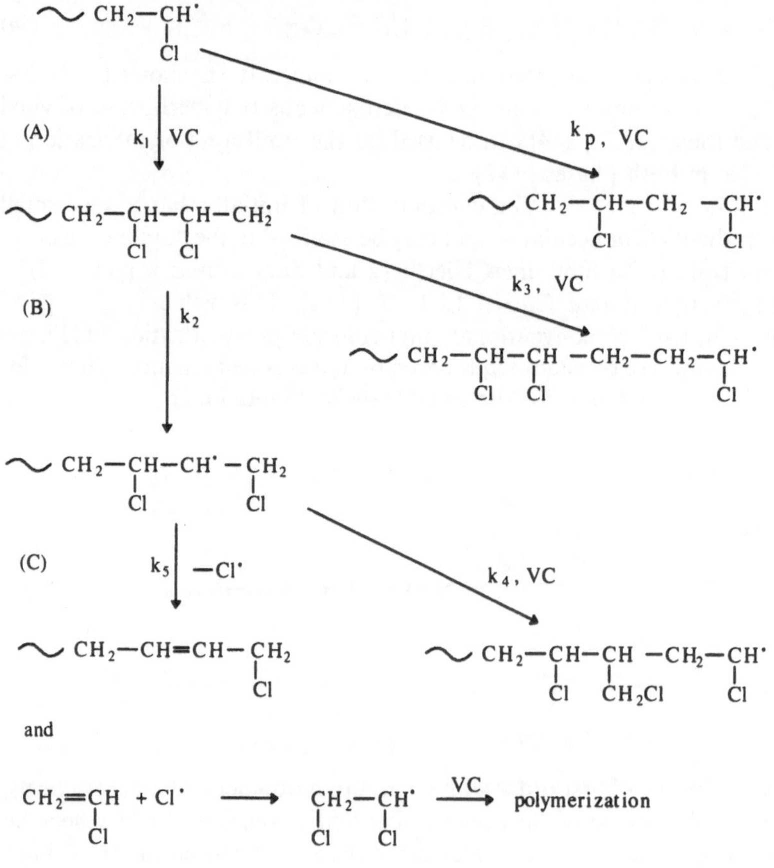

Scheme 2.

with monomer (or polymer). The competing reaction at level C (propagation, k_4) results in further propagation and a chlormethyl branch formation.

The total rate of propagation can, therefore, be approximated as the rate of normal propagation ($k_p[M][P_1^{\cdot}]$) and C_M the ratio between the transfer reaction ($k_5[P_2^{\cdot}]$) and the propagation reaction. From Scheme 2, the following expression for the overall C_M was suggested [114]

$$C_M = (k_5 k_1)/k_p(k_5 + k_4[M]) \tag{55}$$

Even this reaction scheme is very complex and does not describe completely all pertinent reactions, especially not at reduced monomer pressure. It contains the long chain branching (LCB) which proceeds to a higher extent at subsaturation pressures [82]. Additional structural changes or defects were also observed such as: 1) appearance of 1-chloro-2-alkene chain and 1,3-dichloroalkane chain ends [152], enhanced formation of 2,4-dichlorobutyl branches [152], and increased numbers of internal double bonds [153]. The butyl branches are formed by intramolecular backbiting [114, 151, 152], which would have no influence on the molecular weight. Transfer to polymer from macroradicals is a plausible mechanism for the formation of LCB [150, 152]. The reaction Scheme 2 must be extended by a number of these reactions. All reactions leading to LCB will increase M_n. However, formation of a long chain branch after transfer from a chlorine atom implies that an existing macromolecule is enlarged at the expense of the formation of a new macromolecule. In other words, the number of polymer particles would be decreased, which, of course, tends to increase M_n as well. The increased degree of LCB at reduced monomer concentration will thus decrease the order of M_n with respect to $[M]_{eq}$. The decreased termination at reduced monomer concentration should in principle also decrease the influence of $[M]_{eq}$. With decreasing monomer concentration LCB increases [113]. The transfer of a radical activity to polymer from chlorine atom generates the internal double bonds [152, 153].

The double bond content in PVC resins has been found to vary in the range of 1.5 to 4.0 per 1000 VC units. Hildebrand et al. [154] found that the number of double bonds per unit weight of polymer increases as the polymerization temperature increases and is proportional to the number of polymer molecules. Double bond defects can arise from both transfer to monomer and termination by disproportionation. These bonds are called "terminal double bonds" and are located at, or close to the ends of polymer chains. The various anomalous and unsaturated structures found in PVC resins are considered to the primary cause for the low thermal stability of the polymer.

Hjertberg and Sörvik [82, 114] showed that the subsaturation polymerization usually results in polymers with inferior thermal stability. The rate of dehydrochlorination was found to be proportional to $1/M_n$ as well as LCB.

Desorption of radicals from polymer particles refers to the radicals formed in the chain-transfer to monomer. Vinyl chloride monomer is known to be an active chain-transfer agent. The high-water solubility of VC monomer favors desorption of radicals from particle. Exit of radicals from polymer particles refers the radicals generated by the chain transfer and formed from one monomer unit [155, 156]. Desorption of radicals generally decreases the rate of polymerization and the average degree of polymerization. Desorption and reabsorption of radicals can increase termination and decrease the radical activity of polymer particles. Under such conditions, the average number of radicals per particle, n, should be much below 0.5. The value of n for a given polymerization process is determined by the proportion between the rates of initiation, desorption and termination.

Tauer [157] has shown that there exists a different relation between termination and transfer reactions in the aqueous phase than within the latex particles. It was pointed out that the chain transfer events govern the polymerization process in particle (chain transfer exceeds termination). The author showed that termination greatly exceeded transfer in the aqueous solution polymerization. The experimentally determined termination rate constants for the aqueous solution polymerization of VC at the beginning of the reaction are 4 or 5 orders of magnitude higher than those of the polymerization within the monomer-swollen latex particles. Therefore, the oligomers are much shorter than would be expected from calculation using kinetic constants from steady-state emulsion polymerization and its model. The degree of polymerization of VC oligomers is $P = 3 - 10$, which is much lower than the expected one.

3.6 Copolymerization

The kinetic behavior in the emulsion copolymerization of VC and vinyl monomers has not been described till now. The emulsion copolymerization appears to be very complex due to large differences in water solubilities of monomers and polymers, the solubility of polymers in their monomers, the chain-transfer to monomers and the reactivities of monomer and growing radicals.

Capek et al. [158, 159] studied the kinetics of emulsion copolymerization of VC and butyl acrylate (BA) initiated by peroxodisulfate in the presence of an anionic emulsifier (Dowfax 2A1). Copolymerizations were carried out at saturation and subsaturation pressures and conducted to low and high conversions.

Conversion curves for the systems with high BA concentrations are similar to those for the conventional emulsion polymerization (Fig. 9). In systems with high concentration of VC, no acceleration of polymerization is observed. The specific rate of butyl acrylate in the emulsion copolymerization of BA and VC increases with increasing BA concentration ($R_p \propto [BA]^{0.5}$). The rate of emulsion polymerization of BA under the same reaction conditions is proportional to $[BA]^{1.0}$. The reaction order 1.0 is twice that obtained for the fractional rate of BA in the presence of VC. The high reaction order is due to the fall in the monomer concentration in particles with decreasing initial monomer concentration. Thus, at the low initial monomer concentration polymerization proceeds under the monomer-starved conditions and at high monomer concentration under monomer-saturated conditions. In contrast, the specific rate of VC

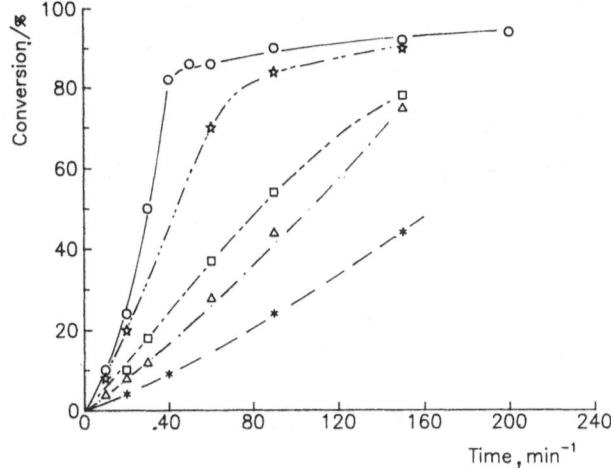

Fig. 9. Variation of the total monomer conversion in the emulsion copolymerization of vinyl chloride and butyl acrylate with the reaction time and with the mole fraction of vinyl chloride, x_{VC} [158]. x_{VC}: 0.34 (○), 0.67 (☆), 0.89 (□), 0.95 (△), 1.0 (∗). Recipe: 300 g water, 200 g monomer, 14.8 g Dowfax 2A1, 0.342 g ammonium peroxodisulfate, 0.075 g NaHCO₃, 60 °C

polymerization decreases with increasing VC monomer concentration. This is caused by a variation of monomer concentration in polymer particles. With increasing BA content in the monomer feed, the equilibrium monomer concentration in particles increases, whereas with increasing VC concentration it decreases the equilibrium monomer concentration. With an increasing number of VC units in copolymer, the swellability of VC/BA polymer particles and the rate of polymerization decrease.

The reactor pressure was found to increase with increasing both the concentration of VC and conversion up to 70–80% [158] (see Fig. 10). This was ascribed to the increase of the mole fraction of VC in vapor and polymer particle phase. Beyond the critical conversion (70–80%), the reactor pressure abruptly decreases which is due to the onset of the gel effect (depletion of VC monomer droplets). With decreasing VC concentration, the critical conversion is shifted to higher conversion. The pressure increase of VC was ascribed to the exotherm which is very often observed in polymerizations of acrylates. The gel effect, which is operative in polyVC/BA particles, increases the pressure of VC due to the temperature increase in the polymer-rich domains.

The size of particles increased with increasing VC concentration (see Fig. 11 and Table 10). At high concentrations of VC, the size of polymer particles increases throughout polymerization. At low VC concentrations, the size of particles increases in the low and medium conversion range and is nearly constant at high conversion.

From the experimental results (R_p and N), the following relationships were obtained

$$R_{ps, BA} \propto [N]^2 \quad \text{and} \quad R_{pT} \propto [N]^{0.7} \tag{56}$$

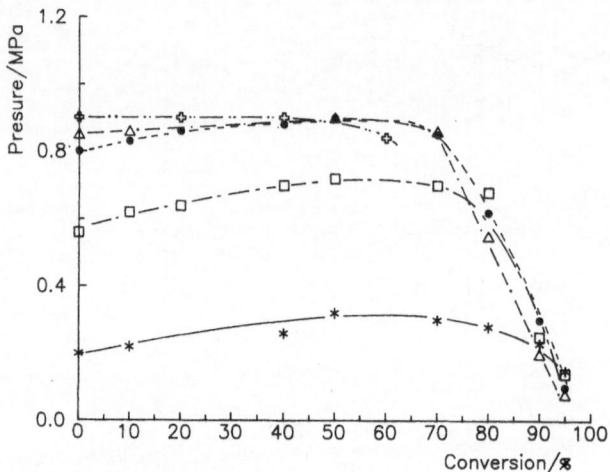

Fig. 10. Variation of the pressure of vinyl chloride monomer in the emulsion copolymerization of vinyl chloride and butyl acrylate with conversion and with the mole fraction of vinyl chloride, x_{VC} [158]. x_{VC}: (∗) 0.34, (□) 0.67, (●) 0.89, (△) 0.95, (⬡) 1.0. Other conditions are given in the legend to Fig. 9

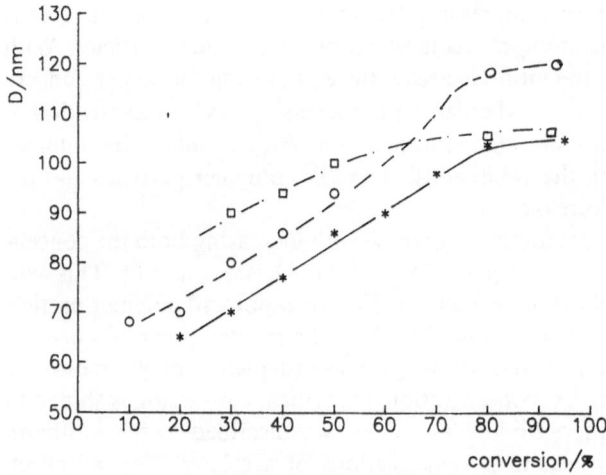

Fig. 11. Variation of the particle diameter in the emulsion copolymerization of vinyl chloride and butyl acrylate with conversion and with the mole fraction of vinyl chloride, x_{VC} [158]. x_{VC}: (□) 0.34, (∗) 0.67, (○) 0.89. Other conditions are given in the legend to Fig. 9

where R_{pT} is the overall rate of polymerization (the sum of both specific rates, $R_{ps, BA}$ and $R_{ps, VC}$).

The decrease of the exponent of [N] with increasing VC concentration was ascribed to decreasing sensitivity of R_p to emulsifier type and stability of polymer particles. The rate of polymerization of BA favors the micellar mechanism, whereas that of VC favors the homogeneous nucleation and particle coagulation.

Table 10. Variation of kinetic, colloidal and molecular weight parameters in the emulsion copolymerization of vinyl chloride with the mole fraction of vinyl chloride, x_{VC} [158]

x_{VC}	$R_p \times 10^4$ [1] $(mol\,dm^{-3}\,s^{-1})$	D [2] (nm)	$N \times 10^{-17}$ [3] $/dm^3$	$[\eta]$ [4] $(cm^3\,g^{-1})$
0.286	–	101	11.4	3.9
0.336	18.4	110	8.3	3.5
0.672	20.2	120	6.0	2.7
0.891	12.7	125	4.9	1.2
0.949	11.1	140	3.4	1.1
1.0	5.1	700	0.03	1.0

[1] The overall (maximum) rate of polymerization, [2] Particle diameter, [3] Number of polymer particles, [4] Intrinsic viscosity (measured in acetone at 30 °C).

As expected, the molecular weights of copolymers increase with increasing numbers of particles and rate of polymerization (with increasing BA concentration). The molecular weights are suggested as being regulated by the chain-transfer to VC monomers, desorption of radicals and formation of trapped radicals (the phase separation or incompatibility of growing radicals with the polymer phase).

At the beginning of polymerization, the copolymer is enriched in BA units. As a reaction advances, the comonomer feed is enriched in VC. At higher conversions, incorporation of VC units to copolymer abruptly increases. The reactivity ratios $r_{VC} = 0.1$ and $r_{BA} = 5.5$ were estimated for low conversions. They show very low ability of both monomers to copolymerize.

At low and medium conversions, the polymer phase is rich in polyBA, whereas at high conversions, PVC is formed. The particles formed consist of a polyBA-rich core and a polyVC-rich shell.

At low and medium conversions at saturation pressure, a strong shift in the copolymer composition is observed. At subsaturation pressure, the concentration of VC units in copolymer increases monotonously with increasing conversion.

Svoboda et al. [160] investigated the emulsion copolymerization and terpolymerization of VC with vinyl accetate, butyl acrylate and/or ethyl acrylate. The polymerizations preceeded under batch and continuous conditions and were initiated by peroxodisulfates. Anionic emulsifiers (sodium dodecyl sulfate, sodium dodecylbenzene sulfonate, . .) and blends of anionic and non-ionic emulsifiers (mostly polyoxyethylene type) were used. Copolymer latexes prepared with emulsifier blends were much more stable than those with an anionic emulsifier. As expected, the copolymers prepared by continuous polymerization gave copolymers with homogeneous composition. In the batch copolymerizations, the shift in the copolymer composition with conversion was observed and particles with broader size distribution were prepared. For example, the batch VC/ethyl acrylate polymer latexes gave particles with a diameter from 180 nm to 320 nm.

The reactivity ratios for the ternary system VC(1)/VAc(2)/BA(3) in the batch conditions estimated from low conversion studies are as follows:

$$r_{12} = 1.68, \, r_{13} = 0.07, \, r_{23} = 0.06, \, r_{21} = 0.23,$$

$$r_{31} = 4.4 \quad \text{and} \quad r_{32} = 3.07$$

They show very low copolymerizability of VC with BA and high reactivity of VC with VAc. VAc copolymerizes slightly with BA but strongly with VC. Combination of all monomers gave copolymers with homogeneous structure and rich in VC units from the beginning of polymerization.

Modification of PVC latexes has been extensively investigated to improve dynamic behavior of plastisols in which the PVC phase is composed of PVC particles [161]. Incorporation of monomers such as vinyl acetate (VAc) and vinylidene chloride (VD) to the PVC particle phase is one of the promising techniques for modifying the properties of PVC particles. Oikawa [162] recently proposed that the dynamic behavior of PVC plastisol was remarkably improved when the original PVC particles were covered with a poly(methyl methacrylate) shell.

Synthesis of various composite particles was performed by Omi et al. [163] employing PVC-base latexes (PVC, PVC-VAc and PVc-VD), as seed polymer particles. They investigated the effects of unsaturated monomers (MMA, St and BMA) of a different water-solubility and compatibility of polymers (PMMA, PSt, . .) with PVC seed polymer or copolymers. MMA is the monomer with a high water-solubility and St or BMA with a low water-solubility. BMA is employed as a polar and styrene as an unpolar monomer.

Typical S-shaped conversion time curves of the emulsion polymerization of MMA, MMA/St or St in the presence of PVC-VAc seed latex (with weight ratio monomer/polymer (M/P) = 2) were obtained. The rate of polymerization of MMA (in interval 2) (R_p = 13.8 g polymer/min) was much higher than that of St (R_p = 2.1 g polymer/min). The comonomer system (MMA/St, 20/30, wt. ratio) revealed an intermediate feature between the two systems (R_p = 5.5 g polymer/min). The same initial rate (R_p = 0.03 g polymer/min) was observed in St and MMA/St runs. In the MMA system, the interval 1 (the low conversion range) was very short (~5 min). In contrast, the interval 1 is very long in polymerizations of St (~ 350 min) and MMA/St (~ 250 min). The secondary nucleation of particles was proportional to the SDS concentration. The secondary nucleation was suppressed when the small seed particles with a smaller monomer/polymer ratio were used. In addition, the choice of a less water-soluble monomer is a key factor in performing complete seed polymerization without nucleation of secondary particles.

The molecular weights of PVC-VAc/PMMA (M_n = 9.8 × 10^4, M_w/M_n = 11.3) were larger than that of PVC-VAc/PS (M_n = 7.6 × 10^4, M_w/M_n = 7.6). The enhanced gel effect and probable formation of the graft linkage between core and shell were supposed to be responsible for these results. The molecular weight distribution of the shell polymer overlapped with that of the core seed polymer (M_n = 6.2 × 10^4, M_w/M_n = 2.5).

Neither the graft linkage nor the miscibility between the core and shell was observed when St and MMA-St were chosen as the shell monomer, while substantial evidence was found between a PVC-VAc core and PMMA.

SEM photographs of composite particles have shown that PVC-VAc/PMMA particles are spherical whereas PVC-VAc/PSt ones are aspherical. This was attributed to the different compatibility of a shell polymer with a seed PVC polymer or copolymer. PMMA is compatible with PVC-VAC and, therefore, the spherical particles or the IPN structure were developed. The incompatible St or PSt led to the formation of aspherical particles (i.e., raspberry-like particles).

4 Conclusions

The foregoing results indicate that the emulsion polymerization of VC is influenced in a complex way by the high water solubility of the monomer and the limited miscibility of polymer and monomer and deviates strongly from the conventional emulsion polymerization. This is discussed in terms of the water-phase polymerization of VC, precipitation polymerization of VC, polymerization in the emulsifier layer zone (a water/particle interphase), desorption of radicals from particles, chain transfer to monomer and polymer, formation of unsaturated and branching structures of PVC, formation of occluded radicals and diffusion of more volatile radical species to the vapor phase.

The monomer-saturated emulsifier layer (or the shell of polymer particles), the high chain transfer constant to VC and the polymerization in the interphase should promote desorption of radicals from particles to the aqueous phase. Desorbed radicals may take part in initiation and termination or re-enter the particles. In both phases, desorbed or re-entered radicals are more efficient in termination. In addition, the mobile radicals irreversibly diffuse to the particle core in which they are immobilized by the polymer phase and/or occluded by propagation.

In the water-phase polymerization (at low conversion) of VC, the reaction order with respect to the initiator concentration is around 0.5. This is interpreted in terms of the high water solubility of VC and bimolecular termination of growing radicals. The independence of the rate of polymerization from the particle concentration is taken as an evidence of water – phase polymerization. In systems with high initiator concentration, the primary radical concentration decreased the reaction order below 0.5 or 0.4. In most cases, however, the reaction order with respect to [initiator] (at low or medium initiator concentration) was found to be above 0.5. This behavior was ascribed to the restricted bimolecular termination of growing radicals and increased contribution of the monomolecular termination events. The long lifetime and large fraction of transferred or occluded radicals favor the monomolecular termination mechanism. Additionally, the degradative chain transfer to VC or unsaturated PVC decreases the radical activity of system which may also be ascribed to the first radical loss.

The reaction order higher than 0.5 or lower than 1.0 is the result of three contributions: a) biradical termination of growing radicals for which $R_p \sim [I]^{0.5}$, b) the primary radical termination for which $R_p \sim [I]^{<0.5}$ and c) the first order radical process for which $R_p \sim [I]^{1.0}$. The reaction order with respect to [I] may be influenced by variation in the initiator efficiency, phase separation, and existence of polymer-rich and monomer-rich domains.

At subsaturation conditions, the rate is proportional to the 0.4th power of the initiator concentration which agrees well with the micellar predictions. This behavior indicates that the monomolecular termination is more suppressed while the bimolecular termination is more favoured. Thus, the reaction order 0.4 should result from a strong contribution of primary radical termination and/or desorption of radicals.

Under the present reaction conditions, the initiator efficiency, f, is expected to be somewhat suppressed. In the rigid micro-surroundings, the polymer wrapping layer becomes very dense, so that diffusion of mobile radicals is restricted. Thus, some of mobile radicals fail to initiate polymerization of VC. In addition, the chain transfer to monomer and polymer generating less stable radicals (due to shielding of radical ends, the phase separation and the resonance stabilization), the high mobility and penetration activity of primary radicals (derived from initiator or generated by chain transfer to monomer) may lead to deactivation of growing radicals and a decrease in f.

The decrease of the rate of polymerization with conversion is ascribed to the consumption of monomer, the decrease of the initiator efficiency and the formation of polymer-rich phase. The unsaturated polymer phase catches the mobile radicals and deactivates them by occlusion and/or resonance stabilization. The fact that the rate of polymerization is inversely proportional to the polymer fraction in the system favors the hypothesis concerning the deactivation role of the unsaturated PVC.

The chain transfer to VC monomer is found to have a strong influence on the polymerization process through the formation of less reactive radicals and their desorption from polymer particles to the aqueous phase. The probability of such events is enhanced because the number of entangled or occluded radicals is high. The high efficiency of chain-transfer to monomer results from the low molecular weights of polymers formed in the emulsion systems and the value of polydispersity index close to 2.0.

The low radical concentration (active centers) in the PVC particles is mostly ascribed to the intensive chain transfer to monomer and polymer and desorption of monomeric VC radicals from polymer particles to the aqueous phase. For these reasons, the average number of radicals per particle (n) is much lower than 0.5 and decreases with increasing particle concentration. The value of n varies in the range from 0.001 to 0.1.

The presence of radical byproducts such as Cl radicals and the formation of unsaturated and defect structures in PVC point to a complex polymerization mechanism. The partitioning of the volatile radicals (monomeric, Cl, . .) between vapor, water and polymer particles may vary the radical concentration in particles and the aqueous phase. Exit of radicals from the particles and the aqueous phase to the vapor phase probably negatively influences the growth events in the system. The low concentration of monomer in the vapor phase favors more termination. The decrease of the chain transfer to monomer with decreasing reactor pressure (at subsaturation pressures) favors the hypothesis that the volatile radicals exit to the vapor phase. The chain transfer is constant at saturation pressure and decreases with increasing pressure at subsaturation conditions.

The chain transfer to VC and/or PVC is responsible for the short and long chain branching and unsaturation structures which is more pronounced at subsaturation pressures.

In the emulsion polymerization and especially with VC a different relation between transfer and termination reactions in the aqueous phase and polymer particle particles is expected. Termination exceeds transfer events in the aqueous phase and transfer events termination in particles. The biomolecular termination is expected to proceed in the aqueous phase while the monomolecular radical loss process in the PVC particles.

The molecular weights of PVC under monomer saturation conditions are nearly independent of conversion whereas under subsaturation pressures increases with conversion up to the total consumption of monomer. Under monomer saturated conditions the concentration of monomer at reaction loci is constant up to ca. 80% conversion. Here, the growth and termination events proceed under stable conditions (the constant monomer concentration and the chain transfer effect) at which the molecular weights are independent of conversion. As the polymerization proceeds at subsaturation pressures, the concentration of monomer decreases and the weight ratio monomer/polymer decreases and so does the growth of polymer chain.

In small particles, polymers of lower molecular weights are formed, whereas in large particles, the polymers formed are of higher molecular weights. This behavior results from radical desorption which is more operative in small particles.

The number of particles is mostly independent of initiator concentration. This indicates that primary radicals do not play a primary role. Thus, the primary particle formation is regulated not ony by f, k_d and initiator concentration but also by other kinetic events such as the particle association, the occlusion of radicals and the chain transfer events.

The rate of polymerization is only slightly affected by the particle concentration. The reaction order reaches a maximum value of 0.2 [164]. This behavior favors the assumption that polymerization of VC takes place in the emulsifier layer zone in which the polymerization seems to be governed by the kinetics of the "homogeneous" polymerization. Desorption of radicals from this layer to the aqueous phase is expected to be operative.

Vinyl chloride monomer is distributed between emulsifier micelles, polymer particles, monomer droplets, vapor and water. Monomer droplets disappear or a pressure drop appears at ca. 80% conversion which clearly indicates that PVC particles are only slightly saturated with monomer. Transition between the monomer-saturated and starved conditions is given by the critical conversion at which the monomer droplets disappear or by the decrease in the weight ratio monomer/polymer in the latex particles. Thus, the extent of saturation of PVC particles by VC monomer is given by the conversion at which monomer droplets disappear or where the concentration is inversely proportional to the "lifetime" of monomer droplets in the system. The presence of VC monomer droplets at high conversion saturates the aqueous phase and favors the homonucleation mechanism. In addition, the precipitation of oligomer radicals from the VC phase favors the nucleation of monomer droplets. The reaction order with respect to [VC] at subsaturated pressures is close to 1.0. The results obtained in these micellar systems are discussed in terms of polymerization under mono-mer-starved conditions (the particle core is only slightly swollen by monomer) and the continuous-phase polymerization. Thus, with increasing monomer concentration the VC concentration at the reaction loci increases and so does the rate of polymerization.

Under saturation conditions, the rate of polymerization of VC is found to be independent of or slightly dependent on the emulsifier concentration. The number of particles as well as the molecular weights of PVC increase with increasing emulsifier concentration. These results indicate that kinetic events typical for the classical emulsion polymerization in particles are not dominant. The polymerization in small particles was more sensitive to emulsifier than that in large particles. This results from the dominant role of the emulsifier layer zone which is more "active" in small particles. Here the particle size distribution increases with increasing [E]. This indicates that the secondary nucleation of particles takes place and/or the nucleation period is proportional to [emulsi-fier].

The reaction order with respect to [E] is known to decrease with increasing water-solubility of monomer. Therefore, the high-water solubility of VC mono-mer depresses the dependence of the rate on [E].

The layers of emulsifier on the particle surface are suggested to act as reaction loci in which the polymerization of VC proceeds homogeneously. The adsorbed layer zone of emulsifier saturated with monomer and oligomer rad-icals is responsible for the growth of polymer chains and particles.

Experimental results show that the solubility of VC increases with increasing emulsifier concentration. Thus, the fraction of VCM in micelles increases and so does the rate of polymerization. Here the initial rate of polymerization increases most intensively with [E] at very low conversion range.

At subsaturation conditions, the rate is proportional to the 0.4th power of the emulsifier concentration which deviates slightly from the micellar predic-tions. This behavior is attributed to the monomer-solubilization role of the emulsifier layer zone.

In the emulsion polymerization of VC, the formation of polymer particles by the coagulative nucleation mechanism is expected to be operative. As polymerization proceeds, the formation of primary particles takes place. In this process, the primary particles undergo coalescence to form large polymer particles and/or they associate with large particles.

More information about the effect of VC monomer on the polymerization process can be obtained from the copolymerization of VC. Conversion curves for the systems without or with a low VC concentration showed a sigmoidal shape typical of the classical emulsion polymerization. The addition of higher amounts of VC leads, however, to deviation from the emulsion copolymerization kinetics, i.e., the conversion curves take a shape similar to that for the solution or precipitation polymerization. These results show that growth events in the VC emulsion systems are somewhat restricted and termination becomes more efficient.

The pressure increase at ca. 80% conversion was ascribed to the exotherm (a gel effect) which is very intensive and often observed in polymerization of acrylate or methacrylates at low conversion. The addition of VC monomer to acrylate or methacrylate systems is accompanied by a shift of a weak gel effect to higher conversion.

The reactivity ratios clearly show a low reactivity of VC in copolymerization with acrylates and also a strong shift in the copolymer composition with conversion is observed. Under such conditions, the particles formed consist of the polyacrylate-rich core and the polyVC-rich shell.

The size of polymer particles increases with increasing amounts of VC in the comonomer feed. The high association activity of VC polymerization initiates the growth of particles by association and therefore large particles are formed. The high chain transfer activity of VC as well as the precipitation of PVC growing radicals decrease the molecular weights of copolymers.

The low activation energy typical for the classical emulsion polymerization was also estimated in the emulsion polymerization of VC.

From the above discussion it is evident that the following factors influence the free-radical emulsion polymerization of vinyl chloride

- high water solubility of VC monomer,
- mutual insolubility of polymer and monomer,
- high chain transfer to monomer and polymer,
- polymerization in the aqueous phase and the emulsifier layer zone,
- formation of unsaturated bonds in PVC backbone and of branched structures,
- desorption of monomeric radicals from polymer particles into the aqueous phase,
- exit of monomeric radicals and other volatile low molecular weight radicals into the vapor phase.

Acknowledgement. The author is indebted to the Alexander von Humboldt Stiftung for financial support.

5 References

1. British patent 1 071344 to Borden Chemical Co.
2. German patent 2 822357 to Wacker Chemie GmbH
3. Japanese patent 59 219380 to Nippon Mectron Co., Ltd.
4. Capek I, Akashi M (1993) J Macromol Sci, Rev Macromol Chem Phys, C33: 369
5. Billmeyr FW (1971) Textbook of polymer science. 2nd edn, Wiley, New York
6. Smith WV, Ewart RH (1948) J Chem Phys 16: 592
7. Harkins WD (1945) J Chem Phys 13: 381
8. Bamford CH, Jenkins AD (1953) Proc R Soc London, Ser A, 216: 515; (1953) Proc R Soc, London, Ser A 220: 228
9. Bamford CH, Jenkins AD, Johnson AD (1957) Proc R Soc London, Ser A, 251: 364
10. Bamford CH, Jenkins AD (1954) J Polym Sci 14: 511; (1956) J. Polym Sci 20: 405
11. Thomas WM (1961) Adv Polym Sci 2: 401
12. Lewis OG, King RM (1969) Adv Chem Ser 91: 25
13. Capek I, Tuan LQ (1986) Makromol Chem 187: 2063
14. Gardon JL (1968) J Polym Sci Part A-1; 6; 623, 6: 643, 6: 665, 6: 687, 6: 2853, 6: 2859
15. Ugelstad J, Hansen FK (1976) Rubber Chem Tech 43: 536
16. Feeney PJ, Napper DH, Gilbert RG (1984) Macromolecules 17: 2520
17. Nomura M, Harada M, Nakagawara K, Guchi WE, Nagata S, J Chem Jpn 4: 160
18. Hansen FK, Ugelstad FJ (1982) In: Piirma I (ed) Emulsion polymerization. Academic, New York
19. Eliseeva VI (1981) Acta Polymerica 32: 355
20. Medvedev SS (1957) Proc Int Symp Macromol Chem, Praha, Pergamon, London (Czechoslovak Academy of Sciences and Czechoslovak Chemical Society), p 174
21. Priest WJ (1952) J Phys Chem 56: 1077
22. Fich RM, Tsai CH (1971) In: Fitch RM (ed) Polymer colloids. Plenum, New York
23. Roe CP (1968) Ind Eng Chem 60: 20
24. Sujov AV, Grizkova IA, Medvedev SS (1972) Kolloid Zh 34: 203
25. Fitch RM, Watson RC (1979) J Colloid Interface Sci 68: 14
26. Lichti G, Gilbert RG, Napper DH (1983) J Polym Sci, Polym Chem Ed 21: 269
27. Dainton FS, Eaton RS (1959) J Polym Sci 39: 313
28. Morton M, Kaiserman S, Altier M (1954) J Colloid Sci 9: 300
29. Hawkett BS, Napper DH, Gilbert RG (1980) J Chem Soc Farday Trans 1, 76: 1323
30. Balke ST, Hamielec AE (1973), J Appl Polym Sci 17: 905
31. Ayrey G, Haynes AC (1973) Eur Polym J 9: 1029
32. Russell GT, Napper DH, Gilbert RG (1988) Macromolecules 21: 2141
33. Noyes RM (1954) J Chem Phys 22: 1349
34. Noyes RM (1955) J Am Chem Soc 77: 2042
35. Stickler M, Panke D, Hamielec AE (1984) J Polym Sci Polym Chem Ed 22: 2243
36. Ballard MJ, Gilbert RG, Napper DH, Pomery PJ, O'Sullivan PW, O'Donnell JH (1986) Macromolecules 19: 1303
37. Capek I, Bartoň J, Tuan LQ, Svoboda V, Novotný V (1987) Makromol Chem 188: 1723
38. Capek I, Bartoň J, Orolínová E (1984) Chem Zvesti 38: 803
39. Wunderlich W, Stickler M (1985) Polym Sci Technol 31: 505
40. Hawkett BS, Napper DH, Gilbert RG (1981) J Polym Sci, Polym Chem Ed 19: 3173
41. Shen JC, Wang GB, Yang ML, Zheng YG (1992) Polymer International 28: 75
42. Capek I, Riza M, Akashi M (1992) Polym J 24: 959
43. Zhu S, Tian Y, Hamielec AE, Eaton DR (1990) Macromolecules 23: 1144
44. Schulz GV (1956) Z Phys Chem 8: 290
45. Smoluchowski M (1918) Z Phys Chem 92: 129
46. Flory PJ (1953) Principles of polymer chemistry. Cornell University Press Ithaca, N.Y.
47. North AM, Reed GA (1961) Trans Fraday Soc 57: 859
48. Buback M, Gilbert RG, Russell GT, Hill DJT, Moad G, O'Driscoll KF, Shen J, Winnik MA (1992) J Polym Sci Polym Chem Ed 30: 851
49. Buback M, Garcio-Rubio LH, Gilbert RG, Napper DH, Guillot J, Hamielec AE, Hill D, O'Driscoll KF, Olaj OF, Shen J, Solomon D, Moad B, Stickler M, Tirrel M, Winnik MA (1988) J Polym Sci Polym Lett Ed 26: 293

50. Tirrell M (1984) Rubber Chem Technol 57: 523
51. Okieimen EF (1983) J Macromol Sci Chem A20: 711
52. Bamford CH, Jenkins AD, Johnstone R (1989) Trans Faraday Soc 55: 1451
53. Deb PC, Meyerhoff G (1974) Eur Polym J 10: 709
54. Deb PC (1982) Eur Polym J 18: 769
55. Deb PC (1975), Eur Polym J 11, 31 (1975)
56. Bamford CH, Barb WG, Jenkins AD, Onyon PF (1958) Kinetics of vinyl polymerization by radical mechanisms. Butterworths, London, p 111
57. Nomura M, Yamamoto K, Horie I, Fujita K, Harada M (1982), J Appl Polym Sci 27, 2483 (1982)
58. Lichti G, Sangster DF, Whang BCY, Napper DH, Napper RG (1982) J Chem Soc Faraday Trans 1, 78: 2129
59. Burnett JD, Melville HW (1950) Trans Faraday Soc 46: 976
60. Thomas WM, Pellon JJ (1954) J Polym Sci 13: 329
61. Nomura M (1982) In: Piirma I (ed) Emulsion polymerization, Academic, New York
62. North AM (1965) Makromol Chem 83: 15
63. Hallensleben ML (1977) Eur Polym J 13: 437
64. Scott GE, Senogles E (1973), J Macromol Sci Rev Macromol Chem C9: 49
65. Park GS, Smith DG (1969) Trans Faraday Soc 65: 1854
66. Rudin A (1982) The elements of polymer science and engineering, Academic, London
67. Fitch RM, Tsai CH (1970) J Polym Sci Polym Lett Ed 8: 703
68. Chatterjee SP, Banerjee M, Konar RS (1978) J Polym Sci Polym Chem Ed 16: 1517
69. Hansen FK, Ugelstad J (1979) J Polym Sci Polym Chem Ed 17: 3069
70. Gerrens H, Fink W, Kohnlein E (1967) J Polym Sci C 16: 2781
71. Gerrens H, Fink W, Köhlein E (1965) IUPAC International Symposium on Macromolecular Chemistry, Prague
72. Bengough WI (1960) Proc Roy Soc A, 260: 205
73. Meehan EJ, Kolthoff IM, Tamberk N, Segal CL (1957) J Polym Sci 24: 215
74. Barlett PD, Nozaki K (1948) J Polym Sci 3: 216
75. Burnett GM, Wright WW (1954) Proc Roy Soc (London), A221: 37, 221: 41
76. Peggion E, Testa F, Talamini G (1964) Makromol Chem 71: 173
77. Schindler A, Breitenbach JW (1955), Ricerca Sci 25: 34
78. Ugelstadt J, Mörk PC, Rangnes P, Dahl P (1969), J Polym Sci Part C No 27: 49
79. Friis N, Hamielec AE (1975) J Appl Polym Sci 19: 97
80. Nilsson H, Silvergren C, Törnell B (1983) Angew Makromol Chem 112: 125
81. Sörvik EM, Hjertberg T (1978) J Macromol Sci Chem A11: 1349
82. Hjertberg T, Sörvik EM (1978) J Polym Sci Polym Chem Ed 16: 645
83. Giskehaug K (1966) In: S.I. Monograph No. 20. The chemistry polymerization process, p 235
84. Burnett GM, Wright WW (1954) Proc Roy Soc (London), Ser A, 221: 28
85. Mork PC, Ugelstad J (1969) Makromolekulare Chem 128: 83
86. Karakas G, Orbey N (1989) British Polym J 21: 399
87. Robb ID (1969) J Polym Sci Part A-1, Polym Chem 7. 417
88. Goebel KH, Schneider JH, Jaeger W, Reinisch G (1982) Acta Polymerica 33: 49
89. Min KW, Ray WH (1974) J Macromol Sci, Part C11: 177
90. Corso C (1961) Mat Plastiche 8: 781
91. Boieshan V (1990) Acta Polymerica 41: 303
92. Neelson J, Jaeger W, Reinisch G, Schneider HJ (1987) Acta Polymerica 38: 1
93. Kolthoff IM, Miller IK (1951) J Am Chem Soc 73: 3055
94. Georgescu CS, Butucea V, Sarbu A, Ionescu A, Deaconescu I, Hagiopol C (1989) Macromol Chem Macromol Symp 29: 329
95. Georgescu CS, Sarbu A, Butucea V (1991) Makromol Chem Makromol Symp 47: 337
96. Boieshan V (1990) Acta Polymerica 41: 298
97. Okieimen EF (1983) J Macromol Sci Chem A20: 711
98. Brit. Pat 984, 487 (1963) (Sicedison S.p.A.)
99. Bovey FA, Kolthof IM, Medalia AI, Meehan AJ (1955) Emulsion polymerization. Interscence, New York, Chap. V.
100. Greth GG, Wilson JE (1961) J Appl Polym Sci 5: 135
101. Testa F, Vianello G (1969) J Polym Sci Part C, No 27: 69
102. Neelson J, Hecht P, Jaeger W, Reinisch G (1987) Acta Polymerica 38: 555

103. Tauer K, Neelsen J, Hellmich C (1985) Acta Polymerica 36: 665
104. Chedron H (1960) Kunststoffe 50: 568
105. Hopff VH, Fakla I (1965) Makromol Chem 88: 54
106. Shepetilnikov BV, Eliseeva VI, Zuikov VI (1978) Vysokomol Soed A20: 2097
107. Goebel KH, Schneider HJ, Jaeger W, Reinisch G (1981) Acta Polymerica 32: 117
108. Ugelstadt J, Pork PC, Dahl P, Rangues P (1965) J Polym Sci Part C, Polym Symp 27: 49
109. Gerrens H, Fink N, Kohlen E (1965) Int. Symposium of Macromolec Chem, Preprint 331, Prague
110. Barriac J, Knorr R, Stabel EB, Stannet V (1975) Polymer Preprints ACS, Div Polym Chem 16: 205
111. Fischer N (1974) Polymer Engn Sci 14: 322
112. Müller R (1974) Chemie Ing Techn 46: 655
113. Hansen FK, Ugelstad J (1977) J Polym Sci Polym Chem Ed 17: 3033
114. Hjertberg T, Sörvik EM (1986) J Polym Sci Part A, Polym Chem 24: 1313
115. Butucea V, Sarbu A, Georgescu CA, Ionescu A, Deaconescu I (1989) Rev Roum Chim 34: 1155
116. Tauer K, Petruschke M (1986) Acta Polymerica 37: 313
117. Heiskanen T (1965) Acta Polytechn Scand Chem incl Metallurgy Ser No. 165
118. Boieshan V, Levitschi D (1991) Acta Polymerica 42: 551
119. Maron S, Elder M (1954) J Colloid Sci 9: 89
120. Ugelstad J (1980) Eur Pat Appl 3: 905
121. Dewald RC, Hart LH, Carroll WF (1984) J Polym Sci Polym Chem Ed 22: 2923
122. Dewald RC, Hart LH, Carroll WF (1984) J Polym Sci Polym Chem Ed 22: 2931
123. Pavljuchenko VH, Ivanchev SS (1983) Acta Polymerica 34: 521
124. Nilsson H, Silvegren C, Törnell B (1978), Eur Polymer J 11: 737
125. Jorgedal A (1967) Paper presented at "Det. 10. Landsmote for Kjemi", Hanko, Norway
126. Schneider HJ, Jaeger W, Dietzel W (1983) Acta Polymerica 34: 574
127. Butucea V, Sarbu A, Georgescu C, Ionescu A, Deaconescu I (1985) Acta Polymerica 36, 389 (1985)
128. Liegeois JM (1977) J Macromol Sci Chem A11: 1379
129. Liegeois JM (1971) J Polym Sci Part C, 33: 147
130. Tulig TJ, Tirell M (1981) Macromolecules 14: 1501
131. Berens AR (1975) Angew Makromol Chem 47: 97
132. Hayden P, Melville H (1960) J Polym Sci 43: 201
133. Hjertberg T (1988) J Appl Polym Sci 36: 129
134. Berrens AR (1975) Angew Makrom Chem 47: 97
135. Hjertberg T, Sorvik EM (1985) In: Klemchuk PP (ed) Polymer stabilization and degradation. American Chemical Society, Washington, DC, p 259
136. Xie TY, Hamielec AE, Wood PE, Woods DR, Chiantore O (1991), Polymer 32: 1696
137. Polymer handbook (1989) 3rd Ed., Brandrup J, Immergut EH (eds) Wiley, New York
138. Neelsen J, Hecht P, Jaeger W, Reinisch G (1987) Acta Polymerica 38: 418
139. Capek I, Bartoň J (1985) Makromol Chem 186: 1297
140. Neelsen J, Jaeger W, Tauer K, Hecht P (1985) Acta Polymerica 36: 694
141. Tauer K, Paulke BR, Müller I, Jaeger W, Reinisch G (1982) Acta Polymerica 33: 287
142. Burnet GM, Wright WW (1954) Proc Roy Soc (London) Ser A, 60: 121
143. Kolthoff IM, Miller IK (1951) J Am Chem Soc 73: 3055
144. Morris CEM, Parts AG (1968) Makromol Chem 119: 212
145. Neelsen J, Jaeger W, Reinisch G, Schneider HJ (1987) Acta Polymerica 38: 1
145. Barton J, Capek I (1993) in: Kemp TJ, Kennedy JF (eds) Radical polymerization in disperse systems, Horwood location
146. Odian G (1981) Principles of polymerization, Wiley, New York, p 227
147. Russo S, Stannet V (1971) Makromol Chem 143: 47
148. Hamielec AE, Gomez – Vaillard R, Marten FL (1982) J Macromol Sci Chem A17: 1005
149. Hjertberg T, Sörvik EM (1982) J Macromol Sci Chem A17: 983
150. Starnes WH, Jr., Schilling FC, Plitz IM, Cais RE, Freed DJ, Hartless RL, Bovey FA (1983) Macromolecules 16: 790
151. Hjertberg T, Sörvik EM, Wendel A (1983) Makromol Chem, Rapid Commun 4: 175
152. Hjertberg T, Sörvik EM (1983) Polymer 24: 673
153. Hjertberg T, Sörvik EM (1983) Polymer 24: 685
154. Hildebrand P, Ahrens W, Brandstetter F, Simak P (1982) J Macromol Sci Chem A17: 1093

155. J. Ugelstad J, Mork PC, Aasen JO (1967) J Polym Sci 5: 2281
156. Nomura M, Harada M (1981) J Appl Polym Sci 26: 17
157. Tauer K (1987) B-Thesis, Academy of sciences of the G.D.R.
158. Capek I, Mrazek Z (1992) Makromol. Chem 193: 1165
159. Capek I, Mrázek Z, Prádová O, Svoboda J (1990) Int Symp on Polyvinylchloride and Polyvinylchloride copolymers, XXV, DNT, Bojnice, CSFR, p 3
160. Svoboda J, Mazanec J, Trgina E (1990) Int Symp on Polyvinylchloride and Polyvinylchloride copolymers, XXV, DNT, Bojnice, CSFR, p 50
161. Hoffman DJ, Saffron PM (1981) ACS Symposium Ser. 165: 209
162. Oikawa S (1986) Japan Patent (Kokai Tokkyo Koko) 185518, 207 418
163. Omi S, Shiiyama E, Aita K, Matsumoto M, Iso M, Sakaya M, Nakano A, Imazawa Y (1990) J Appl Polym Sci 41: 631
164. Ugelstadt J (1977) J Macromol Sci Chem A11: 1281

Editor: Prof. K. Dušek
Received: December 1993

Thermodynamics of Polymer Solutions under Flow:
Phase Separation and Polymer Degradation

D. Jou, J. Casas-Vázquez and M. Criado-Sancho
Departament de Física (Física Estadística), Universitat Autònoma
de Barcelona, 08193 Bellaterra, Catalonia, Spain

The influence of flow on the free energy of polymer solutions is examined from several points of view both on macroscopic and microscopic bases. Application of non-equilibrium chemical potential to the phenomena of flow-induced phase separation and thermodynamically induced polymer degradation is reviewed. Polymer degradation under elongational flow was extensively covered in a recent review in this series. The thermodynamic theory is compared with the dynamical approaches used in the analysis of stability of solutions and it is seen under which conditions the criteria defining the spinodal line under shear (i.e. the limit of the stability region) are the same in both approaches. The thermodynamic analysis may be useful due to its greater simplicity though, in contrast, the details of the phase segregation or homogenisation and the analysis of the dynamical aspects (viscosity, light scattering) are beyond the reach of a strictly thermodynamic method. A short summary of the phenomenology of flow-induced changes in the phase diagram of polymer solutions under flow is given. Perspectives and open problems are pointed out.

List of Symbols and Abbreviations

A	affinity
$\mathbf{A}$	matrix of kinetic constants
a_i	activity of the species i
a_α	the droplet radius
b	length of a segment of the polymer
$\mathbf{B}$	matrix including the hydrodynamic effects or matrix of kinetic constants
c	number of moles per unit volume
$\tilde{c}$	reduced concentration defined as $[\eta]c$
d	slope of the curve $\eta(\dot{\gamma})$ with a change of sign
D	curvilinear diffusion coefficient
D_T	effective diffusion coefficient in the shear direction
D_2	polymer self-diffusion coefficient
$\mathbf{E}$	deformation gradient tensor
f	specific Helmholtz free energy
F	force or thermodynamic force
g	Gibbs free energy per unit volume
$G^*(\omega)$	complex stress-strain coefficient
ΔG_M	Flory-Huggins mixing Gibbs free energy
ΔG_s	non-equilibrium contribution to the Gibbs free energy
H	elastic constant
$J^*(\omega)$	complex compliance
J	steady-state compliance
J_1, J	diffusion fluxes
J^s	entropy flux
k'	entropic elastic constant
k_{ij}	kinetic constant for the breaking of a macromolecule P_j to give P_i
K_{ij}	chemical equilibrium constant defined by the rate k_{ij}/κ_{ij}
l	length of the duct
m	ratio of the partial molar volume of the polymer to one of the solvent
M	molecular weight
M_n	number average for the molecular weight of the polymer
M_w	weight average for the molecular weight of the polymer
M_0	molecular weight of a monomer
n	number of molecules per unit volume (or number density)
n_i	number of moles of the species i
N	number of segments in a polymer
N_A	Avogadro's number
$\tilde{n}_1$	number of moles of the solvent
$\tilde{n}_2$	number of moles of the solute
$N_i^{(\dot{\gamma})}$	number of chains with i monomers under a shear

p	pressure
Δp	pressure difference
$\mathbf{P}^v$	viscous pressure tensor
P_{ij}^v	components of the viscous pressure tensor
P_i	chain with i monomers
q	parameter for defining what kind of averaged molecular weight is used
Q	flow rate
$\boldsymbol{Q}_i$	normal modes of the chain
r	distance to the axis of a duct
R	constant of gases
$\boldsymbol{R}$	the end-to-end vector or the radius of the duct
$\boldsymbol{R}_i$	the vector from bead i to bead $i + 1$
s	specific entropy
T	absolute temperature
T_c	critical temperature
Tr	trace of a tensor
u	specific internal energy
$\boldsymbol{u}$	the local deformation vector
v	specific volume
$\boldsymbol{v}$	velocity vector
$\mathbf{V}$	symmetric part of velocity gradient
x_i	molar fraction of the species i
$\mathbf{W}$	configuration tensor
z	coordination number
α	parameter of the most probable distribution or the friction coefficient
$\dot{\varepsilon}$	the extensional rate
ε_{ij}	component of the strain tensor
ζ	friction coefficient
Γ	gamma function
$\dot{\gamma}$	the shear rate
γ^0	amplitude of the oscillatory shear strain
$\dot{\gamma}^0$	amplitude of the oscillatory shear rate
γ'	infinitesimal displacement gradient in the polymer phase
$\dot{\gamma}_c$	critical shear rate
$\dot{\gamma}_w$	shear rate at the wall
$\gamma_{\alpha\beta}^{(0)}$	interfacial tension in the absence of flow
$\dot{\gamma}_{0.8}$	shear rate for which $\eta(\dot{\gamma})$ is equal to 80% of η_0
η	shear viscosity
η_s	viscosity of the pure solvent
η_0	viscosity for zero shear rate
$\eta^*(\omega)$	complex viscosity
$[\eta]$	intrinsic viscosity
Θ	theta temperature

ϑ	the effective flexibility parameter
κ_{ij}	kinetic constant for the recombination of chains
λ	correction to the chemical equilibrium reaction by effect of shear
λ_i	eigenvalues of the matrix $\mathbf{B}$
μ_i	chemical potential of species i
μ_i^0	reference chemical potential of species i
$\mu_i^{(s)}$	non-equilibrium contribution to the chemical potential of i
μ_{1s}	non-equilibrium contribution to the solvent chemical potential
μ_{p2}	chemical potential under constant P_{12}^v
$\mu_{\dot\gamma2}$	chemical potential under constant $\dot\gamma$
μ_{W2}	chemical potential under constant W
v_i	stoichiometric coefficient of i
π	osmotic pressure
π_ϕ	equilibrium contribution to the osmotic pressure
π_{el}	non-equilibrium contribution to the osmotic pressure
$\boldsymbol{\sigma}$	the stress tensor
τ	relaxation time
τ_d	disengagement time
ϕ	volume fraction of polymer
χ	Flory's interaction parameter
Ψ	the configurational distribution function
Ψ_0	the equilibrium distribution function
$\Psi_1(\dot\gamma)$	the first normal stress coefficient
$\Psi_2(\dot\gamma)$	the second normal stress coefficient
ω	angular frequency
Ω	solid angle
$\boldsymbol{\Omega}$	orthogonal matrix to diagonalize $\mathbf{B}$

1. Introduction

Much interest has been devoted to flow-induced changes in the phase diagram of polymer solutions [1–7]. These phenomena show a very interesting interplay between thermodynamics and hydrodynamics. It is obvious that the usual local-equilibrium thermodynamics must be modified in this case so that the equations of state manifest explicitly the influence of the flow. This, though not sufficient, is certainly a necessary ingredient for the understanding of these phenomena. In our opinion this is, from the point of view of non-equilibrium thermodynamics, one of the most compelling problems at the present moment, because it implies non-equilibrium equations of state which are not known a priori from equilibrium thermodynamics.

The practical importance of this problem is easily understood. Most industrial processing takes place under flow (pumping, extruding, injecting, molding, mixing . . .), in which the polymer undergoes different shear and elongational stresses, depending on the position. Thus a flow-induced change of phase could take place in some positions and not at others, affecting both the rheological and the structural properties of the flow. The fibres formed in these processes would be very sensitive to the extent of the phase transitions (reversible or irreversible) occurring. A wide review of the mechanisms of flow-induced crystallisation has been written by McHugh [8], whereas the importance of the flow-induced changes in the operation of polymer blends and alloys has been analysed by Utracki [9].

Other related fields of interest are the thermodynamically-induced polymer degradation under flow, which may be sensitive in viscous drag reduction or in flows or polymer solutions through packed porous beds [10] as in membrane permeation or flow of oil through soil and rocks. Also, in biological experiments, shear-induced precipitation and aggregation of proteins as horse serum albumin has been observed [10, 11]. Polymer degradation under elongational flow has been studied by Nguyen and Kausch [12], who have extensively and recently reviewed this topic [13] with a special emphasis on the degradation kinetics in the transient elongational regime, which reveals important differences with stagnant conditions in the kinetics of chain scission. The reader interested in this important subject is referred to [13].

A decisive step in the thermodynamic understanding of these phenomena is to formulate a free-energy depending explicitly on the characteristics of the flow. This important problem in non-equilibrium thermodynamics has not yet received all the attention it deserves. It must be noted that several authors have preferred to undertake another way, and to analyse the problem of phase separation or phase homogenisation under shear from a dynamical point of view, i.e. by writing dynamical equations for the behaviour of concentration and velocity fluctuations and analysing the stability of the corresponding set of equations. Of course, the dynamical procedure has a wider range of potentialities than the pure thermodynamic analysis: the latter one may be able to set the spinodal line limiting the region of stability, but it certainly cannot give a detailed view of the features of the processes of segregation of both phases or the changes in viscosity observed during the segregation. However, the existence of both methods is not contradictory; e.g. the dynamical method may describe the instability through the change of sign of an effective diffusion coefficient, but this change of sign is produced at the spinodal line, and this fact is related to the vanishing of the first derivative of the effective chemical potential with respect to the composition. Furthermore, the dynamical analysis cannot avoid the use of equations of state in the dynamical equations; therefore, to find and analyse equations of state in non-equilibrium conditions is of interest even in the case when one does not rely a priori on thermodynamic arguments to obtain the spinodal line. Thus, although there is a common ground for thermodynamical and dynamical analyses, both methods have their own advantages and disadvantages, so that it would be unwise to dismiss a priori any of them.

Here we provide a critical overview of the different efforts that have been made towards a thermodynamic analysis of the phase diagram of polymer solutions under shear. We also review the underlying microscopic basis for the use of a generalized non-equilibrium free energy under shear flow as well as the main criticisms and problems that it has encountered.

In Sec. 2 we start from a macroscopic point of view, or more precisely from the perspective of extended irreversible thermodynamics, rational thermodynamics and theories with internal variables. In Sect. 3, we analyse several microscopic theories and compare their results with the macroscopic ones. We discuss in Sect. 4 some thermodynamic subtleties underlying a proper use and definition of a chemical potential of a material subjected to a shear flow. Section 5 is devoted to several examples of the application of some methods devised to attack this problem, and Sec. 6 is a brief review of the main features of the experimental phenomenology. Section 7 presents some results concerning the thermodynamic degradation of polymers under laminar shear flow, and Sec. 8 compares the main lines of the thermodynamic method with those of the dynamical approaches. To conclude, Sec. 9 gives a view on some of the criticisms directed at the use of thermodynamic analysis of phase changes in solutions under flow and underlines several open problems which may be the basis of future research.

2. Non-Equilibrium Thermodynamics Under Flow

Local-equilibrium thermodynamics assumes that the equations of state keep out of equilibrium the same form as in equilibrium, but with a local meaning [14]. According to this point of view, there is not any proper thermodynamics under flow, since the flow does not change the equations of state. This approach is inadequate to deal with systems with internal degrees of freedom, so that on some occasions [15] one includes some internal variables in the set of thermo-dynamic variables. In this case the flow may influence the thermodynamic equations of state through its action on the internal variables such as, for instance, the polymeric configuration.

In the 1960s, a new non-equilibrium thermodynamic theory was proposed, the so-called rational thermodynamics [16–18]. This theory assumed that the entropy and the (absolute) temperature are primitive quantities. Instead of a local-equilibrium assumption, it was assumed that the entropy, or the free energy, could depend on the history of the strain or rate of strain, thus allowing for an explicit influence of the flow on the thermodynamic analysis. The theory developed a very powerful and elegant formalism to obtain thermodynamic restrictions on the memory functions.

The transport equations of local-equilibrium thermodynamics or of rational thermodynamics do usually lead to infinite speeds of propagation for the signals, but this unphysical fact has not received any remarkable attention. At the end of the sixties, a new approach now called extended irreversible thermodynamics (EIT) [19, 20] was proposed and it has been much developed during the 1980s [21–31]. This theory assumes that the entropy depends on, apart from the classical variables, the dissipative fluxes as the viscous presure tensor. This point of view may be closely related to the theories of internal variables in some systems as polymer solutions. However, in several aspects it is more general than these theories because it is also applicable to ideal monatomic gases where there are no proper internal variables. The direct influence of the viscous pressure tensor and other fluxes, such as heat flux or diffusion flux, on the thermodyn-amic potentials clearly opens a way towards thermodynamics under flow.

2.1 Basic Rheological Quantities

In the study of polymeric systems it is assumed that the viscous pressure tensor depends not only on the velocity gradient but on its own time rate by means of a relaxational term, which is usually written in terms of a frame-indifferent time derivative [16–18, 32, 33], as the corotational derivative or the upper convected derivative. Since the latter is much used in the rheological literature, we will utilize it here. Thus, the evolution equation for the viscous pressure tensor $\mathbf{P}^v$ corresponding to the upper convected Maxwell model [32, 33] has the form

$$d\mathbf{P}^v/dt - (\nabla v)^{\mathrm{T}} . \mathbf{P}^v - \mathbf{P}^v . (\nabla v) = -(1/\tau)\mathbf{P}^v - 2(\eta/\tau)\mathbf{V} , \qquad (1)$$

with $\mathbf{V}$ the symmetric part of the velocity gradient and superscript T indicating transposition. The fluid will from now on be considered as incompressible. This does not mean that $\mathrm{Tr}\,\mathbf{P}^v = 0$, but that $\nabla . v = 0$, so that the linear contribution to $\mathrm{Tr}\,\mathbf{P}^v$, which is proportional to $\nabla . v$, will be zero in this case but, as we will see, second-order nonlinear contributions may appear giving a nonvanishing trace of $\mathbf{P}^v$.

For further purposes it will be convenient to have explicit expressions for $\mathbf{P}^v$ in several steady flows. In a purely shear flow corresponding to $v = (v_x(y), 0, 0)$, the velocity gradient will be

$$\nabla v = \begin{pmatrix} 0 & 0 & 0 \\ \dot{\gamma} & 0 & 0 \\ 0 & 0 & 0 \end{pmatrix}, \tag{2}$$

where $\dot{\gamma}\,(=\partial v_x/\partial y)$ is the shear rate. Introduction of Eq. (2) into Eq. (1) yields, in the steady situation,

$$\mathbf{P}^v = \begin{pmatrix} -2\tau\eta\dot{\gamma}^2 & -\eta\dot{\gamma} & 0 \\ -\eta\dot{\gamma} & 0 & 0 \\ 0 & 0 & 0 \end{pmatrix}. \tag{3}$$

For the planar extensional flow, $v = (v_x(x), v_y(y), 0)$, one has

$$\nabla v = \begin{pmatrix} \dot{\varepsilon} & 0 & 0 \\ 0 & -\dot{\varepsilon} & 0 \\ 0 & 0 & 0 \end{pmatrix}, \tag{4}$$

with $\dot{\varepsilon}\,(=\partial v_x/\partial x = -\partial v_y/\partial y)$ the extensional rate. In the steady state, Eq. (1) yields

$$\mathbf{P}^v = \begin{pmatrix} -2\eta\dot{\varepsilon}(1 - 2\tau\dot{\varepsilon})^{-1} & 0 & 0 \\ 0 & 2\eta\dot{\varepsilon}(1 + 2\tau\dot{\varepsilon})^{-1} & 0 \\ 0 & 0 & 0 \end{pmatrix}. \tag{5}$$

The rheological quantities of interest are the shear viscosity and the first and second normal stress coefficients $\eta(\dot{\gamma})$, $\Psi_1(\dot{\gamma})$ and $\Psi_2(\dot{\gamma})$ respectively, which are defined as

$$\begin{aligned} P^v_{12} &= -\eta(\dot{\gamma})\dot{\gamma} \;, \\ P^v_{11} - P^v_{22} &= -\Psi_1(\dot{\gamma})\dot{\gamma}^2 \;, \\ P^v_{22} - P^v_{33} &= -\Psi_2(\dot{\gamma})\dot{\gamma}^2 \;, \end{aligned} \tag{6}$$

where P^v_{ij}, with $i, j = 1, 2, 3$, indicate components of $\mathbf{P}^v$. In particular, for the upper convected Maxwell model we thus have, according to Eq. (3)

$$P^v_{12} = -\eta\dot{\gamma} \;, \quad P^v_{11} - P^v_{22} = -2\tau\eta\dot{\gamma}^2 \;, \quad P^v_{22} - P^v_{33} = 0 \;, \tag{7}$$

so that the second normal stress is zero and the first normal stress coefficient Ψ_1 is

$$\Psi_1(\dot{\gamma}) = 2\tau\eta \;. \tag{8}$$

For comparison with other authors, it is of interest to write the quantity τ/η in terms of parameters which are well-known and much used in rheology [34]. The shear linear viscoelastic effects are usually summarized in terms of three equivalent complex functions: 1) a complex stress-strain coefficient $G^*(\omega)$; 2) a complex viscosity $\eta^*(\omega)$; or 3) a complex compliance $J^*(\omega)$. These quantities are defined in the following way [30]. Assume that for small deformations the shear stress oscillates as

$$P^v_{12} = Re\,[P^{v0}_{12}\,e^{i\omega t}]\;.\tag{9}$$

Then, G^*, η^*, and J^* are defined as

$$P^{v0}_{12} = iG^*\gamma^0\;,\tag{10}$$

$$P^{v0}_{12} = -\eta^*\dot{\gamma}^0\;,\tag{11}$$

$$\gamma^0 = -iJ^*P^{v0}_{12}\;,\tag{12}$$

where γ^0 and $\dot{\gamma}^0$ are the amplitude of the oscillatory shear strain and of the shear rate respectively. These three functions are closely related to each other, as it stems from their definition, by means of

$$\eta^*(\omega) = (i\omega)^{-1}G^*(\omega),\quad J^*(\omega) = 1/G^*(\omega)\;.\tag{13}$$

The three quantities are often written as

$$J^*(\omega) = J' - iJ'',\quad G^*(\omega) = G' + iG'',\quad \eta^*(\omega) = \eta' - i\eta''\;.\tag{14}$$

Note that $J'(\omega)$ is the strain in phase with stress divided by the stress, so that it is a measure of the energy stored and recovered per cycle. $J''(\omega)$ is the strain $90°$ out of phase with the stress divided by the stress, and it is a measure of the energy lost into heat per cycle.

According to the linear model of viscoelasticity, the viscous pressure tensor is given in general as

$$\mathbf{P}^v = -\int_{-\infty}^{t} G(t - t')\dot{\gamma}(t')\,dt'\;,\tag{15}$$

where G is known as the relaxation modulus, and $\dot{\gamma}$ as the rate-of-strain tensor which is the double of $\mathbf{V}$. In a generalized Maxwell model, for instance, one has for $G(t - t')$

$$G(t - t') = \sum_i (\eta_j/\tau_j)\exp\,[-(t - t')/\tau_j]\;,\tag{16}$$

in such a way that for a small-amplitude oscillatory motion [32] integration of Eq. (15) and the use of the definitions at Eq. (14) of η^* and G^* yield

$$\eta'(\omega) = \sum_j \eta_j[1 + (\tau_j\omega)^2]^{-1}\;,\quad \eta''(\omega) = \sum_j \eta_j\tau_j\omega\,[1 + (\tau_j\omega)^2]^{-1}\;,\tag{17}$$

or

$$G'(\omega) = \sum_j \eta_j \tau_j \omega^2 \left[1 + (\tau_j \omega)^2\right]^{-1} , \quad G''(\omega) = \sum_j \eta_j \omega \left[1 + (\tau_j \omega)^2\right]^{-1} , (18)$$

where η_j and τ_j now correspond to the jth Maxwell element. Furthermore, the steady-state compliance J is defined as

$$J = \int_0^\infty s G(s) \, ds \left[\int_0^\infty G(s) \, ds\right]^{-2} , \tag{19}$$

which for a Maxwell viscoelastic model yields

$$J = \sum_j \eta_j \tau_j \left(\sum_j \eta_j\right)^{-2} . \tag{20}$$

2.2 Extended Irreversible Thermodynamics

We begin our presentation with extended irreversible thermodynamics. We will not deal here with all its details, which have been widely reviewed by Jou, Casas-Vázquez and Lebon [26, 35], García-Colín [27], and Müller and Ruggeri [36], or in some collective works [24, 25, 30]. As we have said, EIT assumes that the entropy may depend, in addition to the classical variables, on the dissipative fluxes. We will neglect diffusion and thermal conduction and will consider that the only non-equilibrium quantity in the space of independent thermodynamic variables is the viscous pressure tensor. To avoid unnecessary formal complications we will use as a variable the whole tensor $\mathbf{P}^v$, instead of splitting it into trace and corresponding traceless parts. In this situation, the generalized Gibbs equation for a simple fluid is, up to the second order in $\mathbf{P}^v$, [26, 27, 35, 36]

$$ds = T^{-1} du + T^{-1} p \, dv - (\tau v / 2\eta T) \mathbf{P}^v : d\mathbf{P}^v , \tag{21}$$

with u and v the specific internal energy and the specific volume, T and p the absolute temperature and the thermodynamic pressure, η the shear viscosity and τ the relaxation time of $\mathbf{P}^v$. In general one could take, instead of a single relaxation time, several relaxation times and could assume that $\mathbf{P}^v$ is the sum of several different contributions $\mathbf{P}_i^v$, each of them with its own relaxation time τ_i. In this case, corresponding to a multicomponent fluid, one would have instead of Eq. (21) [37–39]

$$ds = T^{-1} du + T^{-1} p \, dv - \sum_i (\tau_i v / 2\eta_i T) \mathbf{P}_i^v : d\mathbf{P}_i^v . \tag{22}$$

Thus, the τ in Eq. (21) may be considered as an averaged relaxation time.

Note that in terms of the steady-state compliance (Eq. (20)), one may write the generalized Gibbs equation as

$$ds = T^{-1} du + T^{-1} p \, dv - (v / 2T) J \mathbf{P}^v : d\mathbf{P}^v . \tag{23}$$

Note, furthermore, that Eq. (9) may be written in terms of J as

$$P_{11}^v - P_{22}^v = -2J(P_{12}^v)^2 \ . \tag{24}$$

This relation is not exclusive of Maxwell upper convected fluids, but it is also obeyed by Maxwell corotational fluids or by Rivlin-Ericksen second-order fluids [40].

For shear flow, the generalized Gibbs equation, Eq. (23), reduces to

$$ds = T^{-1}du + T^{-1}pdv - (vJ/2T)[P_{11}^v dP_{11}^v + 2P_{12}^v dP_{12}^v] \ , \tag{25}$$

and since P_{11}^v is of the order of $\dot{\gamma}^2$ whereas P_{12}^v is of the order of $\dot{\gamma}$, for low values of $\dot{\gamma}$ we may neglect the contribution of P_{11}^v and write

$$ds = T^{-1}du + T^{-1}pdv - (vJ/T)P_{12}^v dP_{12}^v \ . \tag{26}$$

For the planar extensional flow in Eq. (7), the entropy would be

$$ds = T^{-1}du + T^{-1}pdv - (vJ/2T)[P_{11}^v dP_{11}^v + P_{22}^v dP_{22}^v] \ . \tag{27}$$

Integration of Eq. (26) leads to

$$s(u, v, \mathbf{P}^v) = s_{eq}(u, v) - (vJ/2T)(P_{12}^v)^2 \ , \tag{28}$$

where subscript eq stands for the equilibrium value. Similarly, for the extensional flow (Eq. (27)) one obtains, up to the second order in $\dot{\varepsilon}$,

$$s(u, v, \mathbf{P}^v) = s_{eq}(u, v) - 2(vJ/T)(\eta\dot{\varepsilon})^2 \ . \tag{29}$$

We now direct our attention to the Helmholtz free energy f, defined as

$$f = u - Ts \ , \tag{30}$$

where we are using values per unit mass. We are interested in the contribution Δf of the flow to the free energy, for which we write

$$\Delta f = \Delta u - T\Delta s \ , \tag{31}$$

with

$$\Delta f = f - f_{eq}, \quad \Delta u = u - u_{eq}, \quad \Delta s = s - s_{eq} \ . \tag{32}$$

Note that for fixed temperature, the internal energy under flow, u, is in general not equal to the internal energy at equilibrium, because the flow may stretch or deform the molecules thus storing internal energy in them.

Up to the second order in P_{12}^v we have, according to Eq. (26),

$$\Delta f = u - u_{eq} - T[s_{eq}(u, v) - (vJ/2T)(P_{12}^v)^2 - s_{eq}(u_{eq}, v)] = (vJ/2)(P_{12}^v)^2 \ , \tag{33}$$

where we have taken into account that $s_{eq}(u) = s_{eq}(u_{eq}) + T^{-1}(u - u_{eq}) +$ higher-order terms.

In a more general situation, when one is not restricted to second order in shear rate, one assumes that the energy contribution, proportional to $(P_{12}^v)^2$, is much higher than the entropic one. We will comment on this point later from a microscopic point of view. One should thus have

$$\Delta f = \Delta u = vJ(P_{12}^v)^2 \ . \tag{34}$$

This is consistent with the meaning of J as stored energy. This expression has been used by many authors [1, 2, 41, 42]. Equation (33), already used by Onuki [5], and Eq. (34) will center our further discussion.

In spite that several attempts to study the thermodynamics of viscoelastic systems have been undertaken [15, 16, 40], the attention of the authors was concentrated on the constitutive equations rather than on the equations of state. This explains the scarce attention received by the point we are studying in this review.

2.3 Rational Thermodynamics

We compare the previous results with the ones obtained in rational thermodynamics [16, 18, 40]. As has been previously said, this theory uses for the free energy an expression which does not coincide with the local-equilibrium one. For a simple fluid the free energy may depend on its classical variables and on the strain history. If ε_{ij} is the strain tensor

$$\varepsilon_{ij} = (1/2)\,[\partial u_j/\partial x_i + \partial u_i/\partial x_j] \ , \tag{35}$$

with u the local deformation vector at point r, one writes for the free energy, in the absence of thermal effects [43–45],

$$F = F_{eq} + \int D_{ij}(t - t')(\partial \varepsilon_{ij}/\partial t')\,\mathrm{d}t'$$
$$+ \tfrac{1}{2}\iint G_{ijkl}(t - t', t - t'')(\partial \varepsilon_{ij}(t')/\partial t')(\partial \varepsilon_{ij}(t'')/\partial t'')\,\mathrm{d}t'\,\mathrm{d}t'' \ . \tag{36}$$

In an isotropic system D_{ij} should be zero. The viscous stress tensor $\boldsymbol{\sigma}^v = -\mathbf{P}^v$ is obtained from Eq. (36) through the relation

$$\sigma_{ij}^v = (\partial(\rho F)/\partial(\partial \varepsilon_{ij}/\partial t'')) = \int G_{ijkl}(t - t', 0)(\partial \varepsilon_{kl}(t')/\partial t')\,\mathrm{d}t' \ . \tag{37}$$

Note that the knowledge of the viscous stress tensor $\boldsymbol{\sigma}^v$ does not allow the knowledge of F, because in Eq. (37) only the modulus $G_{ijkl}(t - t', 0)$ appears whereas in the free energy the dependence of $G_{ijkl}(t - t', t - t'')$ with respect to its second argument is also needed. Note also that rational thermodynamics does not consider the viscous stress tensor as an independent variable, but it uses as an extra variable the velocity gradient $\partial \varepsilon_{ij}/\partial t$, and one aims to obtain a constitutive equation such as Eq. (37) expressing the viscous stress tensor in terms of the history of the rate-of-deformation tensor. Equation (36) may be written by making an explicit distinction between the trace and the traceless part of the rate-of-deformation tensor, but we will keep here the simplest form of it.

In order to deal with the usual viscoelastic systems, one writes [45]

$$G_1(t,t') = \sum_n G_{1n} e^{-(t+t')/\tau_n} \ . \tag{38}$$

Here, G_{1n} are the moduli $G_{1n} = \eta_n/\tau_n$. If we assume now for $\partial \varepsilon_{ij}/\partial t$ a history of the form $(\partial \varepsilon_{12}/\partial t) = \dot{\gamma} h(t)$, with $h(t)$ the Heaviside step function, and we use Eq. (38), Eq. (36) becomes

$$\rho F = \rho F_{eq} + \tfrac{1}{2} \iint G_n e^{-(t+t')/\tau_n} \dot{\gamma}^2 \, dt \, dt'$$

$$= \rho F_{eq} + \tfrac{1}{2} \sum_n \tau_n^2 G_n \dot{\gamma}^2 = \rho F_{eq} + \tfrac{1}{2} \sum_n \tau_n \eta_n \dot{\gamma}^2 \ , \tag{39}$$

which is identical to the free energy obtained from Eq. (22).

2.4 Internal Variables

Finally, one should examine the point of view of theories with internal variables. Indeed, when working with polymer solutions the macromolecular configuration, or the average molecular configuration, of the polymers is incorporated as a supplementary variable into the entropy or the free energy. One usually takes as a description of the configuration the so-called configuration tensor

$$\langle RR \rangle = \int \psi(R) RR \, dR \ , \tag{40}$$

with Ψ the configurational distribution function. More elaborate thermodynamic functions including as variables the whole distribution function have also been considered [46]. Here, R is the end-to-end vector of the macromolecules. Other descriptions are possible in terms of $\langle R_i R_i \rangle$, with R_i the vector from bead i to bead $i + 1$, or the vector related with the i-th normal mode in a Rouse-Zimm description [38, 47].

As will be seen in the next section, and as it is well known, the configuration tensors are directly related to the viscous pressure tensor. Indeed, for the contribution of the i-th normal mode to the viscous pressure tensor one has [38, 47]

$$\mathbf{P}_i^v = -nH \langle R_i R_i \rangle + nkT \mathbf{U} \ , \tag{41}$$

H being an elastic constant characterising the intramolecular interactions and n the number of molecules per unit volume. Thus, introduction of $\mathbf{P}_i^v$ or of $\langle R_i R_i \rangle$ as independent variables into the free energy is essentially equivalent, in the case of dilute polymer solutions (this is not so in the case of ideal gases where there are not internal degrees of freedom). The use of $\mathbf{P}_i^v$ or of $\langle R_i R_i \rangle$ as variables has, in both cases, some characteristic advantages. For the analysis of non-equilibrium steady states, $\mathbf{P}_i^v$ is directly related to the observables. For the microscopic understanding of the macromolecular processes taking place and for the analysis of light-scattering experiments, the use of $\langle R_i R_i \rangle$ is more suitable. The dynamical equations for the configuration tensor thus provide evolution equations for the viscous pressure tensor.

3 Statistical Mechanics of Non-Equilibrium States

A comparison of the results of the previous section with some microscopic descriptions will undoubtedly clarify the notions discussed there. A lot of work on this topic has been carried out for monatomic ideal gases, where it is known that, for low values of the viscous pressure, the expression for the entropy in non-equilibrium states depends on the viscous pressure in the way predicted by the macroscopic theory of extended irreversible thermodynamics. Also, for dilute real monatomic gases, one may obtain from kinetic theory an expression for the entropy where the kinetic viscous pressure and the potential viscous pressure appear in the entropy [48]. For high values of the shear, one may also obtain more complicated dependences of the entropy in terms of the viscous pressure tensor [49]. We will not deal with these systems, as we are focusing the interest of the present review on polymer solutions.

In these systems, the first attempts to take into account the influence of flow on the free energy were undertaken from a microscopic point of view [41, 50–53]. The work by Marrucci [41] was especially influential, and it was used as the starting point of the works by Sun and Denn [51], ver Stratte and Philipoff [42], Rangel-Nafaile et al. [1] and Wolf [2], in spite of several limitations.

3.1 Kinetic Theory of Dilute Polymeric Solutions

The result obtained by Marrucci [41] was that the excess free energy per unit volume related to the flow is

$$\Delta f = -\tfrac{1}{2} \mathrm{Tr} \mathbf{P}^v \ . \tag{42}$$

There are several ways to show this result. The simplest way is to assume a freely jointed chain of N statistically independent segments of length b. Since one supposes that the orientations of the successive segments are independent, the mean value of the end-to-end vector of the chain, R, is given by the well-known result

$$\langle R^2 \rangle = Nb^2 \ . \tag{43}$$

In the Gaussian approximation of the random-flight model, the conformational distribution function of R is [47]

$$\psi(R) = \left(\frac{3}{2\pi Nb^2}\right)^{3/2} \exp\left[-\frac{3}{2b^2N}R^2\right] . \tag{44}$$

In order to obtain the Helmholtz free energy f of a chain with the end-to-end vector R fixed we recall that the equilibrium distribution function ψ can be expressed as

$$\psi(R) \propto \exp\left[-f/kT\right] . \tag{45}$$

Thus, the free energy of the chain can be written as

$$f = const. + \frac{3kT}{2Nb^2}R^2 . \tag{46}$$

One may obtain an expression for the force necessary to modify R by means of the derivative $-(\partial f/\partial R)_T = (3kT/Nb^2)R$. This shows that the freely jointed chain has an elastic constant of entropic origin given by

$$k' = \frac{3kT}{Nb^2} . \tag{47}$$

Then, since there are n independent chains per unit volume, the free energy of the solution per unit volume will be

$$f(T,n) = const. + \frac{3nkT}{2Nb^2}\langle R^2\rangle . \tag{48}$$

The corresponding viscous pressure tensor is given by the Kramers expression [54] for a solution of n Hookean dumbbells per unit volume

$$\mathbf{P}^v = -nk'\langle RR\rangle + nkT\,\mathbf{U} , \tag{49}$$

with k' the elastic constant of the dumbbells and $\mathbf{U}$ the unit tensor. Note that according to Eq. (47), $\mathbf{P}^v$ will vanish in equilibrium, as was to be expected. The flow will change the value of $\langle R^2\rangle$, so that the contribution of the flow to the free energy Δf may be written as

$$\Delta f = \tfrac{1}{2}nk'\langle R^2\rangle_{flow} - \tfrac{1}{2}nk'\langle R^2\rangle_{eq} . \tag{50}$$

On the other hand, since $\mathrm{Tr}\,\mathbf{P}^v = 0$ in equilibrium Eq. (49) can be rewritten as

$$\mathbf{P}^v = -nk'\langle RR\rangle + \tfrac{1}{3}nk'\langle R^2\rangle_{eq}\mathbf{U} , \tag{51}$$

and therefore one will have

$$\mathrm{Tr}\,\mathbf{P}^v = -nk'\langle R^2\rangle_{flow} + nk'\langle R^2\rangle_{eq} . \tag{52}$$

Comparison of Eqs. (50) and (52) yields Eq. (42). This original derivation by Marrucci states very clearly its limitations: 1) the modifications of f are of entropic origin; no change in the internal energy of the chain is considered; 2) the solution must be very dilute so that the chains do not interact; 3) the hydrodynamic interactions between the different parts of the macromolecule are neglected; 4) the internal friction of the molecule is neglected; 5) the linear relationship between elastic forces and end-to-end vectors breaks down when the macromolecules are much extended.

A more complete derivation of the free energy may be obtained from the Rouse-Zimm bead-and-spring model, taking into account the change in the internal energy of the chains, which are now not formed by rigid rods but by elastic springs, and including, in the Zimm model, a preaveraged form of the hydrodynamic interaction between the different parts of the macromolecule. The derivation of the form of the free energy was provided by Sarti and Marrucci [52] and by Booij [53] (see also [33]).

Let $\psi(R_1, R_2, \ldots, R_{N-1})$ be the configurational distribution function giving the probability that the bead to bead vectors R_i are between R_i and $R_i + dR_i$ respectively. It is more convenient to work in terms of the normal modes of the chain, which will be denoted as $Q_1, \ldots, Q_{N-1}$. Thus, the contribution of the chains to the entropy per unit volume is

$$s = - nk \int \psi \ln \psi \, dQ_1 \ldots dQ_{N-1} . \tag{53}$$

The internal energy of the macromolecules may be written as

$$u = -nkT \int \psi \ln \psi_0 \, dQ_1 \ldots dQ_{N-1} , \tag{54}$$

with ψ_0 the equilibrium distribution function

$$\psi_0 = \prod_i \psi_i = (H/2\pi kT)^{3(N-1)/2} \exp\left[-(H/2kT)(Q_1^2 + \cdots + Q_{N-1}^2)\right] . \tag{55}$$

The internal energy of the chain (divided by kT) is the argument of the exponential in Eq. (55) according to the Boltzmann statistics, so that Eq. (54) is indeed the energy, except for an additive function coming from the logarithm of the normalization factor, which will not be important because we are interested in the excess Helmholtz free energy due to the flow rather than in the whole value of the free energy.

According to Eqs. (53) and (54) the expression for the Helmholtz free energy is

$$f = u - Ts = - nkT \int \psi [\ln \psi_0 - \ln \psi] \, dQ_1 \ldots dQ_{N-1} . \tag{56}$$

or

$$f = nkT \int \psi \ln (\psi/\psi_0) \, dQ_1 \ldots dQ_{N-1} . \tag{57}$$

Under non-equilibrium conditions, the distribution function must fulfill the continuity equation

$$\partial\psi/\partial t = -\sum_i \partial(\psi\dot{R}_i)/\partial R_i , \tag{58}$$

where $\dot{R}_i$ are obtained by a force balance over the beads [32, 52, 55]

$$\zeta[\dot{R}_i - (\boldsymbol{\nabla}\boldsymbol{v}).R_i] = -kT\sum_j B_{ij} \partial\ln(\psi/\psi_0)/\partial R_j . \tag{59}$$

The elastic forces are expressed through the function ψ_0, ζ is the friction coefficient of the beads and the matrix **B** may include the effects of the hydrodynamic interactions amongst the beads in the pre-averaged Zimm treatment.

In terms of the normal modes, Eqs. (57) and (58) may be written as

$$-\frac{\partial\psi_i}{\partial t} = \frac{\partial}{\partial Q_i}.\left[\psi_i(\boldsymbol{\nabla}\boldsymbol{v}).Q_i - (H/\zeta)\lambda_i\psi_iQ_i - (kT/\zeta)\lambda_i\frac{\partial}{\partial Q_i}\psi_i\right], \tag{60}$$

with $\psi_i(Q_i)$ the distribution function of the normal mode Q_i. The normal modes are related to the bead-to-bead vectors R_j through

$$Q_i = \sum_j \Omega_{ij}.R_j , \tag{61}$$

where the orthogonal matrix Ω diagonalizes the matrix **B** in Eq. (59). The corresponding diagonal elements of $\Omega.\mathbf{B}.\Omega^T$ are the λ_i, i.e. the eigenvalues of the matrix **B**, to each of whom corresponds a relaxation time $\tau_i = \lambda_i^{-1}$.

A solution of Eq. (60) can be written as [55]

$$\psi_i/\psi_{i0} = (H/kT)^{-3/2}[\det\langle Q_iQ_i\rangle]^{-1/2}\exp[(H/2kT)$$
$$\times\{Q_iQ_i - (1/2)Q_i\langle Q_iQ_j\rangle^{-1}Q_j\}] , \tag{62}$$

with ψ_{i0} the i-th normal mode distribution function at equilibrium.

When Eq. (62) is introduced into eq. (57) one has

$$f = (nkT/2)\left\{\sum_i \text{Tr}[(H/kT)\langle Q_iQ_i\rangle - \mathbf{U}]\right.$$
$$\left. - \sum_i \ln[\det(H/kT)\langle Q_iQ_i\rangle]\right\} . \tag{63}$$

If one takes into account that $\mathbf{P}_i^v$ is given by Eq. (41), one may write Eq. (63) in the form

$$\Delta f = -(nkT/2)\left\{\sum_i \text{Tr } \mathbf{P}_i^{v'} + \sum_i \ln\left[\det\left(\mathbf{U} - \mathbf{P}_i^{v'}\right)\right]\right\} , \qquad (64)$$

with $\mathbf{P}_i^{v'} = (nkT)^{-1}\mathbf{P}_i^v$.

If we use the expression at Eq. (3) in Eq. (64) we obtain

$$\Delta f = (1/2)\left\{2\tau\eta\dot{\gamma}^2 - nkT \ln\left[1 + 2(\tau\eta\dot{\gamma}^2/nkT) - \eta^2\dot{\gamma}^2/(nkT)^2\right]\right\} . \qquad (65)$$

For small values of $\dot{\gamma}$ one may develop the logarithm in Eq. (65) and one finds

$$\Delta f = (1/2)\tau\eta\dot{\gamma}^2 = (1/2)J(P_{12}^v)^2 , \qquad (66)$$

which has precisely the form Eq. (28) predicted by extended irreversible thermodynamics for low values of P_{12}^v. This confirmation of EIT at low values of $\dot{\gamma}$ could also have been obtained without need of the general expression at Eq. (62) but by writing the second-order solution of ψ for a shear flow as [39]

$$\psi_i = \psi_0\left\{1 + (\zeta/8kT)(\boldsymbol{\nabla v}){:}\boldsymbol{Q}_i\boldsymbol{Q}_i\right\} , \qquad (67)$$

and introducing it into Eq. (56).

For high values of $\dot{\gamma}$, the terms in $2\tau\eta\dot{\gamma}^2$ will be more important than the terms in the logarithm, so that Eq. (65) will reduce to

$$\Delta f = \tau\eta\dot{\gamma}^2 = J(P_{12}^v)^2 , \qquad (68)$$

which is the expression following from Eqs. (42) and (5), and that has been used since Marrucci's work [41].

For planar extensional flows, we may introduce the viscous pressure tensor (5) into the free energy as given by Eq. (64), and we arrive at

$$\Delta f = (nkT/2)\left\{-(P_{11}^v + P_{22}^v) - \ln\left[1 + P_{11}^v + P_{22}^v + P_{11}^v P_{22}^v\right]\right\} . \qquad (69)$$

By developing $\ln(1 + x) \approx x - (x^2/2)$ one has, up to the second order in P_{11}^v,

$$\Delta f = (1/4)nkT\left[(P_{11}^v)^2 + (P_{22}^v)^2\right] , \qquad (70)$$

in agreement with the EIT expression at Eq. (29) for low values of P_{11}^v and P_{22}^v.

The microscopic derivation presented here takes into account a pre-averaged form of the hydrodynamic interaction, as in the Zimm model, and takes no account of the intermolecular interactions. It is interesting to have explicit microscopic expressions for the non-equilibrium free energy, as some authors cast doubts on the possibility of obtaining general thermodynamic potentials for polymer solutions under flow, by arguing that in general the friction forces do not come from any potential. Only in the situation when the flow is irrotational, as in planar extensional flow, and when the friction force is assumed to be proportional to the velocity, may the friction come from a potential [3, 56]. However, it has been seen here that an expression for the free energy may be found in more general circumstances. Other microscopic derivations and applications of a free energy under flow have been presented by Doi and Edwards [47].

3.2 Reptation Model

Microscopic expressions for the free energy under flow have also been obtained for concentrated solutions by using very different starting points to the ones mentioned here. Two especially useful models for the analysis of concentrated solutions are the transient network model [57] and the reptation model [47, 58–63].

We will briefly comment here on the results from the reptation model [59–63]. This model incorporates the idea that the main molecular motion in a melt is that of reptation, which is the diffusive wriggling of the molecular chain along its own length. The basic idea is that the transverse motions of each chain are impeded by the meshwork of the strands of other chains around it. If the system is deformed, the cages along which the chain moves are distorted and the chains are carried into new configurations in such a way that stress relaxation proceeds first by a relatively rapid equilibration of chain configurations within the distorted cages (of the order of the Rouse relaxation times), and then by a relatively slow diffusion of chains out of the distorted cages (of the order of the so-called disengagement time).

The relation at Eq. (56) for the free energy in terms of the distribution function may also be used in these models, but with a different meaning of the distribution function, where one defines $\psi(u, s, t)$ as the probability that the tangent vector at s and t is in the direction u. The connection of u with the viscous pressure tensor $\mathbf{P}^v$ is understood when one writes

$$\mathbf{P}^v = -\frac{3nkT}{Nb^2} \left\langle \int_0^{L(t)} ds\, L(t) \left[u(s, t)u(s, t) - \tfrac{1}{3}\mathbf{U} \right] \right\rangle . \tag{71}$$

This formula, that generalizes Eqs. (41) and (49), shows that the viscous pressure tensor is determined by two quantities, the length of the tube $L(t)$ and the orientation of $u(s, t)$.

If one adopts the independent-alignment approximation in the reptation model, one has

$$\Delta f = nNkT \int \psi \ln \left[\psi(u, s, t)/\psi(u_0, s, t) \right] du , \tag{72}$$

where n is the number of polymer chains, N is the number of monomers per polymer and $\psi(u, s, t)$ is the orientation distribution function in the deformed state for a primitive segment located a distance s along the primitive chain; $\psi(u_0, s, t)$ is the corresponding distribution function in the undeformed fluid.

In the limit of small deformations of the polymer, one may consider that the segments are of the same length and number as in equilibrium, and that the deformation only changes its orientation in space. This is called the independent alignment approximation, in which the primitive chain segments are taken to behave as rigid rods in such a way that their distribution function is completely specified in the space of unit vectors u [60–63]. The change in the orientational distribution function $\psi(u)$ due to the deformation of the fluid is obtained in a straightforward way [61] by requiring the continuity condition

$$\psi_0(u_0)\, d\Omega_0 = \psi(u)\, d\Omega \ , \tag{73}$$

with u_0 and u, and Ω_0 and Ω corresponding to the unit vectors and solid angles before and after the deformation. On the other hand, since the continuum deforms at a constant volume, one must have

$$r_0^3 d\Omega_0 = r^3\, d\Omega \ , \tag{74}$$

using r as the magnitude of the position vector r. Then the ratio $\psi(u)/\psi_0(u_0)$ can be written as r^3/r_0^3 so that

$$\Delta f = nNkT \langle \ln(r/r_0)^3 \rangle \ . \tag{75}$$

The ratio r^3/r_0^3 may also be written in terms of the deformation gradient tensor E, defined by means of the relation $r = E.r_0$, as $r^3/r_0^3 = \det(E)$, so that Eq. (75) becomes

$$\Delta f = nNkT \langle \ln[\det(E)]\rangle \ . \tag{76}$$

For a plane shear deformation γ in the z direction, one has $x = z_0\gamma + x_0$ and $z = z_0$, and then

$$r^2 = r_0^2 + \gamma^2 z_0^2 + 2\gamma x_0 z_0 \ . \tag{77}$$

If we have a steady shear flow instead of a static shear deformation, the simplest naive way to proceed would be to take $x_0 = 0$ (i.e. to follow the motion of the fluid) and for the shear deformation the value $\gamma = \dot{\gamma}\tau$, with τ the relaxation time, which in the reptation model may be identified with the reptation or disengagement time τ_d. For small values of the shear rate, writing $\ln(r^2/r_0^2) = \ln(1 + \gamma^2 \cos^2\theta) = \gamma^2 \cos^2\theta$ with $\cos\theta \equiv x_0/r_0$, and taking into account that the mean value of $\cos^2\theta$ is 1/2, we have

$$\Delta f = \tfrac{1}{2}nNkT\dot{\gamma}^2\tau^2 \ , \tag{78}$$

which coincides with the usual expression of EIT for low values of $\dot{\gamma}$. To obtain Eq. (78) we have taken $\ln(r^2/r_0^2)$ rather than $\ln(r^3/r_0^3)$ because we are considering a plane two-dimensional flow.

A more general approximation [59], valid for any arbitrary deformation, is given by

$$\Delta f = nNkT \left\langle \int_{-\infty}^{t} [\partial \mu(t - t')/\partial t] \ln [\det \mathbf{E}(t')] \, dt' \right\rangle , \tag{79}$$

with μ the relaxation function giving the fraction of segments which at time t are still trapped in the deformed tubes. In the reptation model this function has the form

$$\mu(t) = \frac{8}{\pi^2} \sum_{p,odd} \frac{1}{p^2} \exp [- p^2 t/\tau_d] . \tag{80}$$

The general form for $\ln |\mathbf{E}|$ in a shear deformation [59] is obtained by writing the ratio r^2/r_0^2 as $1 + \gamma^2 \cos^2 \theta + 2\gamma \sin \theta \cos \theta \cos \phi$ (here we do not take $x_0 = 0$ a priori), so that integrating with respect to ϕ from 0 to 2π, and setting $\cos \theta \equiv \alpha$, we have

$$\ln [\det (\mathbf{E})] = \tfrac{1}{2} \int_0^1 \ln [\tfrac{1}{2} \{1 + \gamma^2 \alpha^2 + [\alpha^4 (\gamma^4 + 4\gamma^2) - 2\gamma^2 \alpha^2 + 1]^{1/2}\}] \, d\alpha . \tag{81}$$

In a steady shear flow, we would have $\gamma = \dot{\gamma} t$. This would yield a generalization of the simple formula given in Eq. (78), but in fact we will not need it here.

An analogous result can be obtained in the framework of the transient network theory inspired in the molecular network theory of rubberlike elasticity for permanently cross-linked rubberlike solids. Of course, when applied to concentrated polymer solutions and polymer melts one assumes that the junctions are impermanent, being produced and lost during deformation. The processes of creation and loss of segments between junctions thus play an important role in the theory. In Lodge's model [57] the expression for the free energy under a shape history described by the deformation gradient tensor E_{ij} is found to be

$$F(t) = \cos t + \tfrac{1}{2} kTE_{ij}(t) \int_{-\infty}^{t} N^*(t - t') E_{ij}(t') \, dt' , \tag{82}$$

where $N^*(t - t')$ is the concentration at time t of segments of the network created during the interval $(t', t' + dt')$ which is given by

$$N^*(t - t') = \sum_{\kappa, n} L_{\kappa n} \exp [- (t - t')/\tau_{\kappa n}] . \tag{83}$$

Here, $L_{\kappa n}$ and $\tau_{\kappa n}$ are respectively the creation and dissociation rate constants of segments of n free-jointed rigid links of type κ ($\kappa = 1, 2, ...$), the latter being a discrete parameter which may account for different characteristics of the links.

3.3 Phenomenological Derivation

An interesting phenomenological derivation of the excess free energy was given by McHugh [64] for extensional flows, by computing the work done on an element of fluid. In the steady extensional kinematics defined by

$$v = G(x, \; -y/2, \; -z/2) \; , \tag{84}$$

the net force causing extension of the polymer molecules may be obtained from the equations of motion and from the subsequent stress tensor. He considers the extension of a cylinder of fluid in the main body of flow, composed of both polymer and solvent. One assumes that, during the stretching, the dumbbell ends follow the main flow up to the point of their maximum extension. The force needed to extend the cylinder is, in cylindrical coordinates,

$$F = 2\pi \int\limits_0^{R(t)} T_{zz} r \, dr \; , \tag{85}$$

with $T_{zz} = -P_{zz}$ the total stress acting on the moving ends of the cylinder of radius $R(t)$. One may relate the viscous stress σ_{zz}^v as $T_{zz} = -p + \sigma_{zz}^v$, with p the isotropic pressure.

By evaluating T_{zz} from the equations of motion and the velocity field at Eq. (84), one is led to

$$F = \pi R^2(t)(\sigma_{zz}^v - \sigma_{rr}^v) \; , \tag{86}$$

with σ_{rr}^v the viscous stress acting in the radial direction. One has then for the change in the free energy

$$\Delta f = \int\limits_{L_0}^{L} F \, dz \; . \tag{87}$$

Taking into account the continuity equation for the cylinder and using the extension ratio $\alpha = z(t)/L_0$, one has

$$\Delta f = \pi R_0^2 L_0 \int\limits_1^{\alpha} (\sigma_{zz}^v - \sigma_{rr}^v) \frac{d\alpha}{\alpha} \; . \tag{88}$$

Again, in Cartesian rectangular coordinates and dividing Eq. (88) by the volume $\pi R_0^2 L_0$ one has, per unit volume,

$$\Delta f = \int\limits_1^{\alpha} (\sigma_{xx}^v - \sigma_{yy}^v) \frac{d\alpha}{\alpha} \; , \tag{89}$$

with $\sigma_{xx}^v - \sigma_{yy}^v$ the first normal stress difference. For the equivalent extension in shear flow, one needs only to assume that the same computation is valid for the extension occurring along the principal stress directions where the stress tensor is diagonal.

4 Non-Equilibrium Chemical Potential

In this review we are especially interested in the equations of state under flow. Here we refer to the definition of the non-equilibrium chemical potential under flow. We recall that the usual definition of the chemical potential in equilibrium thermodynamics is

$$\mu_i = (\partial G/\partial n_i)_{T,p,n_j} , \tag{90}$$

where G is the Gibbs free energy and n_i the number of moles of the species i in the system. The expression at Eq. (90) can be generalized to non-equilibrium situations by including in G the non-equilibrium contributions. It must be noted that out of equilibrium one must be careful as to which variables should be kept constant during the differentiation. Indeed, one could perform the differentiation either at constant viscous pressure or at constant shear rate, or at constant molecular conformation. In the next paragraphs we will see in a simple example the influence of these parameters on the final expression for the non-equilibrium chemical potential.

To see an example, we will restrict ourselves to the dilute Rouse-Zimm model in theta solvents. Here, the dependence of τ, η and J on the molecular weight and the concentration or number of moles per unit volume, c, is [47]

$$\left.\begin{array}{l} \tau \approx M^2 \\ \eta \approx cM \\ J \approx c^{-1}M \end{array}\right\} \text{(Rouse)} \qquad \left.\begin{array}{l} \tau \approx M^{3/2} \\ \eta \approx cM^{1/2} \\ J \approx c^{-1}M \end{array}\right\} \text{(Zimm).} \tag{91}$$

In fact, there is a whole spectrum of relaxation times. We will take here the average time. Takahashi et al. [65] have seen that the longest relaxation time is usually proportional and close to the weight-average time. In the simple Rouse-Zimm models, all the times follow the same scaling laws.

We will write our expression for the Gibbs free energy per unit volume for small values of P_{12}^v in the form

$$g(T,p,c_i,P_{12}^v) = g_{eq}(T,p,c_i) + (1/2)J(P_{12}^v)^2 . \tag{92}$$

Thus we have

$$\Delta g = (1/2)J(P_{12}^v)^2 . \tag{93}$$

In terms of the shear rate $\dot{\gamma}$ the last equation takes the form

$$\Delta g = (1/2)\tau\eta\dot{\gamma}^2 . \tag{94}$$

One could also refer to the configuration tensor $\mathbf{W}$

$$\mathbf{W} = (1/N)\sum_i [\langle R_i R_i \rangle - (kT/H)\mathbf{U}] , \tag{95}$$

with R_i, the bead-to-bead vector or, in entangled systems, the link-to-link vector. According to Eq. (41) we have

$$P_{12}^v = -cN_AHW_{12} \,, \tag{96}$$

with N_A the Avogadro's number. Thus, Eq. (93) may also be written as

$$\Delta g = (1/2)J(cN_AH)^2 W_{12}^2 \,, \tag{97}$$

We have for the chemical potential

$$\mu_{2,p} = (\partial g/\partial n_2)_{T,p} = (\partial g/\partial c)_{T,p}(\partial c/\partial n_2) = (1/V)(\partial g/\partial c)_{T,p} \,, \tag{98}$$

where V is the volume of the system. We may now compare the three different results for the chemical potential of the polymer that are obtained from the non-equilibrium Gibbs free energy in the Zimm model. At constant P_{12}^v, we have

$$\mu_{P2} = \mu_{eq2} - (1/2)(J/Vc)(P_{12}^v)^2 \,. \tag{99}$$

At constant shear rate $\dot{\gamma}$ we have

$$\mu_{\dot{\gamma}2} = \mu_{eq2} + (1/2)(\tau\eta/Vc)\dot{\gamma}^2 = \mu_{eq2} + (1/2)(J/Vc)(P_{12}^v)^2 \,. \tag{100}$$

At constant conformation W_{12} one would have

$$\mu_{W2} = \mu_{eq2} + (1/2)JcN^2H^2W_{12}^2 = \mu_{eq2} + (1/2)(J/Vc)(P_{12}^v)^2 \,. \tag{101}$$

We have assumed that the elastic constant H of the force between beads does not depend on the concentration. Thus, according to Eq. (99), the chemical potential of polymers is decreased by flow whereas according to Eqs. (100) and (101) it is increased. There is no contradiction in these statements because they refer to different physical conditions. We thus stress the attention that must be paid to these non-equilibrium quantities in the definition and the use of non-equilibrium chemical potential.

Rangel-Nafaile et al. [1] have used (though not directly) a definition at constant P_{12}, whereas Wolf [2] has used a definition at constant $\dot{\gamma}$ and Onuki [5] has used a definition at constant conformation W_{12}. Wolf has concisely justified his use of the potential at constant velocity gradient rather than at constant viscous pressure. According to him, two extreme situations may be envisaged when a phase separation is taking place: 1) the same shear rate is found in both phases or 2) the same shear stress is found in both phases (the same $\eta\dot{\gamma}$). Theoretical considerations concerning macroscopic droplets with sharp phases boundaries point to situation 2 as the most plausible. Wolf, however, argues that at the cloud point the suspended phase consists of extremely small microscopic droplets, and the insignificant interfacial tension leads to diffuse phase boundaries. Since his experiments were performed by keeping fixed the shear rate, Wolf finally takes the potential at constant $\dot{\gamma}$.

Simple scaling laws are also valid in the highly-entangled regime, with [65]

$$\eta \approx M^{3/4} c^b, \qquad J \approx M^0 c^{-2}, \qquad \tau \approx M^{3/4} c^{b-2} ,$$

with η, J, and τ the limiting zero-shear values of $\eta(\dot{\gamma})$, $J(\dot{\gamma})$, and $\tau(\dot{\gamma})$, and b a scaling exponent. In fact, the dependence of J on c given by the simple scaling laws is not always realistic (for instance, instead of the simple dependence $J \approx c^{-1} M$ proposed in Eq. (91), one may have a more complicated dependence on c as it is seen in Eq. (121)). Therefore, the actual situation in the analysis of non-equilibrium chemical potentials is more complicated, because it usually implies dealing with values of η, J, and τ at non-zero shear rates. This makes the dependence of the excess free energy on the density more complicated [1, 2, 66], and this fact precludes from adopting the points raised by Onuki [5] and Rangel-Nafaile et al. [1] regarding the direct use of thermodynamic arguments for studying phase separation out of equilibrium by means of a non-equilibrium chemical potential with J as given by scaling laws. Indeed a naive use of Eq. (91) leads to a decrease of the critical temperature in the presence of a shear flow, in contrast with the behaviour observed in experiments. Rather than a failure of the thermodynamic method this discrepancy should be attributed to the over-simplified forms at Eq. (91).

It is also interesting to show that the condition of material equilibrium of species i between two phases α and β at constant temperature and pressure, which is

$$\mu_i^\alpha(T, p, c_j) = \mu_i^\beta(T, p, c_j) , \tag{102}$$

may be used in the presence of flow. In equilibrium thermodynamics, the condition at Eq. (102) is obtained from the minimization of the Gibbs free energy at constant T and p. Such argument is not directly valid in non-equilibrium situations (though, from the perspective of maximum entropy arguments, it appears legitimate in some circumstances). However, the condition at Eq. (102) may be recovered from a dynamical point of view. Note, indeed, that the rate of variation of G is

$$dG/dt = \mu_i^\alpha(dN_i^\alpha/dt) + \mu_i^\beta(dN_i^\beta/dt) = (\mu_i^\alpha - \mu_i^\beta)(dN_i^\alpha/dt) . \tag{103}$$

We have taken into account mass conservation, according to which $dN_i^\alpha/dt = -dN_i^\beta/dt$, with N_i^α and N_i^β the number of moles of i in phases α and β respectively.

The simplest constitutive equation for dN/dt compatible with a definite decrease of G is

$$dN_i^\alpha/dt = -L(\mu_i^\alpha - \mu_i^\beta) , \tag{104}$$

with L a positive phenomenological coefficient. This is a usual argument in linear irreversible thermodynamics. Thus, the stationary condition $dN/dt = 0$ implies Eq. (102) because of Eq. (103). Since both phases (or one of them) are under flow, it is the actual non-equilibrium chemical potential, including the effect of $\mathbf{P}^v$, which should be used in Eq. (102) instead of the local-equilibrium chemical potential. The use of the generalized μ is thus a reasonable assumption, which will be acceptable if its predictions are experimentally confirmed.

A final comment should be made about the absolute temperature and the pressure in non-equilibrium states. Since, as we have commented, there is a modification in the chemical potential due to the flow, it seems logical to ask whether modifications should also appear in the temperature and the pressure, making them different from the local-equilibrium T and p. The answer is positive [67, 68]. However, one may identify T and p with the temperature and the pressure indicated by a thermometer and a manometer in the usual way, as these instruments read in fact the non-equilibrium values of the temperature and the pressure rather than their local-equilibrium values T and p [69].

5 Application to the Analysis of Phase Diagram Under Shear Flow

The inclusion of non-equilibrium contributions to the Gibbs free energy leads to modifications in the phase diagram of polymer solutions under flow. Such changes have been observed many times, but a definite theory is still lacking. In the next section we briefly summarize the essentials of the phenomenology of these changes. Here we address our attention to the shift of the critical point and the spinodal line corresponding to the separation of two phases in the polymer solution. This problem has been studied explicitly by Rangel-Nafaile et al. [1] and by Wolf [2]. Criado-Sancho et al. [66] studied the influence of the use of several different definitions of the chemical potential in the analysis of the spinodal lines.

It must be noted that the analysis by Rangel-Nafaile et al. [1] for the coexistence lines is satisfactory: the shifts in the coexistence lines or cloud points under shear may be predicted with an error of 3 K. The direct mathematical analysis of the critical temperature, in contrast, is not so satisfactory. However, the latter one is more easily amenable to a closed mathematical formulation, so that we will present it as an illustration.

In the absence of shear, one assumes that the Gibbs free energy of mixing of the solution has the well-known Flory-Huggins form

$$\Delta G_{FH}/RT = n_1 \ln(1\text{-}\phi) + n_2 \ln \phi + \chi(1 - \phi)N\phi \ , \tag{105}$$

with G the free energy of the system (g is the free energy per unit volume), ϕ the volume fraction of the polymer and n_1 and n_2 the number of moles of solvent and of polymer; the volume fraction is defined as

$$\phi = n_2 m(n_1 + mn_2)^{-1} \ , \tag{106}$$

with m the ratio of the partial molar volume of the polymer to the partial molar volume of the solvent. The parameter N is given as $N = n_1 + mn_2$, and the interaction parameter χ is assumed to depend on the temperature as

$$\chi = (1/2) + \varphi[(\Theta/T) - 1] \; , \tag{107}$$

with φ a constant and Θ the theta temperature of the solution.

To obtain the chemical potential of the components we write

$$\mu_i = (\partial \Delta G/\partial n_i)_{T,p} = (\partial \Delta G/\partial \phi)(\partial \phi/\partial n_i) \; , \tag{108}$$

with $(\partial \phi/\partial n_1) = -\phi/N$ and $(\partial \phi/\partial n_2) = (m/N)(1 - \phi)$. The equilibrium chemical potentials have the form

$$(\mu_1/RT)_{FH} = \ln(1 - \phi) + [1 - (1/m)]\phi + \chi\phi^2 \; , \tag{109}$$

$$(\mu_2/RT)_{FH} = \ln\phi + (1 - m)(1 - \phi) + \chi m(1 - \phi)^2 \; . \tag{110}$$

The spinodal line on the plane T–ϕ is given by the condition

$$(\partial\mu_1/\partial\phi) = 0 \; . \tag{111}$$

and the critical temperature, corresponding to the maximum of the spinodal line, is specified by the further condition

$$(\partial^2\mu_1/\partial\phi^2) = 0 \; . \tag{112}$$

From the Flory-Huggins expression Eq. (109) one immediately obtains for Eq. (111) and (112).

$$(1/RT)(\partial\mu_1/\partial\phi) = -(1 - \phi)^{-1} + [1 - (1/m)] + 2\chi\phi = 0 \; , \tag{113}$$

$$(1/RT)(\partial^2\mu_1/\partial\phi^2) = -(1 - \phi)^{-2} + 2\chi = 0 \; . \tag{114}$$

which yields for the coordinates of the critical point

$$\phi_c = [1 + m^{1/2}]^{-1}, \quad \chi_c = (1/2) + \varphi[(\Theta/T_c) - 1] = (1/2)[1 + m^{-1/2}]^2 \; . \tag{115}$$

For very high-N polymers, m is very high and the critical temperature T_c tends towards the theta temperature.

5.1 Non-Equilibrium Effects in Viscoelastic Fluids

To obtain the effects of the stress on the critical temperature, we add to Eq. (105) the non-equilibrium correction

$$\Delta G_s = vNJ(P_{12}^v)^2 \; . \tag{116}$$

An expression for J in terms of ϕ and of P^v_{12} is needed. Explicit expressions for J for polystyrene in dioctyl-phthalate are given by Rangel-Nafaile et al. [1], who obtained them directly from experiments. A different parameterization of their data is given in Criado-Sancho et al. [66]. Note that once J is given there are no free parameters in the theory.

By introducing Eq. (116) in Eq. (108) one finds for the shear contribution to the chemical potential of the solvent

$$\mu_{1s} = (\partial \Delta G_s / \partial \phi)(\partial \phi / \partial n_1) \ . \tag{117}$$

Including this term in Eqs. (111) and (112) one is led, instead of to Eqs. (113) and (114), to

$$- (1 - \phi)^{-1} + (1 - m^{-1}) + 2\chi\phi + (1/RT)(\partial\mu_{1s}/\partial\phi) = 0 \ , \tag{118}$$

$$- (1 - \phi)^{-2} + 2\chi + (1/RT)(\partial^2\mu_{1s}/\partial\phi^2) = 0 \ . \tag{119}$$

The numerical results for polystyrene in dioctyl-phthalate are given in Table 1 for $T_c(\dot{\gamma}) - T_c(0)$, with $T_c(0) = 285$ K, the critical temperature in the quiescent solution. The difference between the theoretical estimations and the experimental results is significant. Rangel-Nafaile et al. [1] attribute them to the sensitivity of the derivatives of G to relatively small errors in the experimental expressions for $J(\phi, P^v_{12})$. These authors, however, have found the coexistence or binodal curve with much less error (only up to 3 K between the experimental results and the theoretical results) using an algorithm based on the modified chemical potentials of both phases.

5.2 Non-Equilibrium Effects in Non-Newtonian Fluids

Up to now, the model which has been presented leads to an increase in the critical point, i.e., to a decrease of the solubility. It could be asked whether this is the general trend. It must be noted, however, that the effects of the shear may be different for different polymers: in some situations the presence of the shear enhances solubility, whereas in other situations it hinders it. The latter effect has been attributed by Wolf [2] to non-Newtonian effects. The model used by this author starts from the hypothesis that the contribution of the shear flow to the

Table. 1 Experimental and theoretical values of the shift in the critical temperature (K) for a solution of polystyrene in dioctyl-phthalate, at different values of P^v_{12}

P^v_{12} (N m^{-2})	100	200	400
Experimental	4	14	24
Rangel-Nafaile et al. [1]	1.7	4.4	10.5
Criado-Sancho et al. [66]	1.6	3.5	9.2

free energy should be included in the thermodynamic calculations in non-equilibrium. He uses for the non-equilibrium contribution the formula

$$\Delta G_s = VJ(\eta\dot{\gamma})^2 \ , \tag{120}$$

with V the molar volume of the system and J the steady-state shear compliance.

Furthermore, when the polymeric solution has a Newtonian behaviour, J may be computed from the Rouse model [34]

$$J = (CM/RTc)\{1 - [\eta_s/\eta(\dot{\gamma})]\}^2 \ , \tag{121}$$

where η_s is the viscosity of pure solvent and C is a constant.

The previous starting equations, together with a semiempirical correlation between $\eta(\dot{\gamma})$ and $\eta(0)$, allowed Wolf [2] to carry out the theoretical study of the system polystyrene/trans-decalin (PS/TD). It is worth saying that the mentioned author proposes for the interaction parameter a dependence not only on the temperature T, as was assumed in Eq. (107), but also on the composition through the expression

$$\chi = (A_0 - A_1) + 2(A_1 - A_2)\phi + 3A_2\phi^2 \ , \tag{122}$$

where A_i are functions of the temperature defined as

$$A_i = (g_{i0}/T) + g_{i1} T \ , \tag{123}$$

where g_{ij} are parameters which may be determined from measurements of the chemical potential of the solvent in different conditions of temperature and composition.

Furthermore, in order to carry out the calculations, it is necessary to complete the previous equations with other equations: 1) to correlate the viscosity with the shear rate and 2) to correlate viscosity with the composition of the mixture. In [2] Wolf uses a correlation $\eta(\dot{\gamma})$ justified from the Rouse model, whereas the dependence of the solvent viscosity with respect to the composition is given by the Martin equation [32].

The phase equilibrium is determined by Wolf by using the double-tangent method. In contrast, Criado-Sancho et al. [70] use the explicit expression for the chemical potential and its derivatives to determine the spinodal curve. Our results are shown in Fig. 1 and they are similar to the results obtained by Wolf, though it must be noted, in contrast with this author, that the effect of the shear is not always to increase the solubility for sufficiently high values of $\dot{\gamma}$, where it is seen that the temperature of the spinodal line is raised or lowered depending on the volume fraction.

When the molecular mass of the polymer is high enough, the non-Newtonian behaviour is enhanced: this is the situation in the PS solutions in trans-decalin reported by Kramer and Wolf [71], where the molecular mass of the polymer is 600 kg/mol and 1770 kg/mol. In these circumstances, the mentioned authors outline the inadequacy of the expressions for ΔG_s and η which have been used

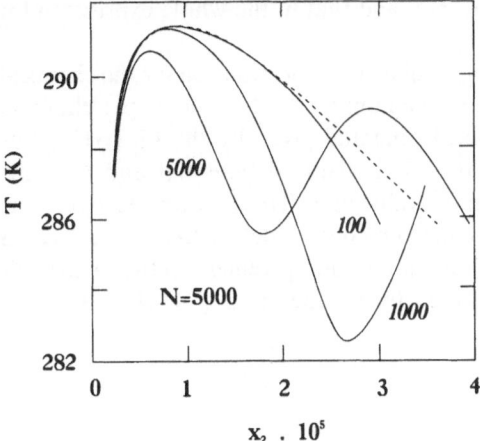

Fig. 1. Spinodal curves for several shear rates when Newtonian but viscoelastic behaviour is assumed. The *dashed curve* corresponds to the Flory-Huggins model

above. As an alternative, they propose for J a new functional relation by following Graessley [72], whereas for ΔG_s they use an expression proposed by Vinogradov and Malkin [73]. The new functional relation is

$$J = C_1 [\eta(0) - \eta_s][\eta(0)^2 \dot{\gamma}_{0.8}]^{-1} , \qquad (124)$$

where C_1 is a constant and $\dot{\gamma}_{0.8}$ corresponds to the shear rate for which $\eta(\dot{\gamma})$ is equal to 80% of $\eta(0)$.

The non-equilibrium contribution to the Gibbs free energy is given by

$$\Delta G_s = VJ[\eta(\dot{\gamma})\dot{\gamma}]^{2(1-d)} , \qquad (125)$$

where the parameter d is the slope of the curve $\eta(\dot{\gamma})$, with a change of sign.

After stating the new basic equations, Eqs. (124) and (125), Kramer et al. [74] follow a method which is not strictly the theoretical one, because both J and ΔG_s must be calculated in accordance with Eqs. (124) and (125) but using in such expressions the values of the viscosity corresponding to the "flow curves" experimentally obtained by these authors.

Later on, Criado-Sancho et al. [70] have shown that it is possible to obtain results analogous to those of [74] for the non-equilibrium contribution ΔG_s by assuming that the dependence of $\eta(\dot{\gamma})$ may be modelled through the following expression

$$\eta = \begin{cases} \eta_0(\dot{\gamma}/\dot{\gamma}_0)^B & \dot{\gamma} > \dot{\gamma}_0 \\ \eta_0 & \dot{\gamma} < \dot{\gamma}_0 \end{cases} , \qquad (126)$$

where in a first approximation the parameter B is directly proportional to the molar fraction of the polymer, and $\dot{\gamma}_0$ is inversely proportional to this quantity. Prior to assuming this functional dependence on composition of $\dot{\gamma}_0$ and B, use has been made of the fact that $B \approx -0.5$, and that in the whole expression for ΔG_s appears the factor $(\dot{\gamma}_0/\dot{\gamma}_{0.8})^{2B+1}$.

Our approximation leads to a non-equilibrium contribution for the chemical potential which is not very sensitive to the number N of segments per chain, in contrast with the equilibrium chemical potential given by the Flory-Huggins model (Fig. 2). As a consequence, the total chemical potential and thus the spinodal curve depend very much on the value of N. Thus, for an effective value of $N = 12\,200$, the spinodal curve predicted by our model for the system TD/PS with molecular mass 1770 kg mol exhibits a good agreement between the calculated values and the experimental data by Kramer et al. [74] (Fig. 3).

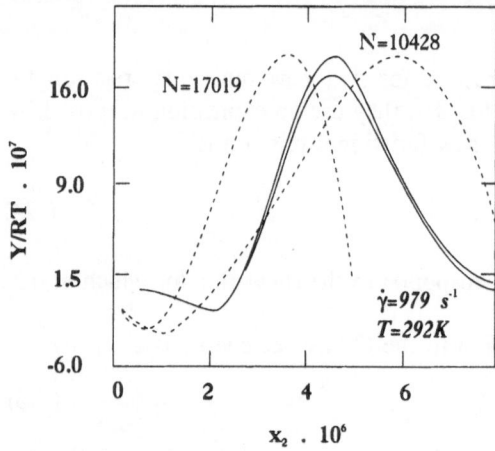

Fig. 2. The *continuous curves* represent the non-equilibrium contribution to the solvent chemical potential for the system TD/PS using two values of N (*upper curve*, $N = 10, 428$, *lower curve* $N = 17, 019$)). The *dashed curves* correspond to μ_1 in the Flory-Huggins model

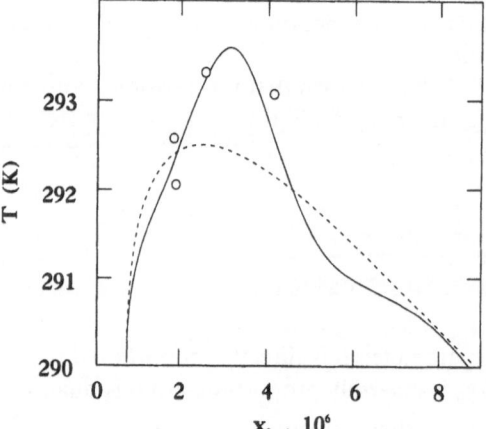

Fig. 3. Spinodal curve for the system TD/PS assuming non-Newtonian behaviour (*continuous curve*) and the one predicted by the Flory-Huggins model. Experimental points are taken from Wolf (1984) and Kramer et al. (1988)

5.3 The Flexibility Approach

A very different approach to the thermodynamics of phase separation under flow has been undertaken by Vrahopoulou-Gilbert and McHugh [75, 76]. The basis of their work is that, since the frictional forces under flow cause the macromolecules to uncoil from their random quiescent conformation to a more extended state of lower conformational entropy, this is equivalent to introduce a degree of rigidity into the polymer coils whose thermodynamic properties then become those of a system of semiflexible chains. This approach (introduction of macromolecular rigidity), and the other one (addition of stored free energy), are not equivalent and they may produce, according to the authors, quite different phase diagrams. The authors have criticized the usual expressions for the stored free energy for being based on solution properties rather than on individual polymer molecules, claiming that no distinction of the stored free energy in both phases was made.

The effective flexibility parameter ϑ of a semiflexible macromolecule stretched by a force F has been proposed to be [75, 77]

$$\vartheta = \frac{(z-2)\exp[-(E/kT)-(Fl/kT)]}{1+(z-2)\exp[-(E/kT)-(Fl/kT)]} , \tag{127}$$

with F the extending force applied to a segment of length l, and z and E respectively the lattice coordination number and intrinsic segment energy relative to the energetically favoured configuration prior to the application of the flow.

The relation between ϑ and the mean-square end-to-end separation of the macromolecule is

$$\langle r^2 \rangle = (x/\vartheta)/l^2(2-\vartheta) \tag{128}$$

for a chain having x segments of length l. For complete flexibility ($E = 0, F = 0$), $r^2 = xzl^2(z-2)$.

By comparing the Gibbs free energy of a system of semiflexible coils in solution containing n_2 polymer molecules and n_1 solvent molecules, one has the following expression for the Flory-Huggins mixing Gibbs free energy

$$(\Delta G_M/RT) = n_1 \ln v_1 + n_2 \ln v_2 + \chi x n_2 v_1$$
$$- n_2 \{\ln x + \ln(z/2e) - (x-2)\ln[(1-\vartheta e)]\} , \tag{129}$$

with e the exponential base.

The Gibbs free energy change associated with the flow is

$$\Delta G_s/RT = (x-2)\ln[(1-\vartheta)(z-1)] . \tag{130}$$

It is zero when $1 - \vartheta = (z-1)^{-1}$, which corresponds to Eq. (127) in the absence of flow.

For the construction of thermodynamic diagrams it is important, according to this view, to make a distinction between the situation when the flexibility of the macromolecules is the same in both phases or when they are different. In the first case, upon equating the chemical potentials in the two phases, the non-equilibrium contributions cancel out. When the flexibility of macromolecules in the concentrated phase, ϑ', differs from that in the dilute phase, ϑ, one obtains changes in the binodal curves. In order that the different flexibility has any effect on the binodal curve it is necessary that $\vartheta > \vartheta'$, i.e. that the flexibility of the chain in the more concentrated phase is less than the random coil flexibility.

When, in contrast to the flexibility approach, one takes an added free energy contribution, Vrahopoulou-Gilbert and McHugh [75] obtain, instead of Eq. (129)

$$(\Delta G_M/RT) = n_1 \ln v_1 + n_2 \ln v_2 + \chi x n_2 v_1$$
$$- n_2 \{\ln(x/2) + \ln[z/(z/1)] + (x-1)\ln[(z-1)/e]$$
$$+ n_2 G_s\} . \tag{131}$$

This expression only coincides with the previous one when G_s is given by Eq. (123), but not in other cases.

5.4 The Droplet Approach

In contrast with a model based on purely quasiequilibrium global thermodynamic analysis, some authors have taken a more local and semimicroscopic point of view by considering the role of the shear on the droplets constituting one phase inside the other one [78, 80]. Usually this approach is taken for the analysis of situations when solubility is enhanced by the flow. The underlying idea is that, since near the critical point the surface tension is very low, the flow may be very effective in producing smaller and smaller droplets. Homogenisation would be achieved when the dimension of the droplets is of the order of the macromolecular gyration radius [80].

Also in the droplet model, the non-equilibrium free energy is needed. Vanoene [79] takes for the flow contribution to the free energy the expression proposed by Janeshitz-Kriegl [50]

$$\Delta F = (1/2)[-\text{Tr}\mathbf{P} + 3p_z] , \tag{132}$$

with $p_z = p_{rr} = -p_{\theta\theta}$ the corresponding components of $\mathbf{P}$. For a Weissenberg fluid in Poiseuille flow, Eq. (132) gives

$$\Delta F = (1/2)[P_{\theta\theta} - P_{zz}] = (1/2)\sigma_2 . \tag{133}$$

When the contribution of the flow free energy is taken into account, as an additional contribution of the droplets one obtains for the interfacial tension under flow $\gamma_{\alpha\beta}$ of a droplet of fluid α in a matrix fluid β

$$\gamma_{\alpha\beta} = \gamma_{\alpha\beta}^{(0)} + (1/6)a_\alpha[(\sigma_2)_\alpha - (\sigma_2)_\beta] \ , \tag{134}$$

with a_α the droplet radius and $\gamma_{\alpha\beta}^{(0)}$ the interfacial tension in the absence of flow.

Vanoene uses $\sigma_2 = -JP_{rz}^2$, as derived by Coleman and Markovitz [40]. He comments that in fact this expression is valid at relatively low shear stress, whereas experimentally one finds that at high stress the normal-stress function becomes proportional to the shear stress rather than to the square of the shear stress.

Vanoene takes for J, the steady state compliance,

$$J = (0.4/RT\rho)(M_z M_{z+1}/M_w^2)M_w \ , \tag{135}$$

with M_w the weight-average molecular weight and M_z and M_{z+1} the so-called z and $z+1$ averages [50]. As a result, he concludes that when initially the domain size is larger than 1 μm the criterion for the formation of a droplet of an arbitrary phase α in a phase β requires that

$$(0.4/RT)[\rho_\alpha^{-1}(M_z M_{z+1}/M_w^2)_\alpha M_{w\alpha} - \rho_\beta^{-1}(M_z M_{z+1}/M_w^2)M_{w\beta}] > 0 \ . \tag{136}$$

From here it follows [79] that: 1) if phase β does not form droplets in phase α, then phase α will form droplets in phase β; 2) if a particular morphology is observed, it should not be influenced by the magnitude of the shear rate except for effects which can be attributed to the hydrodynamic stability of a particular mode of dispersion; 3) the phase with the largest normal stress function will form droplets.

The problem of drop formation may be treated from a thermodynamic point of view [79] by considering the total surface free energy of the dispersion

$$F = \sum A_{ij}\gamma_{ij} \ , \tag{137}$$

with A_{ij} the surface area of droplets i in the matrix j. If the volume fractions of phases α and β are $\phi_\alpha = \phi_\alpha^{(0)} - \phi_\beta^{(0)}$ and $\phi_\beta = 1 - \phi_\alpha^{(0)} + \phi_\beta^{(0)}$, with $\phi_\alpha^{(0)}$ the volume fraction of phase α within the β droplets and $\phi_\beta^{(0)}$ the volume fraction of phase β within the α droplets, then one has

$$F = n_\alpha 4\pi a_\alpha^2 [\gamma_{\alpha\beta}^{(0)} + (1/6)a_\alpha(\Delta\sigma_2)] + n_\beta 4\pi a_\beta^2 [\gamma_{\alpha\beta}^{(0)} - (1/6)a_\beta(\Delta\sigma_2)] \ , \tag{138}$$

if one takes account of the non-equilibrium contribution to the free energy. The droplet approach may be of special interest for the analysis of the segregation in two phases and for the study of the inhomogeneous regions with several domain structures.

6 Solubility and Phase Separation Under Flow: Experimental Results

Changes in solubility of polymers due to the presence of a flow have been reported many times in the literature and have received attention from several points of view. We do not pretend here to be exhaustive on the literature. The papers by Rangel-Nafaile et al. [1] and of Barham and Keller [7] present a wide bibliography on the subject.

The observations are usually made on the basis of turbidity or of solubility. In the first case they are based on the direct observation of a change in the cloud point of the solution under flow as compared with the cloud point of the solution at rest. In other cases the observation stems from changes in viscosity with time at constant shear rate or constant shear stress.

In order to relate the results of the previous section to a concrete example, we assume a polymer solution flowing in a cylindrical duct of radius R. The shear viscous pressure depends on the distance from the axis r as

$$P_{12}^v = -(\Delta p/2l)r ,$$ (139)

with Δp the pressure difference between both ends of the duct and l the length of the duct.

If P_{12}^v is expressed by the usual relation $P_{12}^v = -\eta(\dot\gamma)\dot\gamma$ and remembering that $\eta(\dot\gamma)$ is given by our modelled expression at Eq. (126), we obtain for the velocity of the fluid at a distance r

$$v(r) = A[R^{(B+2)/(B+1)} - r^{(B+2)/(B+1)}] ,$$ (140)

where the parameter A is defined as

$$A = \left(\frac{B+1}{B+2}\right)\left(\frac{\Delta p}{2l\eta_0}\dot\gamma_0^B\right)^{\frac{1}{B+1}} .$$ (141)

The flow rate Q is easily obtained by integration of Eq. (140) over the cross section of the duct. The result is

$$Q = \pi A \frac{B+2}{3B+4} R^{\frac{3B+4}{B+1}} .$$ (142)

While the solution is flowing in the duct, the shear rate is highest at the wall. If the shear rate at the wall $\dot\gamma_w$ is lower than the critical rate $\dot\gamma_c$ obtained by thermodynamic arguments, the fluid remains homogeneous. However, when the shear rate at the wall becomes a little higher than the critical shear rate, the fluid near the wall will decompose into two phases (see Fig. 4). From the definition of

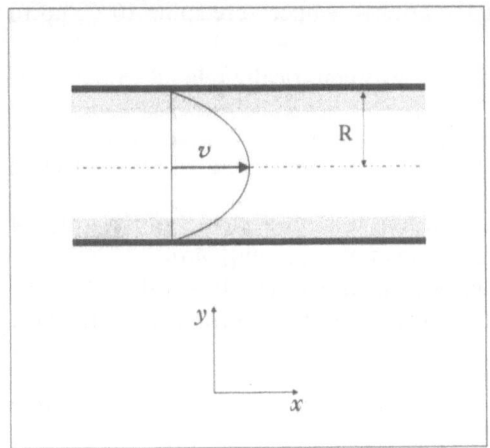

Fig. 4. Polymer solution flows along a cylindrical duct. In the region where the shear rate exceeds the critical values (*dark region*), the homogeneous fluid will decompose into two phases

$\dot{\gamma}$ and the velocity field given by Eq. (140) the value of the shear rate on the wall $\dot{\gamma}_w$ can be written as

$$\dot{\gamma}_w = \frac{B+2}{B+1} A R^{\frac{1}{B+1}} . \tag{143}$$

Elimination of A between Eqs. (142) and (143) allows us to write the following relation for the flux rate in terms of $\dot{\gamma}_w$

$$Q = \pi \dot{\gamma}_w \frac{B+1}{3B+4} R^3 . \tag{144}$$

It is worth noting that when $\dot{\gamma}_c = \dot{\gamma}_w$ Eq. (144) gives the critical flow rate Q_c above which phase separation takes place.

For the system TD/PS we previously considered, one could use the results of Fig. 3 to estimate the correlation between Q_c and the radius of the duct. The critical shear rate will depend on the temperature and the composition. For a composition near to the critical one and a temperature $T = 293.6$ K it follows from Fig. 3 that the critical shear rate is 979 s^{-1} and, as a consequence, one has $Q_c = 615 R^3$ if we take $B = -0.5$. Of course, the value of the numerical coefficient will be a function of temperature and composition of the solution. This is only a simple illustration of how results of practical interest could be obtained from the general results of the previous section.

To our knowledge, the first reports on the influence of flow on the change of polymer solubility are due to Silberberg and Kuhn [81]. These authors studied a solution of two different polymers (polystyrene and ethylcellulose) in a common solvent (benzene). In this system, the critical temperature (at the concentrations studied by the authors) below which phase separation occurs is $T_c = 31.7°$C. The presence of a flow with shear rate $\dot{\gamma} = 200$ s^{-1} was able to reverse the separation and to lower the critical temperature some 10°C, up to 21.7°C.

The change in the critical temperature was empirically related to the shear rate as [78, 81]

$$T_c(\dot{\gamma}) - T_c(0) = -(1/20)\dot{\gamma} , \qquad (145)$$

with $\dot{\gamma}$ expressed in s^{-1}. The phase separation was reversed at $\dot{\gamma}$ of the order of 200 s^{-1}, in laminar flow (the turbulence was not beginning until $\dot{\gamma} = 280 \, s^{-1}$). Silberberg and Kuhn advanced a thermodynamic explanation of this phenomenon. On the one side, the rate at which thermal energy is produced by the field of flow is

$$\text{Rate of thermal energy input} = \eta\dot{\gamma}^2 . \qquad (146)$$

On the other side, the rate at which free energy is consumed by the process of mixing the demixed fluid was obtained according to a formulation by Scott [82]. According to him, the free energy of 1 cm^3 of the system studied by the authors made homogeneous at a temperature ΔT below T_c would be $6 \times 10^{-2}(\Delta T)^2$ erg/cm^3 larger than the free energy of 1 cm^3 of the corresponding stable demixed two-phase system at that temperature. If one assumes that the consumption of this free energy takes place in the time τ_{dif} taken by the coiled polymer to diffuse a distance of the order of its end-to-end distance $(5 \times 10^{-6}$ cm) with a diffusion constant D of the order of 10^{-8} cm^2/s, one may evaluate the rate of free energy consumption as

$$\text{Rate of free energy consumption} = A(\Delta T)^2/\tau_{dif} . \qquad (147)$$

Comparison of Eqs. (146) and (147) yields a stored free energy of the order of

$$\Delta f = A(\Delta T)^2 = \tau_{dif}\,\eta\dot{\gamma}^2 . \qquad (148)$$

Insertion into Eq. (148) of the values of $\eta = 0.11$ Poise, $A = 6 \times 10^{-2}$, and $\tau_{dif} = (\Delta r)^2/2D = 1.25 \times 10^{-3}$ s, and equating Eqs. (146) and (147) yields

$$\Delta T = -[\eta\tau_{dif}/A]^{1/2}\dot{\gamma} , \qquad (149)$$

which numerically corresponds to Eq. (145).

This argument may be given a more accurate form, closer to the usual developments of equilibrium thermodynamics. Inclusion of the effects of the flow into the free energy for the analysis of the phase separation problems was made for the first time by ver Stratte and Philipoff [42] by using Marrucci's formula at Eq. (42) for the free-energy excess. The authors observed that a clear polymer solution, upon passing from a reservoir into a capillary (a region of high shear rate), became turbid in the capillary. After leaving the capillary the solution became clear again. This phenomenon was exhibited only above a certain shear rate or shear stress depending on the polymer, solvent, concentration and temperature. The viscosity of the solution was seen to undergo unusual changes about the turbidity point, passing through a minimum as the shear rate increased. These authors studied polystyrene solutions in two different solvents, decalin and di(-2-ethyl-exil-phthalate). The visual observation of the cloud points was always made in laminar flow conditions. These authors observed an increase in the cloud point temperature, instead of the decrease seen in the two-polymers-one-solvent system of Silberberg and Kuhn, i.e. the cloud point or precipitation curves were shifted to higher temperatures.

Rangel-Nafaile et al. [1] studied solutions of polystyrene in dioctyl-phthalate and they observed a shift of the cloud point curves towards higher temperatures (Fig. 2). The shift in the critical temperature could attain 24°C (at $P_{12}^v = 400$ N/cm^2). They also used a thermodynamic formalism based on Marrucci's formula at Eq. (42) for the excess free energy.

Also in 1984, Wolf used the thermodynamic theory based on Flory-Huggins equation for the mixing free energy and the Marrucci's formula at Eq. (42) for the change in the free energy due to the flow. He studied the phase separation of polystyrene solutions in transdecalin. He observed a decrease in the demixing or cloud point temperatures for relatively low-molecular-weight polystyrene ($N = 5000$) and an increase in the cloud point temperatures for high-molecular-weight polystyrene ($M = 1\,700\,000$) [81]. This behaviour may be understood from what has been said in Sect. 5, as it seems to be due to the non-Newtonian effects arising in high-weight molecular solutions. Furthermore, Wolf found that the demixing curves of the flowing solutions normally exhibited two maxima instead of one [2]. Spinodal curves with two maxima under flow have also been studied by Lingaae-Jorgensen and Sondergaard [84]. Furthermore, Wolf [2] predicts a shear-induced coexistence of three-liquid phases at the temperature at which the branches of the modified critical temperatures cross each other. This point would represent the largest extension of the dissolved state of the polymer than can be achieved by a given shear rate and it was called eulytic point by Wolf. To our knowledge, experimental information on this point is still lacking.

Takebe et al. [85, 86] have studied the shear-induced homogenisation of semidilute solutions of polybutadiene and polystyrene in a common solvent (dioctylphthalate). In this system the presence of a flow enchances solubility, and the decrease in the critical temperature is given by

$$T_c(0) - T_c(\dot{\gamma}) = K\,T_c(0)\dot{\gamma}^{0.50 \pm 0.02} \;, \tag{150}$$

with $K = (2.6 \pm 0.6) \times 10^{-3}$. This expression is different from Eq. (149) and its exponent is almost the same as the one which is found for the changes in solubility of small molecules under shear flow [87, 88]. Such a value of the exponent has been explained by Onuki and Kawasaki [89] by means of the renormalization group approach. However, the value of the prefactor K is four orders of magnitude higher in polymer solutions than in mixtures of small molecules.

Non-equilibrium phase diagrams of PMMA in di-methyl-phthalate, including the coexistence and the spinodal lines, depending on temperature, concentration and shear rate, for the solutions in Couette cylindrical flow have also been studied by Barham and Keller [7], who observed a shift of the coexistence lines towards higher temperatures. The changes may be of the order of up to 30°C.

Here we have referred to upper critical points, i.e. to the maximum temperature below which phase separation is found. Some mixtures present, in contrast, a lower critical point: the solution is miscible at low temperatures but it separates at temperatures high enough. The effect of shear on the lower critical temperature of a solution of polystyrene in poly(vinyl methyl ether) under shear has been explored by Mazich and Carr [90]. Here the flow raises the phase transition temperature of the polymer blend and favours solubility (see Fig. 5).

Another source of study is the one concerning the changes in the dynamical properties such as the shear viscosity. Time-dependent changes of the viscosity under certain conditions were observed and attributed to a network formation induced by the flow. The degree up to which the network is formed would influence the viscosity. The entanglement and disentanglement dynamics would then produce time-dependent changes in the apparent viscosity. Temporary network formation in sheared solutions of poly (methyl methacrylate) in Aroclor were observed by Peterlin et al. [91, 92] and by Matsuo et al. [93].

Barham and Keller [7] have scrutinized possible different explanations for the anomalous flow phenomena observed in solutions. Indeed, a decrease in flow rate at constant stress with increasing time, or an increase in shear stress at constant shear rate with increasing time, could be attributed either to some change of phase induced by the flow or to the formation of an adsorption entanglement layer, the first being a bulk and the second a boundary phenomenon. The origin of turbidity could also be attributed to both phenomena. Furthermore, the presence of a layer on the viscometer is not an unequivocal exclusion of the formation of aggregated phases in the bulk.

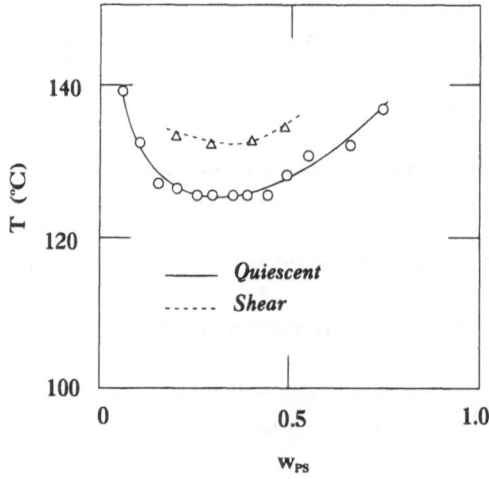

Fig. 5. In solutions with a lower critical point, an increase of the spinodal temperature enhances the solubility

Barham and Keller have proposed experiments under an oscillatory flow as a method to make a distinction between both situations, in which the adsorption layer effects could be much reduced with respect to a steady-state experiment because only a few macromolecules could pass near the layer. In contrast, the shear stress in bulk would be maintained all the time at a relatively high value. In this way, bulk phenomena would not be much affected whereas surface phenomena would be much reduced. In their paper, these authors present a classification of the different experiments according to the kind of explanation (spinodal or nucleated phase separation, or adsorption-entanglement layers) that, in the light of these oscillatory experiments, are the most plausible ones.

7 Thermodynamic Degradation of Polymers Under Flow

The verification that mechanical stresses may produce the degradation of polymers and the first attempts to explain this phenomenon are due to Staudinger [94]. Since then, the phenomenon of degradation under flow has been studied by assuming that only the bonds between monomeric units are broken, and one of the problems has concerned the position of the breaking point along the macromolecule. Some authors in the 1930s such as Freudenberg et al. [95, 96], Kuhn [97] and Ekenstam [98], have considered that bond reactivity cannot be treated as a random process because each bond in a macromolecule depends on its particular chemical environment.

Some experimental results [99, 100] seem to indicate that in some degradation processes all bonds have a similar breaking behaviour, whereas some other studies on *dextran* [101] point out that the breaking of macromolecules due to ultrasound takes place preferentially in a central position along the chain.

The analysis of polymer degradation is usually undertaken through techniques which are typical of chemical kinetics, which is a methodology which implies some specific reaction mechanism as the basis to set the equations which describe the evolution of the system composition with time. From this point of view, the earlier formulations [96, 102] and some of its generalizations [103–105] are from the 1930s and 1940s.

A kinetic model has also been the basis of the studies carried out by Nguyen and Kausch [12, 13] on the mechanical degradation of polymers in transient elongation flow. In elongation flows, degradation is more efficient than in simple shear flow, because in the latter the particle rotates with the flow field and only spends, at each turn, a limited amount of time in high strain-rate regions, whereas in elongational flow the fluid element is in a continued state of dilatation. Nguyen and Kausch [12, 13] have studied the evolution of molecular size distribution during the degradation process starting from the expression of the kinetics describing the rate of bond scission, and incorporated the spatial distribution and the time dependence of strain-rate in transient flow; their results, as well as a detailed comparison with experimental data, have been reviewed [13]. Steady elongational flow could in principle be introduced in the thermodynamic description by using the form given by Eq. (5) for $\mathbf{P}^v$ instead of Eq. (3) corresponding to shear flow; however, the thermodynamic description would not be sufficient for the analysis of the transient behaviour, which requires kinetic mechanisms to be taken specifically into account.

The basic hypothesis underlying any kinetic mechanism for polymer degradation is to assume that each chain is broken only at one point (i.e. only two fragments are produced from each broken macromolecule). The second hypothesis assumes that each elementary reaction

$$P_j \rightarrow P_i + P_{j-i} \tag{151}$$

follows first-order kinetics, i.e.

$$\frac{dn_i}{dt} = k_{ji} n_j \;, \tag{152}$$

where n_i and n_j stand for the number of macromolecules with i and j degrees of polymerization, respectively, and k_{ji} is the kinetic constant describing the breaking of a macromolecule with j monomers to give a macromolecule with i monomers and another one with $j - i$ monomers.

The kinetic model described here is widely used since the work by Basedow et al. [106]. These authors consider a polydisperse macromolecule system, with r the maximum degree of polymerization, in such a way that the coupling of the degradation reactions is given by the following set of linear differential equations

$$\begin{pmatrix} dn_1/dt \\ . \\ . \\ . \\ dn_r/dt \end{pmatrix} = \mathbf{A} \begin{pmatrix} n_1 \\ . \\ . \\ . \\ n_r \end{pmatrix}, \tag{153}$$

where the matrix $\mathbf{A}$ of the coefficients has the form

$$\mathbf{A} = \begin{pmatrix} 0 & 2k_{21} & k_{31} + k_{32} & . & k_{r1} + k_{r,r-1} \\ 0 & -k_{21} & k_{32} + k_{31} & . & k_{r2} + k_{r,r-2} \\ 0 & 0 & -\sum_{j=1}^{2} k_{3j} & . & k_{r3} + k_{r,r-3} \\ . & . & . & . & . \\ 0 & 0 & 0 & 0 & -\sum_{j=1}^{r-1} k_{rj} \end{pmatrix}, \tag{154}$$

according to the reaction mechanism proposed by Simha [103].

Besides numerical complexities, the solution of the above system is only possible through the classical techniques used in the analysis of differential equations, so that the only (and very non-trivial) problem is to know the values of the kinetic constants k_{ij}. In the paper by Basedow et al. [106], three situations are analyzed: (1) the simplest one is to assume that all bonds have the same breaking probability so that the value of k_{ij} is independent of i and j and the macromolecular fragmentation is a random process; (2) a slightly more complicated situation assumes that the kinetic constant depends on the length of the chain which is broken, but not on the length of the ensuring fragments, i.e. the value of k_{ij} depends on i but not on j; (3) the most complicated situation arises when k_{ij} depends on the position of the breaking point; this situation is usually dealt with by assuming that the value of the kinetic constant varies along the chain according to a parabola whose minimum is at the central point of the macromolecule, i.e. one assumes that the bonds near the ends of the chain are more easily broken than the bonds in the central region.

An opposite point of view with respect to the breaking point of the polymer under stress is obtained when one carries out a mechanical analysis taking into account the entanglement of the chains [107]: in this analysis, the central bonds are those preferentially broken. This fact, proposed for the first time by Frenkel [108], is the most important conclusion by Bueche, who found, furthermore, that the breaking probability distribution of monomer-monomer bonds is a Gaussian function which has its maximum at the center of the macromolecule. These conclusions have been experimentally verified in several occasions [109–111].

One may propose the additional hypothesis that the kinetic constant for the breaking of any chain is given by a Gaussian function [112] and consider again the previous mechanism. In this way, one would have a new kinetic model which has been applied to the analysis of polystyrene solutions [12, 13, 113–115], where one assumes that k_{ij} is given by

$$k_{ij} = k(i-1)^X [\sigma_i(2\pi)^{1/2}]^{-1} \exp\{-(1/2\sigma_i^2)(j - i/2)^2\} , \qquad (155)$$

where the standard deviation σ_i is supposed to be proportional to the length of the chain

$$\sigma_i = R(i-1) . \qquad (156)$$

The parameters R and X appearing in Eq. (155) are not known a priori and k is a global kinetic constant which appears in the kinetic equation which describes the breaking of bonds in a global analysis of all the macromolecules in the system, and which may be experimentally determined.

Since a polymer in solution is a polydisperse system, the number of macromolecules of a given length is given by the molecular weight distribution (MWD). From the MWD one may obtain the set of values n_i which are the solution of the previous set of differential equations and, by fitting the subsequent curve with the experimental data, one may obtain the values of R and X.

However, there is an important point: the kinetic mechanism by Simha and Basedow does not include the recombination of macromolecules. Therefore, in order that the experimental data are consistent with the results derived from the theoretical model it is necessary to carry out the measurements under conditions where the recombination of the fragments produced in the breaking of other macromolecules is inhibited. A common technique to achieve this situation is to introduce in the system some scavenger radicals [116] which neutralize the fragments produced in the breaking of the chains. When such recombination inhibitors are not present, the MWD of a sample of polystyrene with a narrow distribution is practically not modified during the degradation, whereas such modification is more conspicuous for wide distributions of the kind of most probable distribution [114, 115]. We will see below that these conclusions are also found by using a purely thermodynamic model [117].

Therefore the kinetic mechanism proposed up to now is not the most general mechanism, and we propose to consider it as a particular case of a degradation-combination scheme of the form

$$P_j \rightleftarrows P_i + P_{j-i} \tag{157}$$

instead of the elementary reaction given in Eq. (151).

When one considers the equilibrium in the scheme at Eq. (157), one must include the new kinetic constant κ_{ji} which corresponds to the recombination of the chains P_i and P_{j-i} and which is related to the chemical equilibrium constant K_{ij} in the usual form

$$K_{ij} = k_{ji}/\kappa_{ji} . \tag{158}$$

Furthermore, the kinetic equations in the degradation-combination mechanism are now the nonlinear set of differential equations

$$dn/dt = [A - B(n_1, \dots n_r)].n , \tag{159}$$

where n is the column vector appearing on the right-hand-side of Eq. (153), A is the matrix of coefficients defined in Eq. (4) and where one has introduced the new matrix

$$\begin{pmatrix} \kappa_{21} n_1 & \kappa_{31} n_1 & . & \kappa_{r1} n_1 & 0 \\ \kappa_{32} n_2 & \kappa_{42} n_2 & . & 0 & 0 \\ . & . & . & . & . \\ \kappa_{r,r-1} n_{r-1} & 0 & . & 0 & 0 \\ 0 & 0 & 0 & 0 & 0 \end{pmatrix} \tag{160}$$

Although in the bibliography the effect of the shear rate on the degradation process is considered [114, 115], it is easy to see in the previous expressions that $\dot{\gamma}$ does not appear in the formalism. Thus the above reaction mechanisms are not satisfactory to formulate a theoretical model for the interpretation of the shear effects on polymer degradation. A way to include the effects of the shear is to consider that the chains are broken when the stored energy due to the shear is applied at some grip points [118] and to use for the stored energy the previous expression used in the analysis of the modification of the phase diagram, where $\dot{\gamma}$ appears explicitly with other viscoelastic properties of the system. Another way would be to assume that the kinetic constants depend on the shear rate, as some calculations in the kinetic theory of gases seem to suggest [119]. An analysis based on strain-rate or shear-rate dependence of the kinetic constants [12, 13] is more detailed than a thermodynamic analysis, which is only concerned with the shear-rate dependence of the equilibrium constant rather than with the shear-rate dependence of the whole kinetic mechanisms.

An alternative analysis to the above kinetic models is a strictly thermodynamical formulation [117] based on the formalism of non-equilibrium chemical potentials which leads to a modification of the equilibrium constant at Eq. (157) under shear. The starting point in our theory is to set equal to zero the affinity A when the chemical equilibrium is reached in the system in a steady shear state, i.e.

$$A = \sum_k \nu_k \mu_k = 0 \ . \tag{161}$$

where ν_k refers to the stoichiometric coefficient of the component k. The derivation of this condition from kinetic considerations would be analogous to that in Eqs. (102)–(104).

The generalized Gibbs function under shear will have the form

$$dG' = -S'dT + V'dp + \sum_k \mu'_k dN_k + (V\tau/\eta)\mathbf{P}^v : d\mathbf{P}^v \ , \tag{162}$$

from where is derived the expression for the generalized chemical potential

$$\mu'_k(T, p, n_j, \mathbf{P}^v) = \mu_k(T, p, N_j) + (1/2)\{\partial(V\tau/n)/\partial N_k\}\mathbf{P}^v : d\mathbf{P}^v \ , \tag{163}$$

where μ_k is the usual local-equilibrium chemical potential.

From the expression for the non-equilibrium chemical potential, it is possible to generalize the affinity by using μ'_k instead of μ_k. In this way, by imposing the condition at Eq. (161) one obtains the relation

$$\prod_k a_k^{\nu_k} = K(T, p)\lambda(T, p, N_j, \mathbf{P}^v) \ , \tag{164}$$

where a_k is the activity, defined in the usual form

$$\mu_k = \mu_k^0(T, p) + RT \ln a_k \ , \tag{165}$$

and K is the equilibrium constant given by

$$\ln K = -\frac{1}{RT} \sum_k \nu_k \mu_k^0 \ . \tag{166}$$

Therefore the function λ, which is given by

$$\lambda = \exp\left\{ -\frac{1}{RT}\sum_j v_j[\tfrac{1}{2}\partial(V\tau/\eta)/\partial N_j]\mathbf{P}^v:\mathbf{P}^v\right\}, \tag{167}$$

will act as a correction to the equilibrium constant which takes into account the effects of the viscous shear over the fluid.

Note that the non-equilibrium chemical potential may also be written as

$$\mu'_k(T,p,N_j,\mathbf{P}^v) = \mu_k(T,p,N_j) + \mu_k^{(s)}(T,p,N_j,\dot{\gamma}), \tag{168}$$

where the term $\mu_k^{(s)}$ contains the non-equilibrium contributions to the chemical potential whose explicit expression may be determined by analogous arguments to those used above in the analysis of the modification of the phase diagram of the polystyrene-trans-decalin system under shear [70].

Another hypothesis introduced in our model to study the polymer degradation in solutions under shear is to consider the system as a multicomponent mixture, constituted by the solvent (component 1) and a set of polymeric species P_i (with i the degree of polymerization). In this way, the polydispersivity plays a fundamental role in the theoretical treatment, because the degradation is attributed to the equilibrium constant of the breaking-recombination reactions of polymeric chains expressed in (6.7) due to the modification arising from the effects of the shear.

The consideration of the solute as a mixture of the chemical species P_i leads to a change with respect to the formalism used above in the analysis of the phase diagram. In that case, polydispersivity did not appear explicitly in the formalism, but now it is crucial, of course, to take into account its effects by including the chemical potentials for macromolecules with different lengths

$$\mu'_j = \mu_j + RT\left(\frac{\partial\Delta G_s}{\partial n_j}\right)_{n_k,\dot{\gamma}}, \tag{169}$$

where ΔG_s is the non-equilibrium contribution to the Gibbs free energy as explained in Sect. 6, and j refers to the number of monomeric units in the chain.

Furthermore, the average character of the polymeric molecular mass in the solution leads us to consider M_2 as a function of n_i and to redefine the concept of molar fractions of the solute x_2. Now, instead of a binary mixture, we have a system with $\tilde{n}_1$ mol of solvent and n_i mol of macromolecules of degree of polymerization i. As a consequence, the total number of mol of solute is given by

$$\tilde{n}_2 = \sum_i n_i, \tag{170}$$

and one may define the global molar fraction of solute as

$$x_2 = \tilde{n}_2/\tilde{n}_0 \tag{171}$$

by introducing the new quantity $\tilde{n}_0 = \tilde{n}_1 + \tilde{n}_2$ which accounts for the total number of mol in the system.

From these hypotheses one gets the following results

$$\frac{1}{RT}\mu_j^{(s)} = +\frac{x_1}{x_2}M + \left(M_2 - \frac{V_2\tilde{c}}{[\eta]}\right)N$$
$$+ \left(\frac{1}{M_2} + x_2 N\right)\tilde{n}_0 \left(\frac{\partial M_2}{\partial n_j}\right)_{n_k,\dot{\gamma}}, \tag{172}$$

where V_2 is the molar volume of the polymer, x_1 and x_2 are the global molar fractions and the functions $\tilde{c}$, M and N depend explicitly on the composition, the shear rate and the viscoelastic properties [117].

For the mean molecular mass M_2 a generalized expression of the kind

$$M_2 = \sum_i n_i M_i^q / \sum_i n_i M_i^{q-1}, \tag{173}$$

is used where M_i is the molecular mass of macromolecules with degree of polymerization i and q is a parameter which depends on the kind of average carried out ($q = 1$, number average, $q = 2$, weight average, $q = 3$, z average, etc).

If one applies Eq. (164) to the equilibrium at Eq. (157) and one assumes for simplicity an ideal mixture one may write

$$\frac{N_j}{N_i N_{j-1}} = \frac{N_j^{(0)}}{N_i^{(0)} N_{j-1}^{(0)}} \lambda(i,j)^{-1}, \tag{174}$$

where N_k is the number of chains with k units and the superscript (0) refers to values with $\dot{y} = 0$. The function λ is obtained by comparison with Eq. (164) and is given by

$$\lambda = \exp\{-\tau_1[(j-i)^q + i^q - j^q] - \tau_2$$
$$[(j-i)^{q-1} + i^{q-1} - j^{q-1}] - \delta\}, \tag{175}$$

where τ_1, τ_2 are functions which depend on q, and δ is independent of q.

When one uses the Schulz distribution [120]

$$W(i) = [\Gamma(h+1)]^{-1}(yM_0)^{h+1}i^h\exp(-yM_0 i), \tag{176}$$

where $W(i)$ is the polymeric mass with a degree of polymerisation between i and $i + di$ (i is considered as a continuous parameter in this model), M_0 is the mass of monomer and h and y are parameters which characterize the solution, Eq. (175) leads to

$$[\Gamma(h)/\Gamma(h_0)][(y_0 M_0)^h/(y_0 M_0)^{h_0}] = [j/i(j-i)]^{h_0-h}\lambda(i,j)^{-1}, \tag{177}$$

with h_0 and y_0 the parameters of the distribution in absence of shear.

When one uses a most probable distribution of the kind

$$W(i) = \alpha^2 i (1 - \alpha)^{i-1}, \tag{178}$$

the parameters of the equilibrium distributions without shear and with shear are related as

$$\alpha(1 - \alpha)^{-1} = \alpha_0 (1 - \alpha_0)^{-1} \lambda(\alpha) \tag{179}$$

in the particular case when $q = 1$.

From Eq. (175) one may obtain several conclusions about the degradation mechanism. In the case when one considers the molecular mass averaged over the number ($q = 1$) it is obvious that λ does not depend on i nor on j: this implies that the modification of the equilibrium constant does not depend on the length of the original chain nor on the length of the resulting fragments after the chain

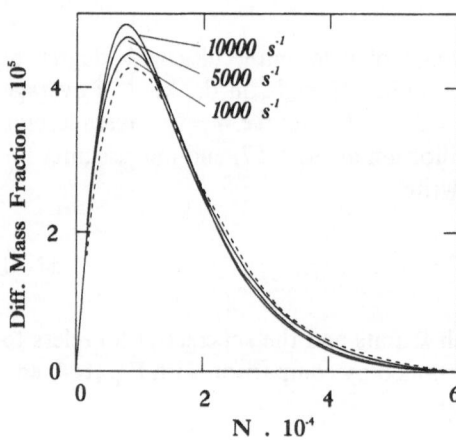

Fig. 6. Differential mass fraction of the system polystyrene-trans-decalin for a molecular weight of 1170 Kg/mol under several shear rates considering satisfied the most probable distribution. *Dashed line* corresponds to an equilibrium situation

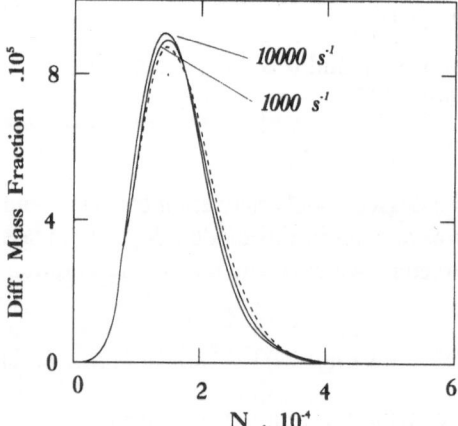

Fig. 7. Differential mass fraction of the system polystyrene-trans-decalin for a molecular weight of 1170 Kg/mol under several shear rates assuming a Schultz distribution with polydispersivity index 1.14 (experimental value). *Dashed line* corresponds to an equilibrium situation

is broken. This result is not very satisfactory from the physical point of view and it corroborates the empirical equations by Grassley [121] where other kind of averages appear, such as the weight average ($q = 2$) or even the z average ($q = 3$) in the expressions for J. In contrast, when one adopts for M_2 a weight average ($q = 2$) it follows that λ has a minimum when the chain is broken at the centre and, therefore, the modification of the MWD is maximum in these conditions, in agreement with the experimental results.

On the other side, the molecular mass before the degradation and after the degradation due to the shear may be related through the following approximation

$$M_w/M_w^{(0)} = \lambda(i,j)^{-1/h}, \tag{180}$$

where, since $|\tau_1| \ll |\delta|$, λ is practically independent of i and j.

When this model is applied to the analysis of a solution of polystyrene of molecular mass 1770 kg/mol in trans-decalin, it is observed that in the degradation process the parameter q has only a very slight influence, whereas the changes in the polydispersivity index, M_w/M_n does influence the results much more strongly (Figs. 6 and 7) in agreement with the experimental observations [114, 115].

8 Dynamical Approach: Non-Equilibrium Chemical Potential and Diffusion Flux

Thermodynamic arguments cannot be blindly extrapolated to non-equilibrium steady states. Thus some authors have argued against the use of thermodynamic arguments in the analysis of the phase diagram of polymer solutions under shear. We will discuss in this section under which conditions dynamical analyses corroborate the conclusions obtained from the point of view of thermodynamics.

Some authors [6, 120] have proposed purely dynamical approaches in which the phase separation under shear is examined as an instability of stationary solutions to some dynamical equations. In Onuki's analysis [6] the chemical potential does not seem to play an explicit role. However, he obtains an effective diffusion coefficient in the shear direction which may become negative under some conditions: this makes the homogeneous state become unstable against fluctuations. It must be noted, however, that up to now Onuki has only been able to obtain an increase in the spinodal temperature under shear, even in the non-Newtonian case. It would be of interest to generalize his theory in order to be able to explain the decrease in the spinodal temperature which is observed in some occasions, as mentioned in Sect. 5.

This dynamical approach is in fact not incompatible with the approach proposed in our paper. To study in detail the situations where both approaches are equivalent we will generalize the Gibbs equation, Eq. (21), to include the effects of diffusion. We will have [22, 123, 124]

$$ds = T^{-1}du + T^{-1}pdv - T^{-1}\tilde{\mu}dc_1 - va_1 J_1.dJ_1 - va_2 \mathbf{P}^v:d\mathbf{P}^v, \quad (181)$$

with c_1 the mass fraction of the solute, $\tilde{\mu} = \mu_1 - \mu_2$ the difference between the specific chemical potentials of the solute and the solvent. According to Eq. (21) the parameter α_2 is equal to $\tau/2\eta T$ and α_1 will be identified below. The corresponding generalized expression for the entropy flux is [2, 123, 124]

$$\mathbf{J}^s = T^{-1}\mathbf{q} - T^{-1}\tilde{\mu}\mathbf{J}_1 + \beta\mathbf{P}^v.\mathbf{J}_1, \quad (182)$$

where β is a coefficient which will appear in the dynamical equations. The dynamical equations for the dissipative fluxes $\mathbf{P}^v$ and $\mathbf{J}_1$ are then

$$\tau_1 \mathbf{J}_1^* + \mathbf{J}_1 = -D'\nabla\tilde{\mu} + D'T\nabla.(\beta\mathbf{P}^v), \quad (183)$$

$$\tau_2(\mathbf{P}^v)^* + \mathbf{P}^v = -2\eta\langle\nabla v\rangle + 2\eta T\beta\langle\nabla\mathbf{J}_1\rangle. \quad (184)$$

The upper star stands for an objective time derivative, as for instance the upper convected time derivative used in Eq. (3). The coefficient D' is related to the diffusivity D through $D = D'(\partial\tilde{\mu}/\partial c_1)_{T,p}$, the angular brackets indicate the symmetric part of the corresponding tensor, and τ_1 and τ_2 are the relaxation times of $\mathbf{J}_1$ and $\mathbf{P}^v$ respectively. The coefficient α_1 turns out to be given by $\tau_1(D'T)^{-1}$.

For situations when the diffusion effects in Eq. (182) may be neglected as compared with those of $\mathbf{P}^v$ (e.g. when the system is subjected to a relatively high shear stress but when the diffusion flux is very low) one has

$$\tilde{\mu}(T,p,c_1,\mathbf{P}^v) = \tilde{\mu}_{eq}(T,p,c_1) + \tfrac{1}{2}T[\partial(v\alpha_2)/\partial c_1]_{T,p}\mathbf{P}^v:\mathbf{P}^v. \quad (185)$$

What we will explore here is under which conditions the usual thermodynamic stability criterion $(\partial\tilde{\mu}/\partial c_1) \geq 0$ may be used to obtain the spinodal line in presence of a shear, as we have assumed in this work.

First of all we will assume that $\beta = 0$. In this case, Eq. (183) reduces the steady state to

$$\mathbf{J}_1 = -D'(c_1,\dot{\gamma})\nabla\tilde{\mu}(c_1,\dot{\gamma}). \quad (186)$$

Note that both $\tilde{\mu}$ and D' may depend on $\dot{\gamma}$. If we focus our attention on a situation with homogeneous shear rate, i.e. with uniform $\dot{\gamma}$, Eq. (186) can be written as

$$\mathbf{J}_1 = -D'(\phi,\dot{\gamma})(\partial\tilde{\mu}/\partial\phi)_{T,\dot{\gamma}}\nabla\phi, \quad (187)$$

where in order to compare our results to those of previous works we have replaced the mass fraction c_1 by the volume fraction ϕ. Thus one may identify an effective diffusion coefficient D_{eff} as

$$D_{eff} = D'(\phi,\dot{\gamma})(\partial\tilde{\mu}/\partial\phi)_{T,\gamma}. \quad (188)$$

When D_{eff} becomes negative, the homogeneous state becomes unstable and the system separates into two or more phases: the inhomogeneities in the solute concentration begin to be amplified. According to the positiveness of entropy production, $D'(\phi, \dot{\gamma})$ is always positive. Thus, for thermodynamically stable states $(\partial \tilde{\mu}/\partial \phi)_{T,\gamma} > 0$ and $D_{eff} > 0$, but when $(\partial \tilde{\mu}/\partial \phi)_{T,\gamma}$ becomes negative D_{eff} will become negative. Then the criterion $(\partial \tilde{\mu}/\partial \phi)_{T,\gamma} = 0$ yields the separation between stable and unstable situations, both from the thermodynamical and the dynamical points of view. Of course, the dynamical approach may predict the kind of phase separation (droplets, percolating domain structures, ...) and the rate of separation, in contrast with the purely thermodynamic analysis.

An example of the use of a generalized chemical potential dependent on the shear rate is provided by Nozières and Quemada [125]. They assumed that the diffusion flux is modified by the shear and described it as

$$J_1 = \alpha(n_1)n_1 F, \tag{189}$$

with $\alpha(n)$ a friction coefficient, F a "thermodynamic force" which for local-equilibrium systems is given by $F = -(\partial \tilde{\mu}_{eq}/\partial y)$ (the fluid is assumed to flow along the x-direction, and the velocity v_x changes along the y-axis), and n_1 is the number density of the solute. In the presence of a shear, they assumed that F may be generalized by including a hydrodynamic lift force F_{lift} which is taken in their work as $F_{lift} = (\chi/2)(\partial \dot{\gamma}^2/\partial y)$, with χ a coefficient which is not specified in [125]. Then, they write

$$F = -(\partial \tilde{\mu}_{eq}/\partial y) - (\chi/2)(\partial \dot{\gamma}/\partial y). \tag{190}$$

If χ is a constant, one may introduce it into the derivative and write a generalized chemical potential of the form

$$\tilde{\mu}' = \tilde{\mu}_{eq} + (\chi/2)\dot{\gamma}^2, \tag{191}$$

so that in this case the non-equilibrium chemical potential $\tilde{\mu}'$ is related to the diffusion flux as

$$J_1 = -\alpha(n_1)n_1(\partial \tilde{\mu}'/\partial y). \tag{192}$$

Nozières and Quemada introduced this flux in the evolution equations for the concentration n and the velocity v_x, which are

$$\rho \frac{\partial v_x}{\partial t} = \eta \frac{\partial \dot{\gamma}}{\partial y} + \dot{\gamma} \frac{\partial \eta}{\partial n_1} \frac{\partial n_1}{\partial y}, \tag{193}$$

$$\rho \frac{\partial n_1}{\partial t} = \frac{\partial}{\partial y} \left[\alpha n_1 \left(\frac{\partial \tilde{\mu}_{eq}}{\partial n_1} \frac{\partial n_1}{\partial y} + \chi \dot{\gamma} \frac{\partial \dot{\gamma}}{\partial y} \right) \right]. \tag{194}$$

It follows from a linear stability analysis that the homogeneous state is unstable when $\dot{\gamma}$ is higher than the critical value $\dot{\gamma}_c$ given by

$$\dot{\gamma}_c^2 = \eta \left(\frac{\partial \tilde{\mu}_{eq}}{\partial n_1} \right) \left[\chi \left(\frac{\partial \eta}{\partial n_1} \right) \right]^{-1}. \tag{195}$$

This result coincides with the purely thermodynamical result based on the requirement $(\partial \tilde{\mu}'/\partial n_1) = 0$ with $\tilde{\mu}'$ given by Eq. (191) and the derivative calculated at constant $\mathbf{P}^v$. Thus in the model by Nozières and Quemada the dynamical and thermodynamical criteria for the spinodal line leads to the same result.

When the coefficient β is not zero, the evolution for the diffusion flux is given by Eq. (183) in which the latter term may be interpreted as the viscous reaction to the differential swelling produced by the change of composition of the mixture due to the diffusion.

In the steady state and in a shear flow with velocity distribution of the form $v_x(y)$, Eq. (183) for the component of the diffusion flux along the shear direction, J_y reduces to

$$J_y = -D'(n_1, \dot{\gamma}) \frac{\partial}{\partial y} [\tilde{\mu}(n_1, \dot{\gamma}) - T\beta P^v_{yy}], \tag{196}$$

where we have taken into account that the quantities appearing in Eq. (183) do not change along the direction of the flow.

This equation is at the basis of the Thomas-Windle model for Case-II diffusion [126]. These authors identify $-T\beta = \bar{v}$, $\bar{v}$ being the partial molar volume of the solvent [123, 126]. In this way, one may identify a generalized chemical potential $\tilde{\mu}''$ as

$$\tilde{\mu}''(n_1, \dot{\gamma}) = \tilde{\mu}(n_1, \dot{\gamma}) + \bar{v} P^v_{yy}, \tag{197}$$

with $\tilde{\mu}(n_1, \dot{\gamma})$ the chemical potential used in Eq. (186). According to [123, 126], the normal stress P^v_{yy} may be interpreted as a supplementary osmotic pressure acting on the system. Then Eq. (196) can simply be written as

$$J_y = -D'(n_1, \dot{\gamma}) \frac{\partial}{\partial y} \tilde{\mu}''(n_1, \dot{\gamma}). \tag{198}$$

In this case, the stability condition, i.e. the condition of a positive effective diffusion coefficient, is

$$\frac{\partial \tilde{\mu}''}{\partial \phi} > 0 \tag{199}$$

instead of the condition $(\partial \tilde{\mu}/\partial \phi) > 0$ found in Eq. (188). In situations where the normal viscous stress P^v_{yy} along the direction of the shear is zero (as for instance in Maxwell upper convected derivative models, where the normal stress along the x-direction is different from zero but the normal stress along the y and z directions are zero) $\tilde{\mu}''(n_1, \dot{\gamma})$ coincides with $\tilde{\mu}(n_1, \dot{\gamma})$. Therefore, the results in the previous sections are valid.

To illustrate this section, we comment on the analysis of phase separation in polymer solutions under shear by Onuki [5, 6]. This author examines the linear stability of a homogeneous shear flow of a semidilute polymer solution against perturbations of the volume fraction ϕ and the velocity v_x, under constant shear stress. Onuki also considers a non-equilibrium free energy which includes the effects of the flow. However, he does not directly use the thermodynamic criterion $(\partial\tilde{\mu}/\partial\phi) \geq 0$ for the analysis of the stability. His hesitation is logical because equilibrium thermodynamic criteria cannot be a priori extrapolated to non-equilibrium steady states.

The equations describing the evolution of v and ϕ are, according to Onuki,

$$\rho_0 \partial v/\partial t = -\nabla.[\mathbf{P}_p^v + \pi\mathbf{U}] - \nabla p - \eta_0 \nabla^2 v, \tag{200}$$

$$\partial\phi/\partial t + \nabla.(\phi v) = -\nabla.\mathbf{J}, \tag{201}$$

with $\mathbf{P}_p^v$ the polymer shear viscous stress, π the osmotic pressure, η_0 the shear viscosity and J the diffusion flux given by [122]

$$\mathbf{J} = -(\phi/\zeta)[\nabla(\pi_\phi + \pi_{el}) + \nabla.\mathbf{P}_p], \tag{202}$$

where ζ is a friction coefficient, π_ϕ and π_{el} are the equilibrium and the non-equilibrium contributions to the osmotic pressure and stand, respectively, for $\pi_\phi = (\phi\partial/\partial\phi - 1)f_{eq}$, $\pi_{el} = (\phi\partial/\partial\phi - 1)f_{el}$, with f_{el} being the elastic contribution of the flow to the Helmholtz free energy f. This equation yields for the component of the diffusion flux along the shear J_y the result

$$J_y = -D_T(n, \dot{\gamma})\frac{\partial\phi}{\partial y}, \tag{203}$$

with $D_T = \zeta^{-1}[\phi\partial(\pi_\phi + \pi_{el} + P_{yy}^v)/\partial\phi].$ \tag{204}

Notice that Eq. (202) may be obtained from Eq. (196) i.e. the normal viscous stress is included into the osmotic pressure as an additional contribution. Onuki writes Eq. (202) following a model by Doi [127] and uses as a stability criterion the positive character of D_T. It is interesting to note that this result is equivalent to the criterion at Eq. (199), as one has $\phi d\pi_{eff} \propto d\tilde{\mu}''$, with $\pi_{eff} = \pi_\phi + \pi_{el} + P_{yy}^v$, so that $\partial\tilde{\mu}''/\partial\phi \propto \phi\partial\pi/\partial\phi$.

An important difference between Onuki's analysis and ours is that we have used experimental data for f_{el} and for the dependence of the shear viscosity on the concentration and on the shear rate, whereas he uses a theoretical scaling relation of the form

$$\eta(\phi, P_{12}^v) = \phi^p \Phi(\tau_d \dot{\gamma}), \tag{205}$$

with p a scaling exponent depending on the model and τ_d a relaxation time. The function $\Phi(x)$ behaves as $\Phi(x) \sim x$ for $x \ll 1$ and $\Phi(x) \sim x^{0.2}$ for $x \gg 1$ and p is 3 for theta solvent though there is the theoretical possibility that $p = 2$. Onuki writes the Flory-Huggins equation in the form

$$f = (kT/a^3)\{N^{-1}\phi\ln\phi + [(1/2) - \chi]\phi^2 + (1/6)\phi^3 + \dots\}, \tag{206}$$

and he concludes (in our notation) that

$$[(1/2) - \chi]_{spinodal} = -(1/2)\phi - (1/2)(a^3/kT)(\partial^2 f_{el}/\partial \phi^2), \qquad (207)$$

or, in a more explicit form, he finds for the change in the spinodal temperature

$$(\Delta T)_s \sim N^{1/2}(T_\theta - T_c)\phi^{p-2}(\dot{\gamma}\tau_d)^\beta \qquad (208)$$

with N the polymerization index, T_θ and T_c the theta temperature and critical temperature, and β an exponent which describes the behaviour of the first normal stress difference N_1 as $N_1 \sim \phi^p(\tau_d\dot{\gamma})^\beta$. Since according to usual scaling theories $\tau_d \sim N^{3.4}\phi^{4.5}$ in theta semidilute regions, $(\Delta T)_s$ increases very strongly with increasing N and ϕ.

To obtain the coexistence line, Onuki assumes as the simplest picture that at the instability a planar interface to the y axis separates a dilute region and a semidilute region. The balance of normal pressure then requires

$$\pi_\phi + \pi_{el} + P^v_{yy} = 0, \qquad (209)$$

and this condition may be expressed in terms of χ as [6]

$$[(1/2) - \chi]_{coexistence} = -(1/6)\phi - (a^3/kT)\phi^{-2}\pi_{el}. \qquad (210)$$

What we have seen in this section is that the instability condition defining the spinodal region is the same in the thermodynamic analysis as in the dynamical analysis when $P^v_{yy} = 0$, so that there is no contradiction between them as long as one limits oneself to a phase stability analysis. In contrast, a dynamical approach is necessary to determine the form of the dependence of the diffusion constant on the shear rate, because such a coefficient has not only a thermodynamic contribution coming from the non-equilibrium chemical potential but also a purely kinetic contribution which is not accessible to thermodynamic arguments.

9 Some Perspectives and Concluding Remarks

Both thermodynamics and hydrodynamics should play a role in the analysis of the stability of solutions under shear. The point of view adopted in this paper is that the hydrodynamical analysis of the system under given boundary conditions determines the velocity and the viscous pressure at each point of the solution. Thermodynamic analysis has then a local meaning; if the thermodynamic stability conditions under shear are not satisfied at a given point, the homogeneous solution at this point will separate into two phases.

It is worthwhile to point out several criticisms or alternatives to our point of view. In the first place, one could discuss the existence of non-equilibrium thermodynamical potentials, even at a local level. Some main criticisms in this line are as follows.

1) Since at low Reynolds number the frictional forces are proportional to the velocity, it follows that they will only be derivable from a potential when the velocity itself is derivable from a potential, *i.e.* for irrotational flow fields [3, 128]. Since shear flow is not irrotational, there have been some doubts about the possibility of their thermodynamic treatment. We think that this objection refers to a condition to use the standard methods of equilibrium statistical physics, by adding to the potentials of the intermolecular forces a potential describing the non-equilibrium frictional forces, rather than pointing to a fundamental general problem. Indeed, we may recall that: a) from a microscopic point of view, as for instance from kinetic theory, several expressions for the free energy under shear flow have been obtained, as we have pointed out in Sect. 3; b) from the macroscopic point of view, one may define a non-equilibrium entropy even in situations when the forces do not derive from a potential; the procedure is to isolate a small part of the system and let it reach equilibrium: then one may define the (initial) non-equilibrium entropy as the (final) local-equilibrium entropy minus the entropy produced during the decay to equilibrium: there is no need that the forces derive from a potential.

2) The hypothesis that the steady state corresponds to the minimization of the free energy is generally considered to be incorrect and, in some examples, the dilute solution kinetic theory [129] gives results in contradiction to the predictions of the free-energy approach [130]. In contrast with this point of view, we recall that the maximum-entropy formalism has been useful in the description of non-equilibrium steady states by including the prescriptions about the dissipative fluxes which keep the system out of equilibrium. Furthermore, we have shown in Sect. 8 that the thermodynamic criteria leading to the stability region have also a dynamic justification under some rather general conditions, so that one does not in fact need extremal criteria for the stability analysis.

3) Whether one should simply add the flow contribution to the non-equilibrium expressions for the free energy, as we have done in this review, or whether some more intimate changes should also be produced by the flow on the local-equilibrium part of the free energy (as, for instance, a modification in the theta temperature) is an important open question. Since the theta temperature is related to the vanishing of the second virial coefficient for the osmotic pressure, it would not be difficult to analyse experimentally the osmotic pressure under shear and to compare the macroscopic predictions derived from the additive non-equilibrium thermodynamical potential with the experimental results. In fact, it follows from the predictions that, in spite of the fact that the theta temperature appearing in the Flory-Huggins equation is not changed, the effective theta temperature obtained from the vanishing of the virial coefficients for the osmotic pressure will be affected by the shear.

Up to now we have commented on the criticisms raised in connection with a non-equilibrium chemical potential. In the next paragraphs we tentatively comment on other perspectives of the problem of stability of polymer solutions under shear. Tirrell [3] gave some years ago a panoramic view of the perspectives on these problems.

1) It is evident that much experimental work is still needed. The coexistence curves should be examined in detail for a higher variety of flows and materials. The change in a few cloud points is not enough to have sufficiently conclusive information. The composition of the individual phases should also be measured. Spectrographic techniques rather than the simple visual observation of the turbidity could be used in order to improve the precision of the data.

2) One needs criteria systematizing the information about shear-induced solubility or shear-induced phase separation. In what conditions and for what systems is the solubility enhanced and when is it decreased? Tirrell [3] has proposed that this question could be related to the reversibility or irreversibility of the changes induced by the flow. In Sect. 5 we have analysed the conditions under which a decrease in the spinodal temperature could appear. For the moment it is difficult to see any connection of this phenomenon to the reversibility or irreversibility of the changes produced by the shear.

3) A dynamical analysis of the phenomena leading to the phase separation after a quench into the separation region (spinodal or nucleational) would be of interest. Such an analysis could be based on equations such as Eq. (191). Much experimental information is still lacking on this point. According to Onuki [6], for deep quenches one should expect gel-like balls in a less-viscous medium, moving relatively in shear but connected by entanglements: this would lead to a considerable decrease in viscosity in the two-phase regime.

The phenomenology of the transition from an inhomogeneous state to a homogeneous state has been studied in detail by Takebe et al. [85, 86] by means of the scattering profile in the plane perpendicular to the shear flow. They have analysed the shear-induced homogenisation of semidilute solutions of polybutadiene and poly-styrene with nearly critical composition, and they have found a rich variety of details, grouped in five different regimes which are present at increasing shear rate: a) a macroscopic phase separation into two coexisting solution phases; b) the macroscopic interface is broken to the shear in percolated domain structure or nonpercolated droplet structure essentially isotropic; c) the shear produces an elongation of the structures along shear direction; d) the amplitude of concentration fluctuations diminishes and e) there is reached a shear-induced homogenized state. The phenomenology could be different for different systems, and also depends on the composition and the quench depth $T_c(0) - T$.

4) One should try to obtain a better understanding of the free energy in inhomogeneous systems with the different structures pointed out by Takebe et al. [85, 86]. The contribution of the surface tension should be taken into account. A combination of the different microscopic or thermodynamic procedures pointed out in Sect. 3 should probably be combined to have sufficiently thorough information about the non-equilibrium contribution to the free energy.

5) Whereas the spinodal curve under shear seems within the reach of thermodynamic analysis, the criteria defining the coexistence curve may be more difficult to obtain. The applicability of the Maxwell rule to non-equilibrium steady states is not clear. Maybe a dynamical criterion related, for instance, to the conditions of mechanical equilibrium and mechanical stability of the domain walls, is unavoidable, or maybe a generalized Maxwell criterion could be directly found from thermodynamics. This is an open question [125].

6) The generalized chemical potential should be used to study in detail the effects of a shear on the diffusivity. We have also mentioned this problem in connection with the dynamical models of phase separation. One should be able to predict, for instance, under which conditions the shear will enhance the diffusion of a solvent towards the wall of a cylinder or diminish it. This problem may be of interest in physiological analyses of the vascular system, where the flow of some materials towards the walls may lead to pathological consequences.

7) The transition from the homogeneous phase to the inhomogeneous phase will yield changes in the viscosity. One should be able to study not only the stability region but also the details of the separation and their correlation with changes in dynamical properties as the viscosity or the light-scattering spectrum. It is clear that these aspects are beyond a purely thermodynamic analysis.

8) To establish a wide thermodynamic theory of fluids under flow, encompassing many different systems, would be of great interest. Indeed, not only polymer solutions exhibit the influence of the flow on their phase changes, but also low-molecular-weight mixtures [88, 89], concentrated systems of soft spheres [132, 132], or liquid crystals [128, 133, 134]. Thus, the understanding of the stored free energy should aim not only at some particular cases of dilute polymer solutions, but it should at least cover real gases, where the pair-correlation function may be distorted due to the flow, thus also leading to changes in the free energy [131, 135]. A single formalism unifying these fields, such as the one proposed in extended thermodynamics, would be very useful to afford perspectives in this direction.

Acknowledgements. We acknowledge the financial support of the Dirección General de Investigación Científica y Técnica of the Spanish Ministry of Education and Science under grant PB 90/0676. DJ also acknowledges a grant of the DGICyT under the program Movilidad del Personal Investigador, which made possible his stay with Dr. M. Grmela in the Centre de Recherche Appliquée sur des Polymères (CRASP) of the Ecole Polytechnique de Montréal, in May–July 1990, where this work was initiated. MC-S acknowledges a fellowship of the Comunidad Autónoma de Madrid which has supported his stay in the UAB from December 1991 – February 1992. Our group has also benefited from EC Human Capital and Mobility program under contract ERB CHR XCT 920 007.

264   D. Jou et al.

10 References

1. Rangel-Nafaile C, Metzner AB, Wissbrun KF (1984) Macromolecules 17: 1187
2. Wolf BA (1984) Macromolecules 17: 615
3. Tirrell M (1986) Fluid Phase Equilibria 30: 367
4. Silberberg A (1988) Physico Chem Hydrodyn 10: 693
5. Onuki A (1989) Phys Rev Lett 62: 2472
6. Onuki A (1990) J Phys Soc Japan 59: 3423, 3427
7. Barham PJ, Keller A (1990) Macromolecules 23: 303
8. McHugh AJ (1982) Polym Engn Sci 22: 15
9. Utracki LA (1990) Polymer alloys and blends. Thermodynamics and rheology. Hanser Publishers, Munich
10. Tirrel M (1978) J Bioengn 2: 183
11. Barbu E, Jolly E (1953) Discuss. Faraday Soc 13: 77
12. Nguyen TQ, Kausch H-H (1989) Makromol Chem 190: 1389
13. Nguyen TQ, Kausch H-H (1992) Adv Polym Sci 100: 73
14. De Groot SR, Mazur P (1962) Non-equilibrium thermodynamics. North-Holland, Amsterdam
15. Meixner J (1954) Z. Naturforschung a 4: 594; 9: 654
16. Coleman BD (1964) Arch Rat Mech Anal 17: 1
17. Coleman BD (1965) Arch Rat Mech Anal 17: 230
18. Truesdell C (1971) Rational thermodynamics. McGraw-Hill, New York
19. Nettleton RE (1959) Phys Fluids 2: 256
20. Müller I (1967) Z Phys 198: 329
21. Jou D, Casas-Vázquez J, Lebon G (1979) J Non-Equilib Thermodyn 4: 349
22. Lebon G, Jou D, Casas-Vázquez J (1980) J Phys A 13: 275
23. Eu BC (1980) J Chem Phys 73: 2158
24. Casas-Vázquez J, Jou D, Lebon G (eds) (1984) Recent developments in non-equilibrium thermodynamics. Springer, Berlin (Lecture Notes in Physics, vol 199)
25. Müller I, Ruggeri T (eds) (1987) Symposium on kinetic theory and extended thermodynamics. Pitagora editrice, Bologna
26. Jou D, Casas-Vázquez J, Lebon G (1988) Rep Prog Phys 51: 1105
27. García-Colín LS (1988) Rev Mex Física 34: 344
28. García-Colín LS, Uribe FJ (1991) J Non-Equilib Thermodyn 16: 89
29. Jou D, Casas-Vázquez J, Lebon G (1992) J Non-Equilib Thermodyn 17: 383
30. Sieniutycz S, Salamon P (eds) (1992) Extended thermodynamic systems. Taylor and Francis, New York (Advances in Thermodynamics, vol 7)
31. Lebon G, Jou D, Casas-Vázquez J (1992) Contemp Phys 33: 41
32. Bird RB, Armstrong RC, Hassager O (1977) Dynamics of polymeric liquids. Volume 1: Fluid mechanics. Wiley, New York
33. Bird RB, Curtiss CF, Armstrong RC, Hassager O (1977) Dynamics of polymeric liquids. Volume 2: Kinetic theory. Wiley, New York
34. Ferry JD (1971) Viscoelastic properties of polymers. Wiley, New York
35. Jou D, Casas-Vázquez J, Lebon G (1993) Extended irreversible thermodynamics. Springer, Berlin
36. Müller I, Ruggeri T (1993) Extended thermodynamics. Springer, Berlin
37. Lebon G, Pérez-García C, Casas-Vázquez J (1986) Physica A 137: 531
38. Lebon G, Pérez-García C, Casas-Vázquez J (1988) J Chem Phys 88: 5068
39. Camacho J, Jou D (1990) J Chem Phys 92: 1339
40. Coleman BD, Markovitz H (1964) J Appl Phys 35: 1
41. Marrucci G (1972) Trans Soc Rheol 16: 321
42. ver Stratte G, Philipoff W (1974) J Polym Sci: Polym Lett 12: 267
43. Breuer S, Onat ET (1964) Z Angew Math Phys 15: 184
44. Gurtin ME, Herrera I (1965) Quart Appl Math 23: 235
45. Christensen RM (1971) Theory of viscoelasticity. An introduction. Academic Press, New York
46. Grmela M (1986) J Rheol 30: 707
47. Doi M, Edwards SF (1986) The theory of polymer dynamics. Clarendon Press, Oxford
48. Jou D, Micenmacher V (1987) J Phys A 20: 6519
49. Santos A, Brey JJ (1991) Physica A 174: 355

50. Janeschitz-Kriegl H (1969) Adv Polym Sci 6: : 170
51. Sun Z-S, Denn MM (1972) AIChE J 18: 1010
52. Sarti GC, Marrucci G (1973) Chem Engn Sci 28: 1053
53. Booij HC (1984) J Chem Phys 80: 4571
54. Kramers HA (1944) Physica 11: 1
55. Lodge AS, Wu YJ (1971) Rheol Acta 10: 539
56. Bhattacharjee SM, Fredrickson GH, Helfand E (1989) J Chem Phys 90: 3305
57. Lodge AS (1968) Rheol Acta 7: 379
58. de Gennes PG (1971) J Chem Phys 55: 572
59. Durning CJ, Tabor M (1986) Macromolecules 19: 2220
60. Doi M, Edwards SF (1978) J Chem Soc Faraday Trans II 74: 918
61. Marrucci G, Grizzutti N (1983) J Rheol 27: 433
62. Currie PK (1982) J Non-Newtonian Fluid Mech 11: 53
63. Doi M, Edwards SF (1978) J Chem Soc Faraday Trans II 74: (a) 1789, (b) 1802, (c) 1818
64. McHugh AJ (1975) J Appl Polym Sci 19: 125
65. Takayashi Y, Noda I, Nagasawa M (1990) Macromolecules 22: 242
66. Criado-Sancho M, Jou D and Casas-Vázquez J (1991) Macromolecules 24: 2834
67. Casas-Vázquez J, Jou D (1989) Acta Physica Hungarica 66: 99
68. Jou D, Casas-Vázquez J (1990) Physica A 163: 49
69. Jou D, Casas-Vázquez J (1992) Phys Rev A 45: 8371
70. Criado-Sancho M, Jou D, Casas-Vázquez J (1993) J Non-Equilib Thermodyn 18: 103
71. Kramer H, Wolf BA (1985) Makromol Chem, Rapid Comm 6: 21
72. Graessly WW (1964) Fortschr Hochpolym-Forsch 16: 1
73. Vinogradov GV, Malkin A (1980) Rheology of polymers. Springer, Berlin
74. Kramer H, Schenck H, Wolf BA (1988) Makromol Chem 189: 1627
75. Vrahopoulou-Gilbert E, McHugh AJ (1984) Macromolecues 17: 2657
76. Vrahopoulou-Gilbert E, McHugh AJ (1986) J Appl Polym Sci 31: 399
77. Elyasevitch G (1982) Adv Polym Sci 43: 205
78. Silberberg A, Kuhn W (1954) J Polym Sci 13: 21
79. Vanoene H (1972) J Colloid Interface Sci 40: 448
80. Wolf BA (1980) Makromol Chem Rapid Comm 1: 231
81. Silberberg A, Kuhn W (1952) Nature 170: 450
82. Scott RL (1949) J Chem Phys 17: 279
83. Wolf BA, Kramer H (1980) J Polym Sci Polym Lett Ed 18: 789
84. Lyngaae-Jorgensen J, Sondergaarrd K (1987) Polym Engn Sci 27: 351
85. Takebe T, Sawaoka R, Hashimoto T (1989) J Chem Phys 91: 4369
86. Takebe T, Fujioka K, Sawaoka R, Hashimoto T (1990) J Chem Phys 93: 5271
87. Beysens D, Gbamadassi M, Boyer L (1979) Phys Rev Lett 43: 1253
88. Beysens D, Gbamadassi M (1980) Phys Lett A 77: 171
89. Onuki A, Kawasaki A (1980) Prog Theor Phys 63: 122
90. Mazich KA, Carr SH (1983) J Appl Phys 54: 5511
91. Peterlin A, Turner DT (1965) J Polym Sci B 3: 517
92. Peterlin A, Quan C, Turner DT (1965) J Polym Sci B 3: 521
93. Matsuo T, Pavan A, Peterlin A, Turner DT (1967) J Colloid Interface Sci 24: 241
94. Staudinger H (1932) Die hochmolekularen organischen Verbindungen. Springer, Berlin
95. Freudenberg K, Kuhn W, Durr W, Boltz F, Steinbrunn G (1930) Ber 63: 1510
96. Freudenberg K, Blomquist G (1935) Ber 68: 2070
97. Kuhn W (1930) Ber 63: 1502
98. Ekenstam A (1936) Ber 69: 549, 553
99. Senti F, Hellman N, Ludwig N, Babcock G, Tobin R, Gass C, Lambe B (1955) J Polym Sci 17: 527
100. Antonini E, Bellelli M, Bruzzesi M, Caputo E, Chiancone E, Rossi-Farelli (1964) Biopolymers 2: 27
101. Basedow AM, Ebert KH (1975) Makromol Chem 176: 745
102. Mark H, Simha R (1940) Trans Faraday Soc 36: 611
103. Simha R (1941) J Appl Phys 12: 569
104. Montroll E (1941) J Am Chem Soc 63: 1215
105. Montroll E, Simha R (1940) J Chem Phys 8: 721
106. Basedow AM, Ebert KH, Ederer HJ (1978) Macromolecules 11, 774
107. Bueche F (1960) J Appl Polym Sci 4: 101

108. Frenkel Yu I (1944) Acta Physicochim. URSS 19: 51
109. Bestul AB (1956) J Chem Phys 24: 1196
110. Ceresa RJ, Watson WF (1959) J Appl Polym Sci 1: 101
111. Mullins L, Watson WF (1959) J Appl Polym Sci 1: 245
112. Glynn PAR, van der Hoff BME, Reilly PM (1972) J Macromol Sci Chem A6: 1653
113. Ballauff M, Wolf BA (1981) Macromolecules 14: 654
114. Ballauff M, Wolf BA (1984) Macromolecules 17: 209
115. Ballauff M, Wolf BA (1988) Adv Polym Sci 85: 2
116. Henglein A (1956) Makromol Chem 18: 37
117. Criado-Sancho M, Jou D, Casas-Vázquez J (1994) J Non-Equilib Thermodyn 19: 137
118. Wolf BA (1987) Makromol Chem Rapid Comm 8: 461
119. Cukrowski AS, Popielawski J (1986) Chem Phys 109: 215
120. Kurata M (1982) Thermodynamics of polymer solutions. Harwood Academic Publishers, Chur
121. Graessley WW (1974) Adv Polym Sci 16: 1
122. Helfand E, Fredrickson GH (1989) Phys Rev Lett 62: 2468
123. Jou D, Camacho J, Grmela M (1991) Macromolecules 24: 3597
124. Casas-Vázquez J, Criado-Sancho M, Jou D (1993) Europhys. Lett. 23: 469
125. Noziéres P, Quemada D (1986) Europhys Lett 2: 129
126. Thomas NL, Windle AH (1882) Polymer 23: 529
127. Doi M (1990) in: Onuki A, Kawasaki K (eds) Dynamics and pattern in complex fluids: New aspects of physics and chemistry interfaces. Springer, Berlin
128. See H, Doi M, Larson R (1990) J Chem Phys 92: 792
129. Aubert JH, Tirrel M (1980) Rheol Acta 19: 452
130. Tirrell M, Malone MF (1977) J Polym Sci: Polym Phys Ed 15: 1569
131. Hanley HJM, Evans DJ (1982) J Chem Phys 76: 3225
132. Hanley HJM, Rainwater JC, Clark NA, Ackerson BJ (1983) J Chem Phys 79: 4448
133. Olmsted PD, Goldbart P (1990) Phys Rev A 41: 4578
134. Grmela M, Ly C (1987) Phys Lett A 120: 281
135. Grmela M (1987) Phys Lett A 120: 276

Editor: Prof. H. H. Kausch
Received: March 1994

Author Index Volumes 101-120

Author Index Vols. 1-100 see Vol. 100

... ...

... ...

... ...
... ...

Subject Index

Springer-Verlag
and the Environment

We at Springer-Verlag firmly believe that an international science publisher has a special obligation to the environment, and our corporate policies consistently reflect this conviction.

We also expect our business partners – paper mills, printers, packaging manufacturers, etc. – to commit themselves to using environmentally friendly materials and production processes.

The paper in this book is made from low- or no-chlorine pulp and is acid free, in conformance with international standards for paper permanency.